P9-DGF-506
THIS LAND
IS YOUR LAND

Canada
Pacific Ocean
Pacific Ocean
Mexico
ROCKY MOUNTAINS
SIERRA NEVADA
Vancouver
Victoria
Calgary
Puget Sound
Olympic Nat'l Park
Seattle
Tacoma
WA
Spokane
COLUMBIA RIVER
Portland
Glacier Nat'l Park
MISSOURI RIVER
Helena
BLACKFOOT RIVER
Butte
MT
Billings
ND
Bismarck
Eugene
OR
Boise
ID
Yellowstone Nat'l Park
Sheridan
Grand Teton Nat'l Park
Pierre
Rapid City
SD
Idaho Falls
WY
Smoke Creek Desert
Reno
NV
Salt Lake City
Provo
Cheyenne
NE
Sacramento
San Francisco Bay
Oakland
San Francisco
San Jose
Yosemite Nat'l Park
UT
Boulder
Denver
CO
COLORADO RIVER
Fresno
CA
Sequoia & Kings Canyon Nat'l Park
Zion Nat'l Park
Glen Canyon Nat'l Park
Pueblo
KS
Las Vegas
Grand Canyon Nat'l Park
Santa Barbara
Santa Monica Bay
Los Angeles
AZ
Santa Fe
Amarillo
Albuquerque
NM
San Diego
Phoenix
Tijuana
Wichita Falls
Roswell
Tucson
El Paso
San Angelo
TX
RIO GRANDE RIVER
Honolulu
HI
Hawaii
Chihuahua
San Anton
Arctic National Wildlife Refuge
AK
Monterrey
Anchorage
Juneau
Tongass Nat'l Forest

James Bay
Quebec
PENOBSCOT RIVER
Lake Superior
MN
Duluth
Montreal
ME
Acadia Nat'l Park
VT
Augusta
Ottawa
Lake Huron
Montpelier
Kingston
NH
Buzzard's Bay
St. Paul
Minneapolis
Concord
HUDSON RIVER
WI
Lake Ontario
Albany
Boston
Toronto
MA
MI
NY
Providence
Buffalo
Hartford
RI
Madison
Lake Michigan
Lansing
CT
Long Island Sound
Detroit
Milwaukee
Lake Erie
Waterloo
Newark
New York City
IA
Chicago
Des Moines
Davenport
PA
Philadelphia
Gary
Cleveland
Trenton
Harrisburg
Pittsburgh
Ft. Wayne
NJ
OH
Dover
Atlantic City
Peoria
Baltimore
DE
Columbus
IL
Indianapolis
Washington DC
Annapolis
Springfield
Cincinnati
IN
WV
MD
Kansas City
Shenandoah Nat'l Park
Chesapeake Bay
Topeka
St. Louis
East St. Louis
Richmond
Charleston
Frankfort
Louisville
VA
MO
Roanoke
KY
Springfield
Raleigh
Knoxville
NC
Nashville
Tulsa
TN
Charlotte
Fort Smith
Memphis
Greenville
MISSISSIPPI RIVER
Columbia
AR
Little Rock
Atlanta
SC
Birmingham
Augusta
Charleston
Atlantic Ocean
MS
AL
GA
Jackson
Savannah
Montgomery
Shreveport
Meridian
Albany
LA
Jacksonville
Mobile
Tallahassee
Baton Rouge
New Orleans
FL
Mississippi River Delta
Orlando
Tampa
St. Petersburg
Fort Lauderdale
Miami
Gulf of Mexico
Everglades Nat'l Park
Key West
Cuba

About the Authors

Jon Naar is an internationally acclaimed environmental author and lecturer. He is environmental consultant to The Creative Coalition, Shomrei Adamah (Keepers of the Earth), the Global Youth Forum, and other organizations. A cofounder of the Solar Coalition, he is a director of the North East Sustainable Energy Association and member of the New York Open Center's Advisory Council and of the Founding Advisory Board of the World Peace Prayer Society. His *Design for a Livable Planet* won the 1990 American Library Association award for best nonfiction book for young adults. In addition to his books, which are listed below, he has written and photographed articles on ecology and the environment for numerous magazines including the Natural Resources Defense Council's *Amicus Journal*, the *New York Times Magazine*, *Interiors*, and *Northeast Sun*.

Alex Naar attended the National Audubon Society's Expedition Institute and has a B.S. degree in environmental science from Lesley College. He has worked for the Environmental Law Foundation, was president of the Environmental Law Society at Golden Gate University, and was an associate editor of *Ecology Law Quarterly*. He was a firefighter and emergency medical technician for nine years. An active Morris dancer, he can often be found exploring the wilds of North America. In addition to coauthoring *This Land Is Your Land*, he worked with Jon Naar on *Design for a Livable Planet*.

Previous Books by Jon Naar:

- *Design for a Livable Planet*, New York: Harper & Row, 1990
- *The New Wind Power*, New York: Penguin, 1982
- *Your Space: How to Put It Together for Practically Nothing* (with Mary Ellen Moore), New York: St. Martin's, 1979
- *Design for a Limited Planet* (with Norma Skurka), New York: Ballantine, 1976
- *Living in One Room* (with Molly Siple), New York: Random House, 1976
- *The Faith of Graffiti* (with Norman Mailer and Mervyn Kurlansky), New York: Praeger, 1974
- *Christopher Columbus*, New York: Picture Progress, 1955

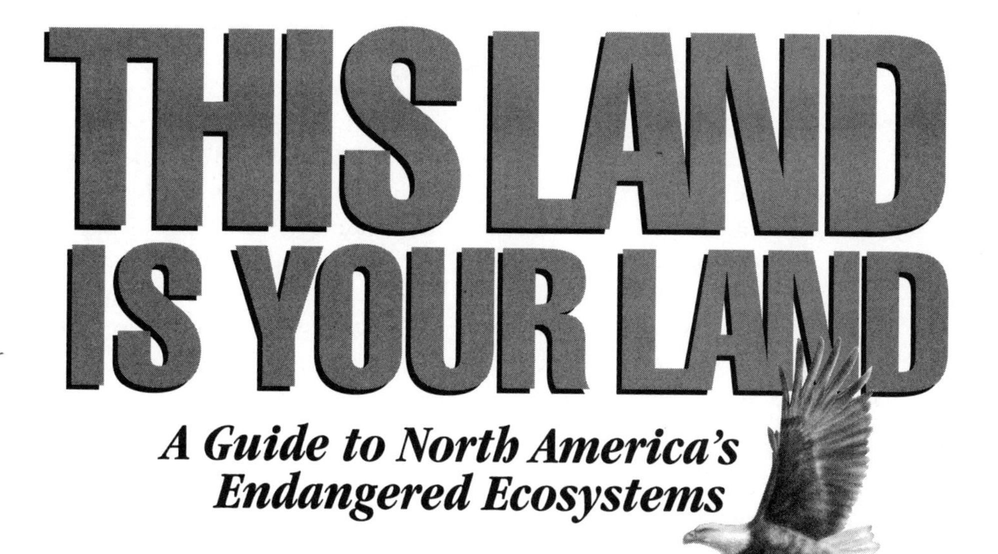

THIS LAND IS YOUR LAND

A Guide to North America's Endangered Ecosystems

JON NAAR
and Alex J. Naar

authors of Design for a Livable Planet

HarperPerennial
A Division of HarperCollins*Publishers*

Dedication

We dedicate *This Land Is Your Land* to Bald Mountain, Muir Woods, the Adirondacks, Walden Pond, the Arctic National Wildlife Refuge, the northern spotted owl and the California gnat catcher, the Xerxes butterfly and the Plymouth gentian, the Pacific yew, and the grandness of the Grand Canyon. To a towering Douglas fir somewhere in Oregon and a gnarled Juniper tree high in the Sierra, may they continue to be there in two hundred years. And finally, to the thousands of other unique and special places and species in the hope that they will be preserved for our children and following generations to know and learn from.

HarperCollins books may be purchased for educational, business, or sales promotional use. For information please write: Special Markets Department, HarperCollins Publishers, Inc., 10 East 53rd Street, New York, NY 10022.

FIRST EDITION

Designed by Joseph L. Santoro

Library of Congress Cataloging-in-Publication Data

Naar, Jon.
This land is your land : a guide to North America's endangered ecosystems / Jon Naar and Alex J. Naar. — 1st ed.
p. cm.
Includes bibliography references and index.
ISBN 0-06-096882-6 (paper)
1. Biotic communities — United States. 2. Environmental protection — United States — Citizen participation. I. Naar, Alex J. II. Title.
QH104.N23 1992
333.7'2'0973 — dc20 91-5058

---

93 94 95 96 97 RRD 10 9 8 7 6 5 4 3 2 1

This Land Is Your Land is printed on recycled paper.

CONTENTS

PREFACE

This land is your land, this land is my land
From California to the New York Island

— Woody Guthrie

We took our title from Woody's song because it captures the magnificent sweep of this continent. His words invoke the incredible richness of this land's natural resources — Yosemite, Yellowstone, the Grand Canyon, the Rocky Mountains, the Mississippi, the Great Lakes, the Adirondacks, the Appalachians, and thousands of other places we can know, where we can find solitude and connect with Nature. We take pride in the song because it invokes our history, the spirit of We the People, the great populist tradition of Paine, Jefferson, Lincoln, and the F.D.R. of the New Deal.

This land is ours in several ways. One-third of the more than 3.5 million square miles of total United States land area is federally or state owned. We own it constitutionally, yet it is not ours to use indiscriminately. As the saying goes, we didn't inherit this land from our ancestors, we have borrowed it from our children. We hold it in trust for future generations. We have this responsibility as the only species that can *knowingly* determine the fate of the land. Through our everyday actions — the way we live, eat, work, travel, and play — we powerfully influence the equilibrium of its intricate natural systems.

We wrote this book as an inventory of the lands and waterways that Guthrie celebrated, to describe the condition they are in today and to show how they can be protected and restored. We focus on ecosystems because they explain the vital links between the water, air, and land communities on which we are dependent. We also examine the important connection between Nature and ourselves.

Like *Design for a Livable Planet, This Land Is Your Land* emphasizes that we live in a society with a long tradition of grass-roots activism where the individual can make a difference. We recognize the important groundwork achieved before us by ordinary citizens, working people, teachers, and children as well by community leaders, conservationists, scientists, legislators, and others. Thanks to their efforts, there exists in North America a foundation of knowledge and experience upon which effective future action can be built.

The threats to our continent's ecosystems are great and time is running out, but we can still act. We can support those who are already out there showing us the way. These are the people and the organizations we describe in the case histories in each chapter, in the "What You Can Do" sections, and in the Directory at the end of the book. Find out who they are, what they stand for, and *join* them in their vital work. It's an exciting prospect. After all, this land is *your* land.

INTRODUCTION

How the Earth Works[1]

"We are Nature, long have we been absent, but now we return."

— Walt Whitman

*E**cology*, from the Greek *oikos* "habitat" and *logos* "the study of," is the science of how living organisms relate to their environment. What are the forms of life in a drop of water, a forest, or a desert? How do they get energy and matter to stay alive? How do they interact with one another and other living and nonliving things? How do they respond to natural and human-induced changes that take place in this environment? What are ecosystems? And what can we do to save and restore them? These are some of the questions ecologists try to answer and which we shall explore in this book.

In physics and chemistry, **matter** (anything that has mass or weight) is usually identified in a spectrum ranging from subatomic particles to the entire universe. Ecology is concerned primarily with the part of the spectrum that includes **organisms, species, populations, communities, ecosystems,** and the **biosphere**. Organisms are any individual life form capable of reproduction, ranging in scale from the tiniest microorganisms such as **phytoplankton** (one-celled, floating plants) to giant sequoia trees and African elephants. A species comprises all organisms of the same kind. Globally there are estimated to be as many as fifty million different species of plants and animals, of which perhaps thirty million are insects. A population is any group of individuals of the same species living in a given area and sharing common features. A community is a group of populations coexisting interdependently in a given area or region; examples might include all the plants and creatures found in an aquarium, a desert, or a forest. Ecological systems (ecosystems) are any self-regulating set of plants and animals that interact with one another and their nonliving environment. Although no two ecosystems are exactly alike, they can be identified generally by the similarity of their plants and animals and the way these organisms interact with one another and their environment. An ecosystem can be as small as a microbial community or as large as an ocean. Major ecosystems such as rivers, wetlands, grasslands, deserts, forests, and oceans are called **biomes**.

The biosphere consists of those parts of the **lithosphere** (earth's crust), **hydrosphere** (water), and **atmosphere** (air) where **biota** (living organisms) can be found. Sometimes referred to as "the fragile skin of life," the biosphere contains all the minerals, water, oxygen, and other nutrients that living things need. Within the biosphere everything is interdependent: Air, water, and soil keep plants and animals alive; plants sustain animals and help renew the soil and air, which in turn help purify water and the earth.

The First Law of Energy

In many respects, the earth is a closed system, and the amount of matter within it is finite. Therefore, the constant recycling of matter is critical to the continuity of life. The principle of conservation of energy in nature, known also as the first law of energy or thermodynamics, is sometimes expressed in the saying "You can't get something for nothing." In other words, the energy gained or lost in any living or nonliving system must equal the energy gained or lost outside that system. For example, the earth's radiant energy gained from the sun is eventually reradiated into space. Left undisturbed, an ecosystem naturally recycles its nutrients as they flow through plants and animals and return to the soil, air, and

water, keeping the ecosystem balanced. In this process some of the energy and nutrients are lost, but nature helps to maintain a balance. Much of the time, the carbon dioxide released naturally into the atmosphere by the respiration and decay of plants is absorbed through the leaves of other growing plants through photosynthesis. However, when carbon dioxide produced by burning fossil fuels is released into the atmosphere, it is added to the amount already there and begins to overload the natural balance of energy.

As we shall see in the following chapters, energy overload comes from many sources, including the lavish agricultural use of nitrates and other chemicals (as fertilizers, pesticides, etc.), which are carried by wind or water into rivers, grasslands, and other ecosystems. This upsets the balance in two ways, taking away too much energy from a field, for example, and adding too much to a forest or a body of water, often with serious ecological consequences.

The Flow of Energy and Matter

The living part of an ecosystem **(biotic)** consists of organisms (plants and animals); the nonliving part **(abiotic)** is made up of physical components (terrain, sunlight, shade, temperature, precipitation, etc.) and chemical components, which are the elements and compounds needed by organisms to live. The type, amount, and variation of these factors determine what and how many plants and animals can exist there, and they, in turn, affect the environment they inhabit. Complex interactions occur both within and between biotic and abiotic parts of the ecosystem. In chapters 5 and 6, for example, we shall see how grasses and trees help form the soil, buffer the wind, hold water, and moderate the climate, and how animals scatter seeds and pollinate plants, and how many other interactions between plants, animals, and their environment take place.

The living and nonliving parts of an ecosystem are connected by a constant exchange of materials through cycling of nutrients driven by energy from the sun. As you can see from the simplified chart on the next page, the major components of an ecosystem are: **solar energy; producers** (plants); **consumers** (of plants and animals); **decomposers** (bacteria, fungi); and **nutrients** (carbon dioxide, oxygen, nitrogen, minerals) important for growth.

Producers are chlorophyll-bearing plants such as algae, grass, and trees, so called because they produce their own food. Through the process of **photosynthesis** they take energy from sunlight and nutrients from soil, water, and air, manufacturing organic compounds that they use to build their tissues and store as chemical energy. The myriad interconnected chemical changes in this process can be summarized as follows: Solar energy + carbon dioxide + water ➜ glucose + oxygen. A portion of the glucose (and other more complex, carbon-containing molecules) is then converted back into carbon dioxide by **cellular respiration**: Glucose + oxygen ➜ carbon dioxide + water + energy. The carbon dioxide (CO_2) is returned to the atmosphere and water, where it can be reused by producers. Photosynthesis and cellular respiration form carbon and oxygen cycles through which plants both live and produce food needed by animals as well as absorb carbon dioxide given off by animals. **Consumers** are organisms that feed on producers and other consumers. They consist of **herbivores** (plant eaters), **carnivores** (animal eaters), and **omnivores** (which eat both). **Decomposers** constitute the

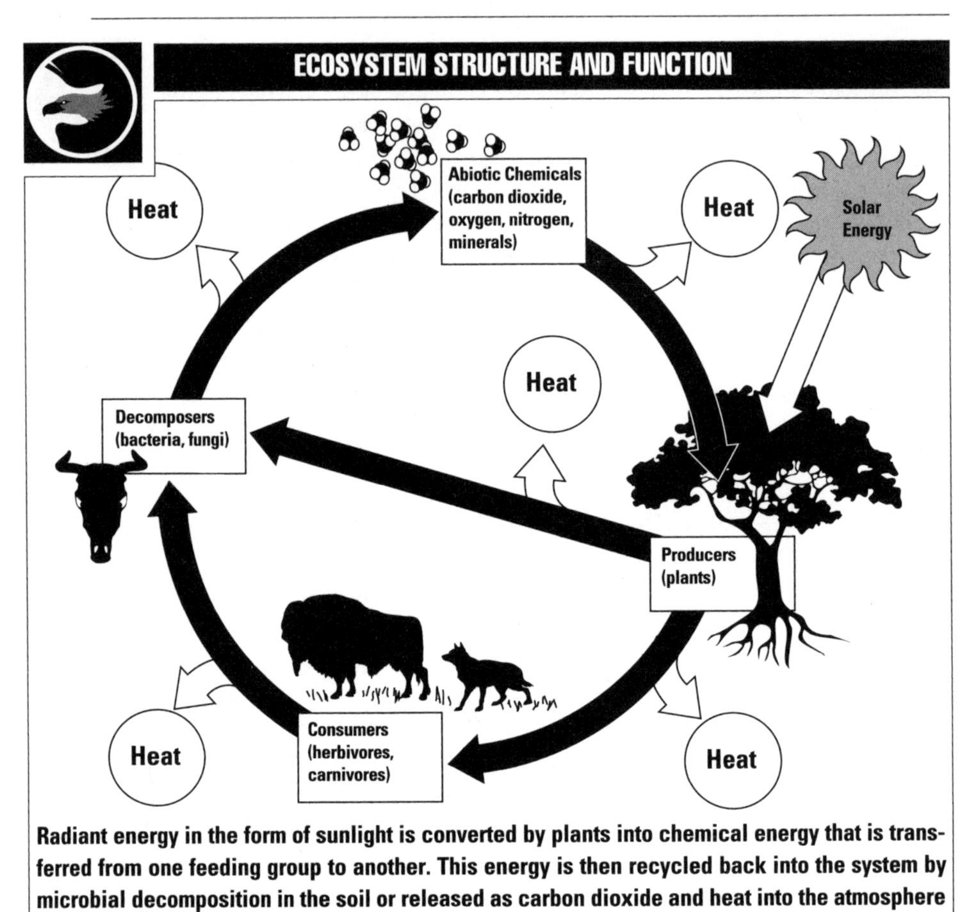

Radiant energy in the form of sunlight is converted by plants into chemical energy that is transferred from one feeding group to another. This energy is then recycled back into the system by microbial decomposition in the soil or released as carbon dioxide and heat into the atmosphere and water for reuse by producers.

numerically largest and perhaps most important group. These are microconsumers — bacteria, fungi, and some insects and worms — that get their food from breaking down complex molecules in the wastes and dead bodies of other organisms and convert them into simpler molecules, most of which can then be returned to the soil and water for reuse by producers. It is decomposition that prevents the earth from running out of raw material. Some idea of how the main components of an ecosystem relate to one another can be seen in the diagram of the edge of the forest on the next page.

Diversity and Distribution of Species

In simple communities with few niches to occupy, such as a drop of water, there is little diversity of species, although the *numbers* of the species existing there are generally very great. In more complex communities such as a forest there is greater species diversity with more specialized (and sometimes less numerous) occupants. The robustness of a community is often seen as a function of its diversity. In general, the more diverse a community is, the better it is able to recover from natural or human disturbances. You can get an idea of how some species are distributed and the role they play by looking at the diagram on the opposite page.

Defining Community

Because no organism can survive completely on its own, populations of plants and animals form natural communities with other populations. Some of them — ponds, tidal beaches, stands of trees within a forest, for example — have clearly defined boundaries. Others blend into one another with overlapping species of plants and animals. The area at the boundary between two communities is the **ecotone** or edge. Within a community each species has an **ecological niche**, representing the specific part of a habitat occupied by an organism, and the role it plays in an ecosystem — i.e., where it lives, what it eats or decomposes (its food niche), how it reproduces, and its physical and chemical requirements such as temperature, humidity, and shade, the chemicals it can tolerate, and its impact on the nonliving parts of its environment. Depending on the extent of its habitat, the sources of its food, and its tolerance to physical and chemical conditions, the niche of an organism may be specialized, as with giant pandas, which get 99 percent of their food from bamboo plants, or generalized, as with cockroaches, rats, and humans, who eat a wide variety of food.

MAIN COMPONENTS OF A FOREST-EDGE ECOSYSTEM

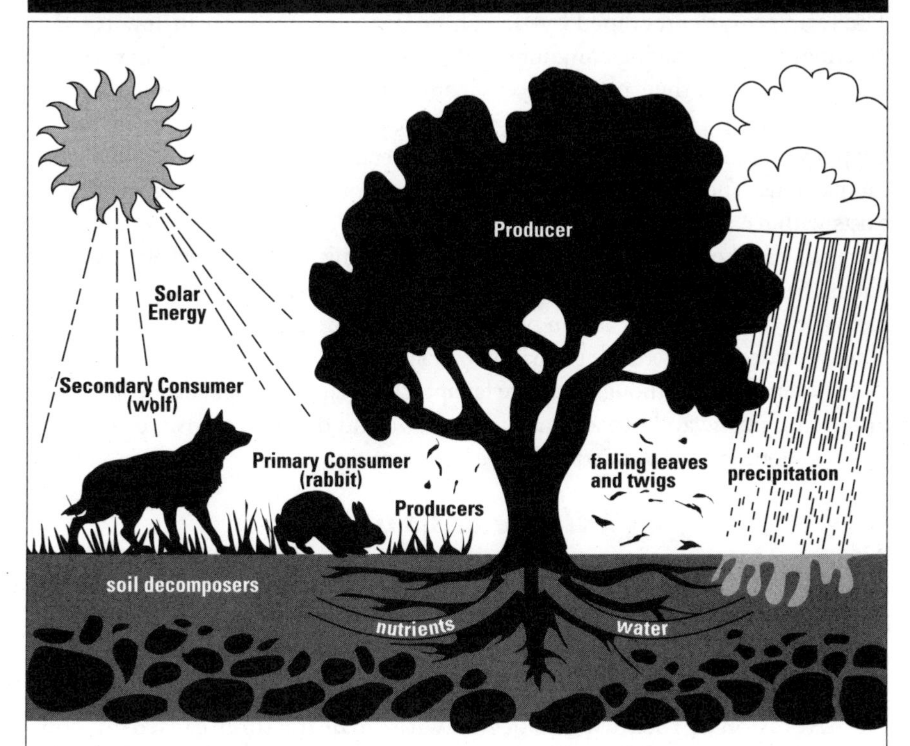

Producers (green plants) are eaten by a primary consumer (rabbit), which is then eaten by a secondary consumer (wolf), which, when it defecates or dies, passes on organic matter that is decomposed back into the soil along with fallen leaves from the tree under which the plants are growing.

At the bottom of the forest you find a subterranean layer of soil, bacteria, worms, insects, and moles interacting with the extensive root systems of herbs, shrubs, and trees. Next is the **floor** or ground layer of mosses, lichens, and grasses, which acts as an insulator, slowing up soil erosion, and stores and recycles foodstuffs such as fallen, decaying leaves, rotting tree stumps and logs, twigs, nuts, flowers, fallen fruit, and the feces and decomposing bodies of animals. The next layers — herbs, shrubs, and trees — are formed unevenly because they grow at different rates. They shelter and feed herbivorous insects, birds, mammals, and the predators that feed on them. At the top is the **canopy** forming the forest roof that modifies weather, light intensity, and other outside influences as well as providing habitat for plants and animals. The distribution of wildlife follows a similar pattern to that of plants. As you might expect, communities with the greatest variety of layers usually have the most species. They may be regularly spaced, clumped (e.g., seeds germinating under a parent plant), or, as is most common, distributed apparently at random.*

How Ecosystems Work

The term *ecosystem*, coined by British botanist Sir Arthur Tansley in 1935, describes a self-regulating community of plants and animals interacting with one another and with their nonliving environment.[2] Ecosystems are inhabited by microorganisms, plants, and animals, which are dependent on one another and on the physical world of terrain, light, shade, temperature, climate, water, and chemical materials. The quality, quantity, and variation of these factors determine what exists within an ecosystem and how it has been able to adapt. The bitter-cold arctic regions are hostile to all but the most hardy organisms such as lichens, dwarfed plants, well-insulated mammals such as caribou and polar bears, and year-round birds such as snowy owls and snow buntings. The intense heat and dryness of the Southwest deserts determine the nature of soil, plants, and animals that can survive there. Interestingly, although they are far apart geographically and exposed to very different climatic and other conditions, arctic and desert regions are arid, and plants in both regions must adapt to the lack of water.

Soil: The Basis of Life

Air, water, and soil are the main media that support life. Soil, in which all land vegetation grows, is the combined product of the weathering of the earth's crust (rocks and clay minerals) and the actions of microorganisms and plants. The creation of soil is an extremely slow process; it can take thousands of years to build up a few inches of topsoil.

Abrasion of rock usually begins with small fractures caused by constant heating and cooling. Smaller particles are created by the action of water, ice, and wind. For example, as rain falls through the air, it picks up carbon dioxide gas to form carbonic acid, which dissolves the surface of the

* Recent studies suggest there may be an underlying pattern in what we perceive as chaos.

rock. If the rock is chemically soft, the acid triggers the release of bicarbonate ions.* Some of these are washed into rivers and oceans, providing the material from which marine creatures form their shells. Other ions are attracted to stable soil particles and help retain calcium, potassium, and other important chemicals in the soil. Vegetation accelerates the process. Lichens gain a foothold on bare rock, creating small crevices in which plants can grow. As their roots die, bacteria and fungi release carbon dioxide and carbonic acid, further dissolving the rock and also adding nutrients to the soil.

Depending on the original rock type, the soil is formed with specific chemical and organic properties that determine which plant communities can become established there. These plant communities in turn begin modifying the soil as well as attracting a specific community of animals, which through burrowing, trampling, and defecating also influence its texture and composition. Through time these modifications in soil, plants, and animals lead to natural changes in the ecosystem. As we shall see in the following chapters, the interactions between the living and the physical worlds are important to the survival of organisms and the ecosystems of which they are part.

Dimensions of Space and Time

Ecosystems exist in space, with measurable dimensions of width, height, and depth. They also evolve over time, with discernible stages of development. Left undisturbed, a giant redwood forest, for example, will last thousands of years. On the other end of the scale, a microbial community may have a life span of less than twelve hours (one half-millionth that of the redwood forest). One can in certain circumstances observe the birth of a pond or the death of a forest.† Knowing the dimensions and characteristics of an ecological community is vital to understanding how it works and how it may react to increasingly changing conditions. For example, if you examine a bay only at its deepest point, you will find nothing about its shallow-water components. Similarly, a sample of a tundra region taken in winter is quite different from one done in summer, and in a young ecosystem samples taken ten to twenty years apart will show significantly different stages of development.

Responses to Stress

Ecosystems are so complex that it is sometimes difficult to understand how they remain stable. Yet most of them have the ability to adapt to moderate and even severe changes in environmental conditions. Hurricanes, volcanic eruptions, and other naturally occurring events often devastate vegetation and soil, but even after an apparent catastrophe, biological activity usually resumes within a season or two, sometimes sooner. And in some cases where flooding washes soil down into river valleys or deltas or where wind deposits soil on grasslands or prairies, fertility may actually be improved. It is interesting to

* Ions are electrically charged atoms or groups of atoms.

† See chapter 2, page 48, and chapter 6, pages 178–179, respectively.

note how the same kind of phenomenon that can devastate one type of community might benefit another. Fire is a case in point. In an eastern deciduous forest, fire burns trees, leaving in its path a field which over time is repopulated by shrubs and eventually trees. This series of changes from one community type to another is known as **succession.*** On the grasslands of the Midwest, fire prevents succession and maintains a stable prairie community by keeping trees from establishing themselves. In redwood forests, fire benefits the ecosystem by clearing the forest floor of **duff** (decomposing organic matter) and enabling the trees' seeds to reach the soil. It also recycles nutrients by turning debris on the ground into fertilizing ash. Fire suppression by humans over the last 150 years has interrupted these natural cycles, with negative effects on the ecosystems. Thus, depending on the system, an individual factor such as fire can drive succession, maintain community stability, or allow for regeneration of a community through time.† The response of ecosystems to different types of natural and human-induced stress is discussed at greater length throughout this book.

Major Land Ecosystems

The biosphere is a mosaic of countless ecosystems, which may be divided into two broad categories — aquatic or water ecosystems, consisting of oceans, rivers, streams, lakes, and ponds, and terrestrial or land ecosystems, consisting primarily of forests, grasslands, tundra, and deserts.

In this book we use the following biome classifications, which generally correspond with familiar geographical regions:

Tundra Stretches from northwest Alaska across the Northern Territories of Canada; Mountain and Alpine Tundra is located at high elevations in the western United States and Canada and in New England. These ecosystems will be covered in chapter 5.

Coniferous Forest Extends south of the tundra in a wide belt from New England across northern New York State and southern Canada through the Rocky Mountains to the Pacific Northwest. It also covers a large region south from Quebec through the Appalachians to Georgia and Alabama; see chapter 6.

Deciduous Forest Occupies most of the eastern United States with the exception of northern New England and New York, Appalachia, and southern Florida; see chapter 6.

Grassland Covers a huge area west of the deciduous forest that includes much of the Midwest, the Great Plains, and parts of north central and southwest Mexico; see Chapter 5.

Chaparral A relatively small but important biome located on the southern coast of California and in Arizona, New Mexico, and Nevada; see chapter 5.

Desert Located between the Rocky Mountains and the Sierra Nevada in

* This is a process in which plant and animal species in a given area are replaced over time by a series of different and usually more complex communities.

† For more on the ecological role of fire, see references in the index.

California, Oregon, Nevada, Utah, Arizona, New Mexico, Texas, and northern Mexico; see chapter 7.

How to Use This Book

This Land Is Your Land is divided into: Part 1: "The Water"; Part 2: "The Land"; Part 3: "The People"; and Resources; A/Directory of Organizations, B/Pending Legislation, Glossary, and Index. For ease of reference all chapters and appendices are coded on every page with the following symbols:

▲ Chapter 1: Rivers

▲ Chapter 2: Lakes, Ponds, Wetlands

▲ Chapter 3: The Edge of the Sea

▲ Chapter 4: Oceans

▲ Chapter 5: Grasslands, Chaparral, Tundra

▲ Chapter 6: Forests

▲ Chapter 7: Deserts

▲ Chapter 8: Public Lands

▲ Chapter 9: Laws of the Land

▲ Chapter 10: Restoring the Earth

▲ A: Directory of Organizations

▲ B: Pending Legislation

At the end of each chapter is a "Resources" section referring you to a network of agencies, organizations, groups, associations, research centers, books, periodicals, and other sources that will help you take action to save or restore endangered ecosystems. Immediately after "Resources" are the "Notes," which give specific references to the material provided in the chapter. To avoid duplication, addresses and telephone numbers of the main agencies and organizations* cited throughout the book are listed in the Directory, pages 331-355.

The "thumb-up" symbol directs you to the "What You Can Do" sections of *This Land Is Your Land*.

* Except for certain regional or local organizations whose coordinates are provided on the page where they are described.

RESOURCES

The following books were particularly helpful in the writing of this Introduction:

— *Basic Ecology*, Eugene P. Odum. Philadelphia: Saunders, 1983.
— *Communities and Ecosystems*, R. H. Whittaker. New York: Macmillan, 1975.
— *Concepts of Ecology*, Edward J. Kormondy. Englewood Cliffs, N.J.: Prentice-Hall, 1984.
— *Ecology*, Paul A. Colinvaux. New York: Wiley, 1986.
— *Ecology*, Charles J. Krebs. New York: Harper & Row, 1978.
— *Ecology and Field Biology*, Robert L. Smith. New York: Harper & Row, 1980.
— *Ecology of Man*, Robert L. Smith. New York: Harper & Row, 1980.
— *Ecology and Our Endangered Life-Support Systems*, Eugene P. Odum. Sunderland, Mass.: Sinauer, 1989.
— *Ecology of Natural Resources*, François Ramade. New York: Wiley, 1984.
— *The Economy of Nature*, Robert E. Ricklefs. New York: Chiron Press, 1986.
— *Elements of Ecology*, Robert L. Smith. New York: Harper & Row, 1985.
— *Energy and Ecology*, David M. Gates. Sunderland, Mass.: Sinauer, 1985.
— *Environment, Power and Society*, Howard T. Odum. New York: Wiley, 1971.
— *Evolutionary Biology*, D. J. Futuyma. Sunderland, Mass.: Sinauer, 1979.
— *Gaia: A New Look at Life*, James E. Lovelock. Oxford, England: Oxford University Press, 1979.
— *Geographical Ecology*, R. H. McArthur. New York: Harper & Row, 1972.
— *Global Ecology*, Charles Southwick. Sunderland, Mass: Sinauer, 1985.
— *Global Ecology: Towards a Science of the Biosphere*, Mitchell B. Rambler, Lynn Margulis, and René Fester, editors. New York: Academic Press, 1989.
— *Living in the Environment*, G. Tyler Miller, Jr. Belmont, Calif.: Wadsworth, 1989.
— *The Living Planet*, David Attenborough. Boston: Little, Brown, 1984.
— *The Machinery of Life: The Living World Around Us and How It Works*, Paul R. Ehrlich. New York: Simon & Schuster, 1986.
— *Man and Earth's Ecosystems*, Charles Bennett. New York: Wiley, 1975.
— *Man and Nature*, George Marsh. New York: Scribner's, 1864.
— *Man's Responsibility for Nature*, John Passmore. New York: Scribner's, 1974.
— *Natural Ecosystems*, W. B. Clapham, Jr. New York: Macmillan, 1984.
— *A Sand County Almanac*, Aldo Leopold. New York: Oxford University Press, 1987.
— *Stress Effects on Natural Ecosystems*, G. W. Barrett and R. Rosenberg, editors. New York: Wiley, 1981.
— *The Web of Life*, John H. Storer. New York: Mentor, 1953.
— *Why Big Fierce Animals Are Rare*, Paul A. Colinvaux. Princeton, N.J.: Princeton University Press, 1978.

NOTES

1. For the organization of the Introduction we are indebted to *This Land's* scientific consultant, Dr. Richard A. Orson, botanist, ecologist, and geomorphologist, assistant professor of environmental sciences at William Paterson College and visiting research scholar with Rutgers University Institute of Marine and Coastal Sciences. We also greatly appreciate the input of Dr. Paul G. (Eric) Smith, ecologist, educator, and Morris dancer.

2. "... the whole *system* (in the sense of physics) including not only the organism-complex, but also the whole complex of physical factors forming what we call the environment of the biome. We cannot separate them (the organisms) from their special environment system with which they form one physical system... . It is the system so formed which [provides] the basic units of nature on the face of the earth... . These *ecosystems* ... are of the most various kinds and sizes." Cited in Robert E. Ricklefs, *The Economy of Nature* (New York: Chiron Press, 1983), p. 13.

Part 1
THE WATER

"All is born of water; all is sustained by water." — Goethe

From the perspective of space, entire continents are mere islands in a vast interconnected body of water. A vital component of all ecosystems, both aquatic and terrestrial, water makes possible the existence of all living organisms. Oceans, and to a lesser extent rivers and lakes, help regulate the earth's atmospheric chemistry, climate, and temperature. Every body of water, even the smallest rainwater puddle, is either an ecosystem itself and/or part of a larger system, providing habitats for countless plants and animals. Aquatic ecosystems are classified broadly as **freshwater** and **marine** or **saltwater**. Freshwater systems are subdivided into **running water systems** (rivers, streams, brooks, rivulets, springs), covered in chapter 1, and **standing water systems** (lakes, ponds, and freshwater wetlands), covered in chapter 2. Marine ecosystems are subdivided into the **coastal/estuarine zone** (estuaries, coastal wetlands, the beach or shore, embayments, intertidal pools, the offshore zone, barrier islands), covered in chapter 3, and the **open sea**, covered in chapter 4.

1
RIVERS

"Rivers are a natural and free-flowing part of the environment, the heart of many ecosystems, they are the arteries and the veins."

— Kevin Coyle, president, American Rivers

Springs, rivulets, brooks, creeks, streams, and rivers are characterized by continuously running water that flows by gravity downhill. Streams may begin as runoff from springs and seepage areas or as outlets of ponds and lakes. As water drains from its source, it flows on a course determined by the lay of the land and the underlying rock foundations. It concentrates in rills that grow into gullies and then forms a stream, which usually feeds with other streams into a river. As it moves downstream, water carries a load of debris that is eventually deposited within or alongside the stream as sand, silt, or mud. In times of heavy rain or melting snow, the water often overflows the river's banks, depositing the debris onto level land and forming a floodplain. When a river flows into a lake or the sea, its velocity is checked and its sediment is deposited at the point of inlet, forming an estuary or a delta, marked by numerous channels, small lakes, marshes, and swampy islands.

Going to the Source

Rivers get their water directly from rain and snow, runoff from precipitation that falls on land too saturated to absorb it, meltdown from glaciers, and via springs or wells from aquifers. These are underground reservoirs providing some 30 percent of the volume of inland waters; in the Mississippi River aquifers provide an estimated half of its flow. Lakes and ponds may also serve as sources of streams and rivers because they are often connected. A river system is composed of tributary streams that converge to form the main trunk. Each stream has its own drainage basin or watershed that funnels its runoff and groundwater downhill into the stream channel. Each watershed is bounded by a ridge or divide. Like their stream counterparts, river tributaries join with those of equal or larger size to form fewer, longer, deeper, wider systems. The Mississippi, for example, is the result of ten stages of tributary convergence.

Rivers shape the terrain through which they flow, carving out valleys, carrying rocks and soil downstream, constantly reworking their own channels as they become clogged with sediment. The Mississippi moves some 450 million tons of silt, clay, and sand annually, enough to push its delta 6 miles seaward every hundred years.* Running water reacts chemically with the rocks it touches, dissolving and carrying off a solution of mineral salts in the form of ions — electrically charged parts of a molecule. These contain calcium, phosphorus, iron, and other nutrients that are vital to plant and animal life along the river and in adjoining ecosystems.

As might be expected, the biological and chemical makeup of running- and standing-water environments are quite different. In the upper reaches of running waters you find organisms that can adhere to rocks and other exposed surfaces: algae; larvae of blackflies, midges, and other insects; and planaria (flatworms). Downstream, floating vegetation appears along with clams, burrowing mayfly nymphs, and other invertebrates. Still farther downstream one comes across crayfish, darters, salmon, and other cold-water fish, then catfish, carp, and other warm-water species. The upper reaches of running-water

* For a report on the Mississippi Delta, see chapter 3, pages 103-105.

ecosystems are rich in oxygen, which decreases as the water flows more sluggishly downstream. Nutrient levels are usually higher downstream because of the steady addition of organic material en route. In small streams nutrients come mainly from trees, plants, and other habitats along the banks.

As a river moves downstream, it often overflows its banks, leaving a wide, flat belt of alluvium (deposited sediment), which forms the basis of floodplains. These vary in size depending on the extent of the flooding. One of the largest, stretching from the confluence of the Ohio and Mississippi rivers at Caro, Illinois, all the way south to New Orleans — a distance of roughly 500 miles — and measuring 125 miles across at its widest point, embraces 30,000 square miles of silt-rich farmland. Floodplains are ideal for wildlife, trees, and certain kinds of farming. They act as temporary storage reservoirs, lowering downstream floods and allowing water to soak through the soil to underground streams that seep slowly back to the river. When people build houses along a river, they compete directly with its natural flow and sooner or later become the victims of high water and flooding. When they plow fields or strip-mine along riverbanks, the soil erodes.

Living with the River

Communities of plants and animals depend on river systems. Floodplain forests need the water and alluvial soil. Silver maple in the East, cottonwood on the Plains, cedar in the Northwest, and cypress in the South grow at the river's edge, as do sycamores and hemlocks. Most other trees die if their roots are soaked in water for more than a few days. Grizzly bears eat salmon and raccoons eat crayfish. Before the nineteenth century otters lived in rivers throughout most of this country, but trapping, shooting, channelization, and dams have driven them away from their original habitats. More than half of North American bird species nest on or near waterways, including thirty-five species of American duck, heron, kingfisher, yellowleg, sandpiper, osprey, and water ouzel.

Ponds, lakes, streams, and rivers teem with algae and thousands of species of insects that feed salmon, trout, pike, and muskellunge in the cold waters of the northern states and perch, bass, sunfish, carp, and catfish in the warmer southern waters. The speed at which a stream or river flows has a critical influence on the wildlife found there and the survival strategies they have evolved. As a river tumbles down from a mountain, the force of its water pouring over the rocks and pebbles leaves sediment in the riverbed. This is where you find salmon, who are strong enough to swim against the current, and creatures that cling to rocks. In the lower, more sluggish reaches of a river, sediment carried down from the mountains makes a nutrient-rich bed in which plants can flourish. Here the fish are more numerous, as are the birds and animals that prey on them.

Changing the River Systems

Rivers are the pulse of many ecosystems. The free-flowing character of a river tells us that the ecosystem is healthy, providing vital habitat for innu-

merable plant and animal species and helping support an intricate network of other interrelated systems touched by the river. It is sad to relate that most rivers in North America suffer from pollution, uncontrolled development, channelization, and other ailments. Of 3.5 million miles of rivers in the U.S., 600,000 miles (17 percent) are dammed, causing enormous and permanent ecological damage. Many others are sucked dry by diversions, changed hydrologically by channelization, and contaminated by uncontrolled development. For generations the human use of flowing water seemed useful and good. It helped settle the West and turn the California drylands into a farm region supplying almost one quarter of the nation's food. The Ohio River serves more commerce than the Panama Canal. In the U.S. 26,000 miles of waterways are channeled for shipping, 60 million acres are irrigated, and hydroelectric dams generate more power than do all the nuclear plants. These huge water projects have become an important feature of our landscape and our way of life. But we are now beginning to see that this "progress" has been bought at a painful environmental price.

Dammed If You Do

Fifty thousand dams straddle the rivers of North America — sixty on the Missouri River alone and twenty-five on the Tennessee. They have achieved such a transformation of the landscape that hydrologists no longer call them rivers but "regulated waterways." Los Angeles, Phoenix, and many other cities could not exist without dams. The far-reaching ecological consequences of these projects are extremely serious. To make way for reservoirs, entire communities of people, animals, and plants must be uprooted or destroyed. The creation of a reservoir radically changes the river's ecosystem. Fish that depend on free-flowing water are replaced by species adapted to the environment of a lake. Although initially these may thrive on nutrients released from submerged soils, the inundated vegetation begins to rot, using up the oxygen in the water. The effect of this process in Glen Canyon in Utah was described by Edward Abbey in his classic *Monkey Wrench Gang*: "Instead of a river he looked down on a motionless body of murky green effluent, dead, stagnant, dull, a scum of oil floating on the surface. On the canyon walls a coating of dried silt and mineral salts, like a bathtub ring, recorded high-water mark."[1]

> *"Rivers are vital to preserving the nation's capacity for wild genetic exchange. When we segment a river by placing dams in it, we cut off important biological exchange. We disassociate habitats from one another."*
> — W. Kent Olson, former president, American Rivers

In such a stinking morass few fish can survive the invasive waterweeds that block out the sunlight. For migratory fish such as salmon and sturgeon, dams are disastrous, grinding up huge quantities of young fish in their turbines.* Another effect of dams is to trap silt, which reduces the capacity of the reservoir behind the dam, often leading to its premature closing.[2]

* For a report on endangered salmon in the Columbia and Snake river system, see pages 20-22.

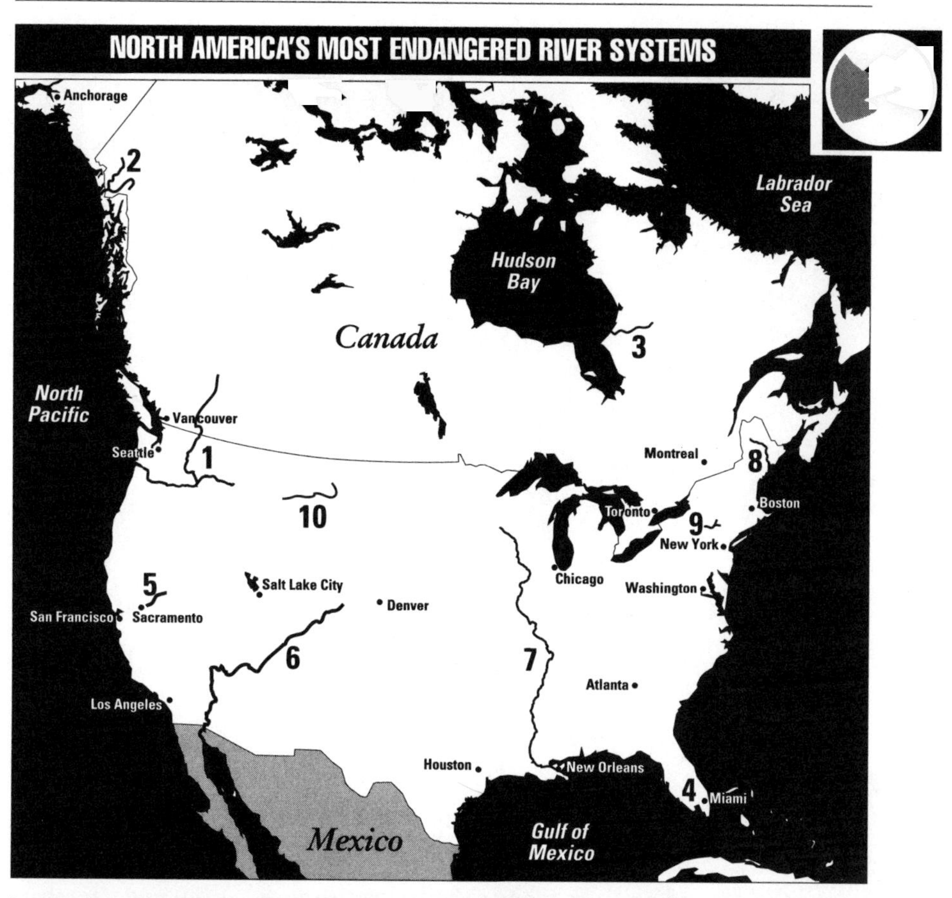

In 1992 American Rivers, North America's leading river conservation organization, listed the following river systems as most endangered in North America:

1. **The Columbia and Snake river system (Pacific Northwest)**
2. **The Alsek and Tatshenshini (British Columbia and Alaska)**
3. **The Great Whale (Quebec)***
4. **The Everglades (Florida)†**
5. **The American (California)**
6. **The Colorado (Colorado, Arizona, California, Baja California)**
7. **The Mississippi (Louisiana, Mississippi, Tennessee, Arkansas, Missouri, Illinois)‡**
8. **The Penobscot (Maine)**
9. **The Beaverkill and Willowmoc river system (New York)**
10. **The Blackfoot (Montana)**

"Rivers today in North America are in worse shape than they've ever been," said Kevin Coyle, announcing the 1992 most endangered list. In addition to being threatened by dams, they are devastated by daily runoff from agriculture, cities and new development, mining, mining wastes and logging-caused erosion.[3]

* For a report on the Great Whale River, see chapter 5, pages 166-169.
† For a report on the Everglades, see chapter 2, pages 64-70.
‡ For a report on the Mississippi River and the Mississippi Delta, see chapter 3, pages 103-105.

Threatened Rivers

In addition to the endangered rivers, American Rivers listed 15 more waterways as highly threatened:

River / State	River / State
Animas / Colorado	Ouachita River / Arkansas & Louisiana
Clavey / California	Passaic / New Jersey
Elwha / Washington	Rio Conchos & Rio Grande / Chihuaha, Mexico & Texas
Gunnison / Colorado	Savannah / Georgia
Illinois / Oregon	Susquehannah / Pennsylvania
Klamath / Oregon	Verde / Arizona
New / North Carolina	Virgin / Arizona, Nevada, & Utah
Ohio River / Ohio, Pennsylvania, & West Virginia	

The Columbia and Snake River System

Draining a region the size of Texas, the Columbia River rolls for 1,243 miles from its headwaters in Columbia Lake, British Columbia, and empties more water into the Pacific than any other river in the Western Hemisphere. Its waters and those of the Snake River, its largest tributary, support a variety of uses — irrigation, hydroelectric power generation, nuclear reactor cooling, recreation, and fisheries. Yet despite its size, the Columbia contains too little water to satisfy all these needs at the same time. In the resulting competition it is the fish and those who depend on them who are suffering the most serious and damaging losses. Traditionally the Columbia's *anadromous** fish (particularly salmon and steelhead trout) have enjoyed one of the largest runs in the world. Today they are totally extinct upriver in British Columbia and Nevada and reduced to "remnant population" in Idaho and parts of Oregon and Washington. Their precarious condition has prompted the sockeye and other strains of Pacific salmon to be considered for listing under the Endangered Species Act, triggering a process that "could affect electricity rates down the West Coast, economic prosperity in Seattle and Portland and growers of some of the nation's most bountiful crops." [4]

Once the passageway for more than fifteen million salmon a year, the Columbia River is backed up by sixteen dams, which have transformed the wild, rushing waterway into a series of sluggish lakes in order to provide the cheapest electricity in the U.S.

The massive dam walls impede or completely block adult salmon swimming upstream, and reservoirs behind them have drowned thousands of acres of fish spawning and nursery habitat. But it is the young salmon, or smolts, heading downstream who suffer most severely; up to 95 percent of them are killed each year before they reach the sea.

Some smolts are simply ground up as they pass through the turbines. Others die because the dams have changed the very nature of the river

* Anadromous fish are marine species that migrate upriver to spawn and breed.

ecosystem — temperature, timing, and velocity — to which salmon are biologically adapted. Cool, swift waters have become warm and sluggish. Without a strong current to flush them downstream, smolts lose their bearings, are exposed to predators and higher temperatures that encourage disease. Delayed too long, the smolts' biological timeclocks run out; they lose the urge to migrate and never reach the sea. In 1990 the American Fisheries Society (AFS) found 214 separate stocks (runs) of salmon to be endangered, threatened, or of special concern.[5]

Efforts to restore the beleaguered salmon have been underway for more than a decade. But despite a billion dollars worth of fish ladders, turbine screens, and programs to truck fish around the dams, salmon populations have continued to plunge. To compensate for the declining populations, many salmon are now bred in hatcheries, but they are found to be genetically inferior. They are much more prone to disease than the wild salmon, which, some studies show, are nine times as likely as hatchery fish to survive and return to the Columbia or Snake rivers for spawning.[6]

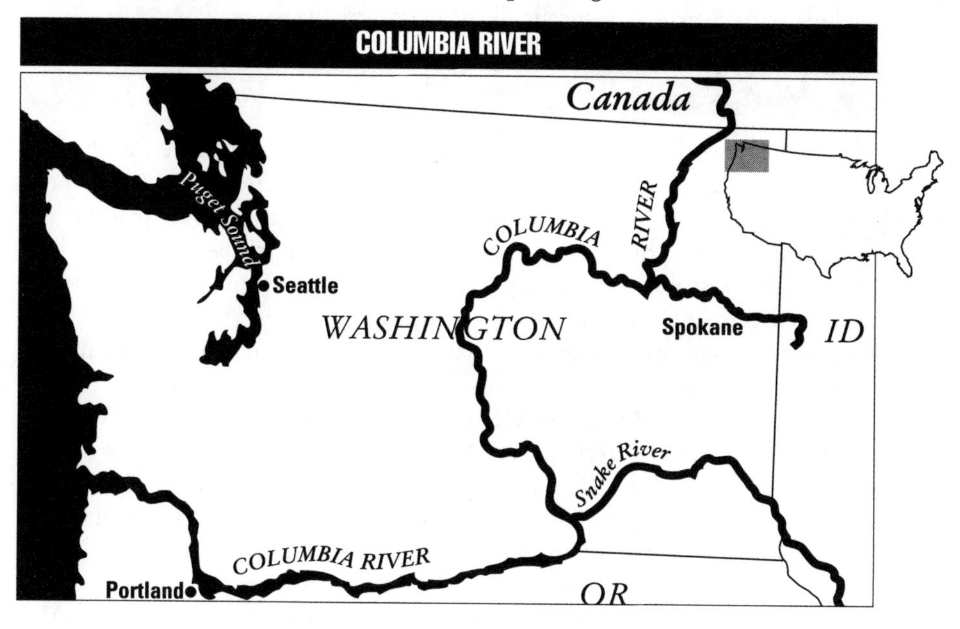

Saving the Salmon

With the Northwest salmon on the verge of extinction, efforts have been initiated to protect the fish and restore the ecosystems of the Columbia and Snake rivers. In November 1991, at the urging of American Rivers, Oregon Trout, the Oregon Natural Resources Council, and other organizations, the National Marine Fisheries Service (NMFS) listed the Snake River sockeye as endangered under the federal Endangered Species Act and proposed listing certain Snake River runs as threatened. The listings pave the way for the development of a comprehensive salmon recovery plan in the Pacific Northwest that might become a model for a national program. This one has designated four fish species — the Snake River spring, summer, and fall chinook salmon and the lower Columbia coho salmon — as threatened or endangered species.

Although the verdict on recovery programs is still unresolved, the consensus among conservationists and ecologisists is that to be successful these programs must emphasize not only safe passage around dams, but also habitat restoration and management of the Columbia more closely to resemble a natural river system. This means, in part, releasing more water downstream, especially during spring and early summer when salmon need it most, rather than in winter when cities demand more electricity.

Unlike the spotted owl controversy,* which focuses on a single industry, the salmon dispute affects many interests. The Columbia and Snake dams supply half of the Pacific Northwest's electric power. Water from the Columbia irrigates more than 3 million acres of farmland. The rivers are a major regional waterway and recreational area. And despite the wild salmon's decline, commercial fishing is still an important part of the region's economy. Once a species is designated federally as endangered, it must be protected without regard to economic consequences.† Should this happen to the salmon in the Pacific Northwest, utility companies, farming and fishing interests, recreationists, and other affected groups might have to change dramatically the way they use the rivers.

What You Can Do to Protect the Columbia/Snake River System

1. Support the efforts of the Washington, D.C.-based American Rivers, which recently opened a Northwest regional office in Seattle to help salmon recovery efforts through intervention in hydroelectric licensing, taking part in other administrative and judicial proceedings, and advocacy in river restoration and management. (For details on American Rivers, see the Directory, page 334.)

2. Contact other citizens' groups that have joined American Rivers in this effort. They include:

— Oregon Natural Resources Council, Yeon Building 1050, 522 SW Fifth Avenue, Portland, OR 97204; (503) 223-9001;

— Oregon Trout, P.O. Box 19540, Portland, OR 97219; (503) 244-2292; American Fisheries Society, 5410 Grosvenor Lane, Suite 110, Bethesda, MD 20814; (301) 897-8616; Idaho chapter (208) 743-6502; Oregon chapter (503) 628-1882.

3. Find out about the work of the following groups as it relates to the Columbia River ecosystem:

— Northwest Renewable Resources Center, 1133 Dexter Horton Building, 710 Second Avenue, Seattle, WA 98104; (206) 623-7361;

— Oregon State Public Interest Research Group, 027 SW Arthur, Portland, OR 97201; (503) 222-9641.

4. Read *Audubon, Sierra, Wild Earth,* and other magazines that discuss the relationship between endangered species and endangered ecosystems such as the Columbia River.

* See the index for further references.

† For more on the designation of endangered species, see chapter 9, pages 286-291.

Alsek-Tatshenshini[7]

The Tatshenshini, or the Tat for short, is North America's wildest river. A major tributary of the Alsek River, it is located in the extreme northwest corner of British Columbia, 1,000 air miles north of Vancouver. The Tat travels 160 miles from the subarctic tundra of the Yukon, through British Columbia's highest ranges, the Fairweather and Saint Elias Mountains, to its mouth on the Gulf of Alaska in the United States. The region contains the world's largest nonpolar ice fields and countless glaciers, some of which are up to 10 miles wide and shear off as icebergs as they encounter the river's edge. A remnant from the geologic era when mountains were reemerging from the Ice Age, the Tat is the habitat of grizzlies, wolves, Dall sheep, mountain goats, moose, and eagles. The river supports major salmon runs of chinook, sockeye, coho, pink, and chub, along with steelhead and rainbow trout. Never roaded, logged, mined, or settled in all its public tenure, its 2.8 million acres of wilderness are without question British Columbia's last frontier. The Alsek-Tatshenshini is a river system that knows no boundaries. In 1992 it was listed by American Rivers as tied for second (with the Great Whale River) of its top ten most endangered rivers in North America.

You may find it hard to believe, but the Tatshenshini is so remote that the first recorded descent by nonindigenous people was in 1972. Since then its spectacular reputation has spread rapidly. The Tat is now rated by the Sierra Club and other organizations as one of the world's wildest rivers, in a class with the Amazon, the Zambezi, and the Colorado. The Tatshenshini watershed encompasses six biogeoclimatic zones: a unique coastal forest system, boreal shrubland, tundra,* alpine tundra, ice, and rock. It also contains the only hedysarum meadows in Canada. Except for one glaring omission, the Tat is relatively well protected. In 1980 its lower reaches were secured by the U.S. government in an expansion of Glacier National Park. Its headwaters in the Yukon are similarly protected in Canada's Kluane National Park. Adjoining the Tatshenshini region to the northwest, the Alaskan territory is preserved as the Wrangell-St. Elias National Park. Only in British Columbia is the Tat unprotected, despite the fact that it is the heartland of a proposed transboundary park complex and provides the sole accessible corridor through the lofty coastal mountains.

Windy Craggy Threat

The pristine Tatshenshini environment is threatened by a proposal to dig two open pits — one of them a mile long, a third of a mile wide, and a thousand feet deep, the other underneath a glacier — on top of Windy Craggy Mountain near the point where the Alsek and the Tatshenshini rivers come together. The Toronto-based Geddes Resources company wants to remove more than 100 million tons of ore containing copper, gold, silver, and cobalt. In a region previously entered only by water or air, a 70-mile access road would cut right into

* For more on the tundra biomes, see chapter 5.

the heart of the wilderness, causing direct habitat loss for grizzly bear, Dall sheep, and juvenile salmon in the river's side channels and tributaries. It would require building eleven bridges, one of them 700 feet long, to span the Tat at its confluence with the O'Connor River. To haul away the ore, 150 giant trucks would drive the road every day starting in 1994 (when the mine is scheduled to open) and continuing for forty or fifty years. On their way to the deep-water port at Haines, Alaska, the trucks would carry toxic dust through the Chilkat Eagle Preserve, where each year 3,500 bald eagles — believed to be the largest winter population of these birds in North America — gather for the late-fall salmon run. With so many vehicles on the road, collisions and animal deaths would be inevitable and the increased access to humans would lead to poaching and disturbance of lambing grounds and migration routes. The presence of a mining camp for six hundred workers and their followers would lead to uncontrolled use of the fragile Glacier Bay backcountry and to dramatic declines in the now undisturbed wildlife populations.

As American Rivers and other conservation groups have warned, the impact of the mining on the Tatshenshini-Alsek ecosystem would be enormous. The copper ore in question has a sulphur content of 35 percent (*six* times greater than that of other British Columbia mines), which, when exposed to the atmosphere, would generate sulfuric acid *as strong as battery acid*. If acid from the associated 200 million tons of waste rock and tailings leaches into streams and rivers, it could literally wipe out the salmon runs, destroying the multimillion-dollar Alaskan commercial fishery downriver and the native-food fishery in the Yukon, as well as the grizzly and eagle populations that depend on the fish in British Columbia and in Glacier Bay National Park. Geddes's plan to store the waste rock underwater behind a 360-foot dam is extremely hazardous, because the site is in Canada's highest-risk earthquake zone, where a major (6.2 on the Richter scale) quake hit in the spring of 1990. Because acid hazards of this type can endure for a thousand years or more, the Tat wilderness would be endangered for centuries after the mine has been abandoned.

Long used by indigenous people to travel from the coast to the interior, the Tat is also their primary salmon fishery and source of food. Based on this traditional use, the Champaigne-Aishihik band has filed a land claim to the area. The 1,200 inhabitants of Haines would also be affected by air and noise pollution, sharply increased traffic, and other impacts from the mining. In a poll conducted by the Borough of Haines in the summer of 1990, 60 percent of the residents opposed the overall project and 69 percent voted specifically against the traffic. In response the president of Geddes warned a community meeting, "A popular vote of the people of Haines, Alaska, and their local government against the mine will not stop us." [8]

The Geddes Maneuver

In the spring of 1990 the Geddes proposal was rejected by the Canadian government on the basis of the acid waste hazard. Nevertheless, the company has pressed ahead with an intensive public relations campaign downplaying the risks. Consistently trying to avoid an environmental impact assessment,

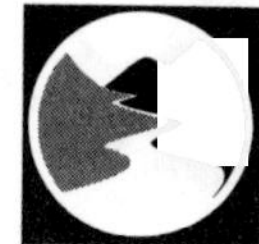

Geddes is moving ahead undaunted. It is widely suspected that the Toronto company plans to get everything in place and then sell out to a larger mining corporation. One scenario is envisaged by American Wildlands: "Since the remote Geddes deposit may be uneconomical to operate, the tragic prospect is that the road could be built into the Tatshenshini wilderness even if the mine never opens. And if approved, the possibility of the project proceeding at any time in the future could loom over this wilderness for a long, long time."[9] In December 1992 Glacier Bay Park and National Preserve, 25 miles downstream from the proposed mine, was declared a World Heritage site by UNESCO, adding great pressure to halt the Windy Craggy project.

What You Can Do to Save the Tat

1. Support the Save Tat Campaign to preserve the Tat wilderness. Spearheaded by Tatshenshini Wild! in Vancouver, British Columbia, (604) 886-8605, this is one of the largest and broadest-based Canadian-American coalitions, involving twenty-six environmental organizations in both countries and representing 2 million people. The campaign includes American Wildlands, American Rivers, the National Audubon Society, the Sierra Club, the Wilderness Society, the Canadian Nature Federation, the Canadian Parks and Wilderness Society, the British Columbia Wildlife Federation, Friends of Yukon Rivers, and the Yukon Conservation Society.
2. Urge your congressional or Canadian elected representatives to press the Canadian government to have comprehensive environmental impact assessments carried out in the region.
3. Write to Dan Robison, U.S. EPA, Box 19, 222 West Seventh Street, Anchorage, AK 99513, expressing your desire to save this important ecosystem.
4. Write to the United Nations Environmental Program, 2 United Nations Plaza, New York, NY, 10017, asking for information about international river protection including that of the Alsek-Tatshenshini.
5. For more information, contact the Alaska Conservation Council, P.O. Box 021692, Juneau, AK 99802; (907) 586-6942; and Lynn Canal Association, P.O. Box 964, Haines, AK 99827; (907) 766-2240.

The Colorado River

The Colorado is neither the biggest, longest, nor, despite its cutting through the Grand Canyon, the most scenic of American rivers. But it has more people, more industry, and a more significant economy dependent on it than any river in the world. Its river system provides over half the water of greater Los Angeles, San Diego, and Phoenix; it grows much of America's domestic production of fresh winter vegetables and generates electricity for Las Vegas and other cities on its way from the Continental Divide to the Gulf of California in Mexico. As *Time* magazine wrote in a cover story, "Life in much of the American West would be unimaginable in its present form without the Colorado."[10] It made the cover because it had just been awarded the dubious distinction of being named

most endangered river in the United States by American Rivers, the nation's leading river-conservation organization. In terms of annual flow, the Colorado doesn't rank among the top twenty-five in the United States, but in its natural state it was "tempestuous, wilful, headstrong."[11] It is also extremely silty, draining a vast watershed in the Rocky Mountains and producing a sediment volume* almost as great as that of the Mississippi, which carries twice as much water. For these reasons and its enormous potential for irrigation and electricity generation, the Colorado had to be tamed by those who believed a wild river was a wasted river.

> ***"If the Colorado River suddenly stopped flowing, you would have four years of carryover capacity in its reservoirs before you had to evacuate most of southern California and Arizona and a good portion of Colorado, New Mexico, Utah, and Wyoming."***
> — Marc Reisner, *Cadillac Desert*[12]

The Colorado's first diversion channel was cut in 1901 to irrigate a vast, flat region in the Sonoran Desert of southern California and Arizona near the Mexican border then known as the Valley of the Dead and renamed Imperial Valley. Within eight months there were two towns, two thousand settlers, and a hundred thousand acres ready for harvesting. By 1904 the

COLORADO RIVER BASIN

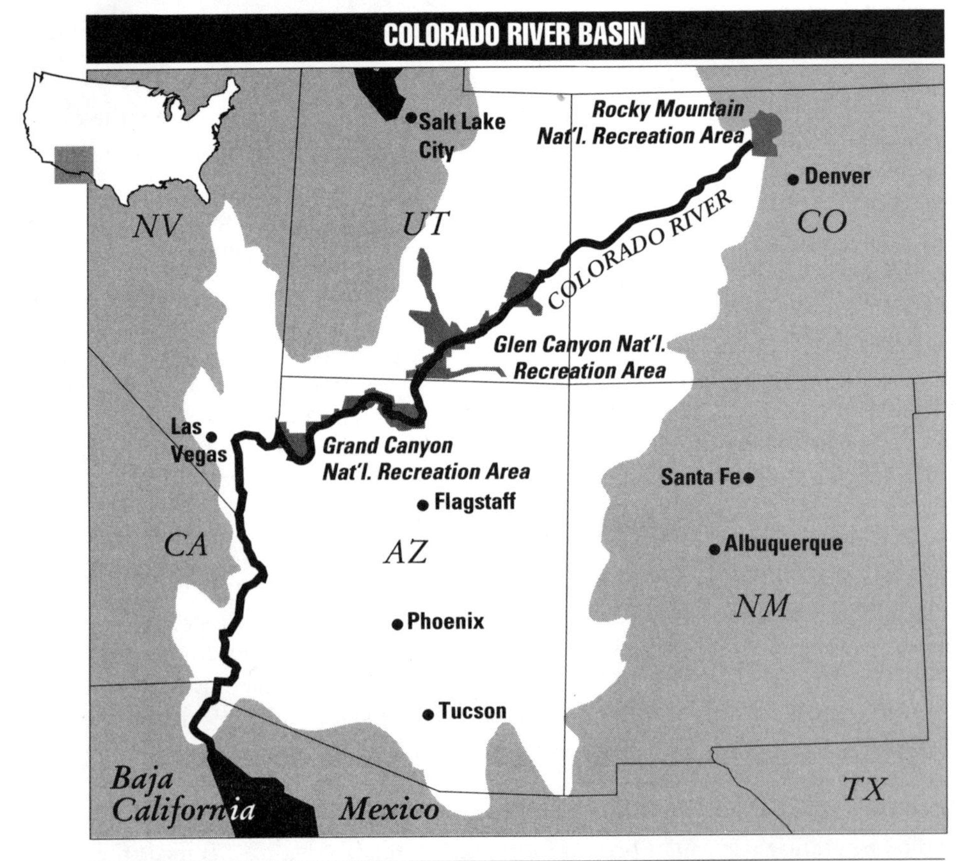

* The name *Colorado* is believed to have come from the river's original red color, from silt.

artificial channel had silted up and a bypass had to be cut. And, as Reisner writes, "It too silted up. Another bypass was cut; it too silted up." [13] Finally, as a temporary expedient, the developers persuaded the Mexican government to cut yet another channel south of the border. In February 1905 the spring floods arrived early, and by summer the Colorado River spilled out of its main channel, washing away the newly created fields, farms, and homes as it found its "phantom" channel, the Alamo River, for the first time in centuries. The only way to control the flooding and the countervailing tendency of the river to dry up was to create a huge dam. Thus was born the project that culminated in the building of Boulder Dam in 1936, the greatest structure the United States had known. Later renamed Hoover Dam, it was the prototype of a multiple-use project, incorporating irrigation, hydroelectricity (nearly one-third of it for pumping water to California), flood control, and urban and industrial development. Lake Mead, its reservoir, can hold the entire flow of the river for two years and is large enough to flood an area equivalent to the state of Pennsylvania 1 foot deep.

The Glen Canyon Story [14]

In 1963 the 638-foot-high Glen Canyon Dam, 16 miles upstream from the Grand Canyon, became the symbol of wild-river and wilderness devastation when it drowned what many people considered the most sublime of all desert canyons. Today its operations drastically affect the Colorado River and the Grand Canyon. The wildly fluctuating water released by the dam as the result of "peaking" operations to enhance hydroelectric power output can change the river's flow below the dam from 3,000 to 30,000 cubic feet per second and cause its level to rise and fall 13 feet in a single day. Another effect of the dam is to trap the river's seasonal runoff of silt in Lake Powell, an artificially created "bathtub" 1,900 miles around, capable of holding a two-year flow of water and visited by more than three million tourists every year. With the silt blocked off, the clear water scours the delicate sandy beaches it once replenished and that form the base of the Grand Canyon's ecosystem. An estimated 16 million tons were eroded in this way between 1983 and 1986. The existence of the dam has lowered the temperature of the river in the Grand Canyon. The release of colder water from Lake Powell has made the Colorado suitable for exotic fish such as rainbow trout but not for several native fish including humpback chub (an endangered species), razorback sucker, and Colorado squawfish. Fish habitats are alternately inundated or exposed by the sharply fluctuating river levels, leaving stranded aquatic creatures to die by the thousands. Beaver disappeared because entrances to their underwater homes in the riverbanks were regularly exposed as water levels rose and fell. The dam's operation also threatens a prized trout fishery and important Hopi and Anasazi archaeological sites.

Archaeological Threats

"In addition to providing significant wildlife habitat and recreational opportunities, the Grand Canyon contains important archaeological sites and shrines. According to Hopi cultural leader Leigh Jenkins, water flowing through the Glen Canyon Dam threatens eight hundred sites, including the Shrine of Emergence, where Hopis believe life began." — American Rivers

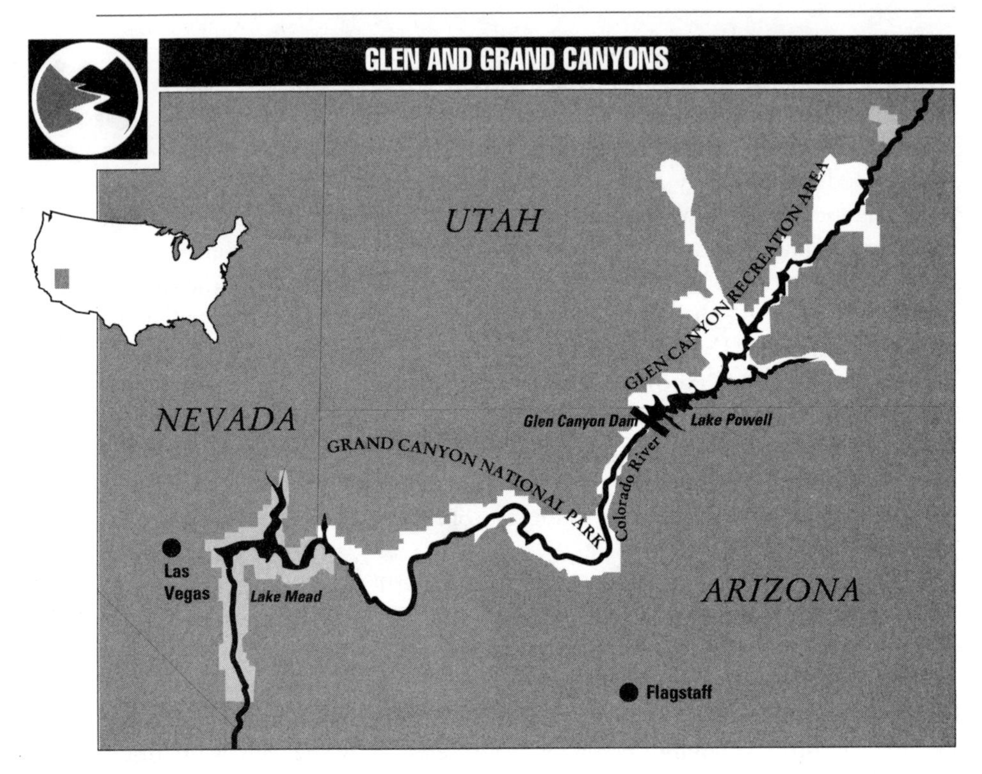

Managing Glen Canyon Dam for power production at the expense of other resources violates the 1968 Colorado River Basin Act (P.L. 90-537), authorizing the generation and sale of electrical power as the "incident of the foregoing purposes" — that is, providing basic outdoor recreation and improving conditions for fish and wildlife, among other priorities. Proposed legislation to improve the operation of Glen Canyon Dam and to protect and enhance downstream environmental, recreational, and cultural resources of the Colorado River in Grand Canyon National Park and the Glen Canyon National Recreation Area has been introduced in the U.S. Congress.* Supported by conservationists, the bills are opposed by a powerful combination of special interests that include western farmers, power companies, and real estate developers. However, as American Rivers asks, "If we are not able to protect the Grand Canyon, a world treasure, how can we hope to protect other wilderness areas?" A legislative mandate to protect the Grand Canyon would set a precedent to modify other environmentally unsound federal dam operations in the entire Colorado River system and throughout the nation.

Drought and Deficit

One key to the river's future may be the revision of a 1922 agreement known as the Colorado River Compact, which divided up the water among upper-basin states — Colorado, New Mexico, Utah, and Wyoming — and lower-basin states — Nevada, Arizona, and California. The fight between the interests of these states and those of other states and Mexico has intensified because of the

* For an update on pending legislation, see Appendix B, pages 359-364.

five-year drought in the West. Until recently the upper-basin states didn't use all the water allotted to them, and California got its water by siphoning off their unused portion and by encouraging conservation at home. But in 1990 Arizona began taking much of its share for the burgeoning populations of Phoenix and Tucson. If the compact is revised, the question will be who gets how much of the Colorado's limited water.

Complicating the dispute is the changing attitude in Washington and the eastern states toward western water. The Bureau of Land Management (BLM), the agency primarily responsible for the damming of the Colorado, is funded by appropriations from a Congress increasingly opposed to the billions of dollars of subsidies that make the river's water available at extremely low prices to western farmers. In the words of California representative George Miller, sponsor of the Grand Canyon bill in the House, "The drought and the deficit have caused people from Pennsylvania, Massachusetts and New York to reassess supporting a bad habit."[15]

What You Can Do to Protect the Grand Canyon and the Colorado River

1. Let the U.S. President know that saving the Grand Canyon and the Colorado River is a major national priority and not just the opportunity for an election campaign "sound bite."
2. Support congressional initiatives to protect the Grand Canyon and improve the operation of Glen Canyon Dam, including support for the Grand Canyon Protection Act.
3. Lobby the secretary of the interior to the same effect.
4. Join and support American Rivers, the Grand Canyon Trust, and other conservation groups in their efforts to protect the Colorado and other threatened river ecosystems.

The Wildlife River[16]

Starting high in the Rocky Mountains, the South Platte River in Colorado and the North Platte in Wyoming join forces at North Platte, Nebraska, to form the Platte River, which crosses Nebraska to join the Missouri. As with many waterways of the West, it flows through a thirsty land where there is never enough water to satisfy human needs. But long before human settlement the river was the springtime rendezvous for half a million sandhill cranes during their long migration to the North. The Platte River Valley is a classic case of an ecosystem on which too many demands are made. In the early 1880s farmers began diverting waters from it for irrigation, a demand that has steadily increased over the years along with demands for additional water to generate electricity and for domestic use. As more and more water was diverted, the Platte shrank from its broad-flowing stream into a system of narrow, braided*

* Rivers can be classified as straight, meandering, or braided, or combinations of the three. A braided river is defined as having multiple channels that flow through shifting sandbars with large seasonal variations in flows.

channels, eliminating the wide open areas sandhill cranes need for their roosting habitat.

In addition to these cranes, some 250 other species of migratory birds use the Platte River habitats to rest, nest, and feed, including 6 species — the Eskimo curlew, whooping crane, bald eagle, least tern, piping plover, and peregrine falcon — federally listed as endangered or threatened and 75 percent of the species on the National Audubon Society's Blue List.* The Eskimo curlew, which is near extinction, formerly gathered in such enormous flocks on the Platte's wet meadows that they were shot by the wagonload. Many of the remaining species, which require a broad, shallow river, are already abandoning reaches of the Platte. Those migrating birds that depend on wetlands are also moving away, because 90 percent of the Platte's wetland habitat has been converted to cropland.[17] Because the Platte is the only spring stopover site for sandhill cranes in the Central Flyway of the U.S., and the largest such site on the continent, loss of this habitat could endanger their entire population. American Rivers named the South Platte as one of the nation's most endangered rivers in 1988 and 1989.

The demand for water from the Platte continues to grow. Although there are many reservoirs along the river, there are plans to build new ones that would claim what remains of the flow. "Without water, you don't raise crops, and without crops you don't raise kids and keep things going," is how one farmer expresses his need.[18] Yet as farmers pump groundwater to irrigate their crops, they lower the water table and drain off more of the vital wetlands. The vicious cycle continues as new plantings require more irrigation water from a shrinking groundwater supply. One possibility, proposed by the U.S. Bureau of Reclamation but seriously questioned by many conservationists, is known as the Prairie Bend Project; it would store water from the Platte in new reservoirs and release it at times of peak flow to replenish groundwater levels for irrigators. The increased water flow, its proponents claim, would dilute the dangerously high nitrate levels (from fertilizer runoff) in the river water used for domestic consumption. The $450 million plan also promises to create three wildlife refuges in the affected area. However, a growing number of farmers and sportspeople believe that such a giant project would further threaten the Platte and doubt that Congress will ever appropriate funds for it because of budget restrictions and lack of public support.

Saving what remains of the Platte River wildlife habitat has become a top priority for conservationists. "Any further diversion of water from the river anywhere is a serious threat. The cranes have lived here for ten million years and we could lose them within our generation," says Peter A. A. Berle, president of the National Audubon Society, which maintains a 2,300-acre sanctuary on the river at Kearny, Nebraska, where sixty thousand cranes came to roost in 1990. Forty-five miles downstream, near Grand Island, the Platte River Whooping Crane Habitat Maintenance Trust, a private nonprofit organization, owns 8,000 acres, which it hopes eventually to triple.

* This is an early warning listing of species not federally listed as threatened or endangered but that are showing local or widespread population declines.

Legal Litmus Test[19]

In December 1990 EPA chief William K. Reilly vetoed the permit to build the Two Forks Dam on the South Platte River. The project, which would have bottled up 1.1 million acre feet of the river's water at a cost of more than $1 billion, was defeated largely through the efforts of American Rivers, Trout Unlimited, Environmental Defense Fund, and other concerned groups. The veto supports the increasingly heard argument that water resources should serve purposes beyond irrigation and municipal needs — fisheries, tourism, wildlife, and even esthetic values, for example. Such a case is now being tested over yet another major threat to the Platte — the Kingsley Dam in central Nebraska. Built in 1937, this dam has destroyed or damaged more than 160 miles of the most critical wildlife habitat on the river. In 1987 the licenses governing Kingsley's operation expired, setting in motion a lengthy legal proceeding over the granting of a new license by the Federal Energy Regulatory Commission (FERC). At that time and subsequently, American Rivers filed motions to intervene in the case, and in late 1990 they were joined by the National Audubon Society and other groups, recommending that FERC attach conditions to the new licenses to increase water conservation, provide more efficient irrigation, and restore wildlife habitat. The case is strengthened by the passage in 1986 of the revised Federal Power Act, which requires FERC to give equal consideration to environmental concerns as well as irrigation and hydropower when it issues new licenses. The outcome of the case will undoubtedly set an important precedent for how FERC implements these new congressional guidelines.

What You Can Do to Protect the Platte

1. Write and call on your legislators to join the fight to save the Platte River as a matter of highest priority.

2. Write to the Platte River Campaign, National Audubon Society, 666 Pennsylvania Avenue, SE, Washington, DC 20003. Ask to have your name added to the Platte River activist mailing list. This will update you regularly on the progress of the fight and alert you to specific action you can take.

3. Subscribe to the *Audubon Activist* to get regular updates on the Platte River and other important environmental issues.

4. Telephone the Audubon twenty-four-hour Actionline — (202) 547-9017 — for a recorded update on fast-moving environmental legislation.

5. At the state level contact the following organizations:

— Nebraska Wildlife Federation, 410 South Eleventh Street, Lincoln, NE 68803;
— Platte River Trust, 2550 North Diers Avenue, Grand Island, NE 68803;
— The Colorado Wildlife Federation, 7475 Dakin Street, #137, Boulder, CO 80221;
— Colorado Environmental Coalition, 777 Grant Street, #606, Denver, CO 80203;
— Trout Unlimited Colorado Council, 1557 Ogden Street, Third Floor, Denver, CO 80218;
— The Wyoming Wildlife Federation, P.O. Box 106, Cheyenne, WY, 82003.

Cleaning Up the Hudson

In 1609 when the English explorer Henry Hudson sailed up the river in his two-masted *Full Moon*, he described its bays, coves, and creeks as "pleasant and proper for man and beast to drink, as well as agreeable to behold."[20] Draining a 13,000-square-mile basin, the 315-mile waterway was navigable for 100 miles inland and impressed the English with its abundance of fish, oysters, and river bank game. Its fertile valleys were inhabited by indigenous tribes that included the Mohicans, whose name means "people of the river that flows two ways," a reference to the Hudson's ocean tides, which reach 165 miles north from the Atlantic Ocean.

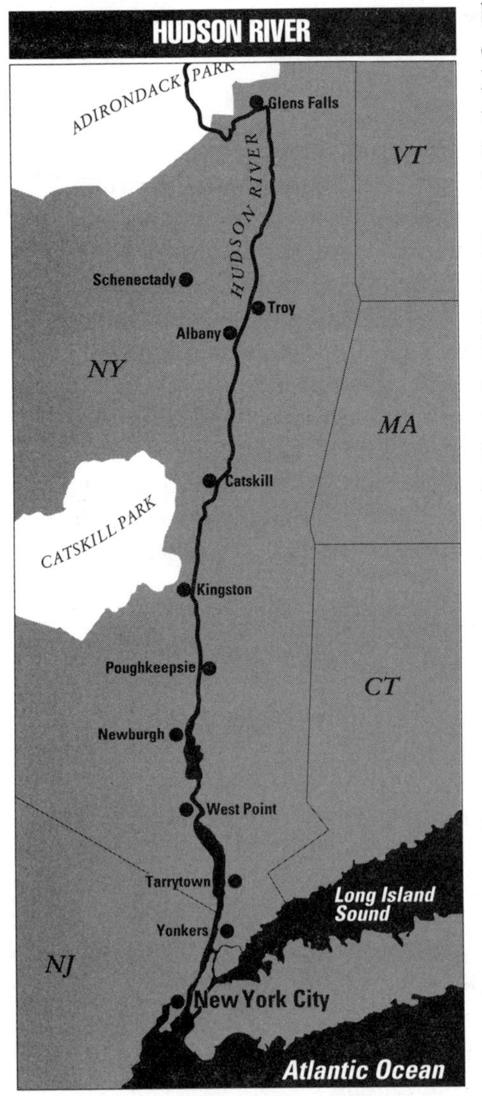

Despite centuries of interaction with human populations, the Hudson has sustained less ecological damage than many other rivers in North America, which have been dammed, diverted, and otherwise altered. Accordingly, it supports many tidally dependent plant and animal communities that were at one time much more abundant in the Northeast. The river and its estuary are a major component of the Atlantic seabord ecosystem, supplying nutrients to and supporting a rich variety of estuarine life and serving as a spawning and nursery ground for ocean fish.[21]

Sometimes called America's first river because of its strategic and commercial importance, the Hudson played an important part in the development of the United States, providing easy access from New York City to upland timber, agriculture, and, through the Erie Canal, to the Great Lakes. By the late nineteenth century its banks were lined with railways and waterfront factories, the building of which filled in the wetlands that were the river's natural cleansers. This created a water-filtering problem that would not be identified for decades.

After World War II there was an enormous growth of paper plants, plastics manufacturing, electric companies, and cement, chemical, and other industries along the Hudson. In the absence of environmental regulation they freely dumped their wastes into the river, as did the cities burgeoning on its shores, from New York City to Troy, north of Albany. In the 1950s there were telltale signs of the Hudson's deterioration. One could see oil slicks from river-borne oil tankers while municipal water supplies became tainted with chemical wastes

and sewage. Even in New York City one could smell the stench coming off the river. By the mid-1960s the entire river had become a sewer. Hazardous chemical pollution was exposed in 1975–76, when a government hearing found that two General Electric plants north of Albany had over a twenty-year period discharged into the river an estimated 1.2 million pounds of PCBs (polychlorinated biphenyls).[22]

The ABCs of PCBs

- Manufactured commercially since 1929, PCBs are a class of more than two hundred different chemical compounds. They are colorless, odorless, and range in consistency from waxy solids to heavy, oily liquids.
- Because they are poor conductors of heat and fire-resistant, PCBs have many industrial applications as insulators in electrical systems and as fillers in paints, adhesives, plastics, and carbonless copy paper.
- Carcinogenic and other health problems associated with PCBs were first suspected in the late 1930s and led to a ban on their production in the U.S. in 1977. Several foreign countries still manufacture PCBs, and products containing PCBs are still imported into the U.S.
- PCB levels of more than two parts per million are considered unsafe for humans.
- PCB-caused closure of the 40 miles of the upper Hudson to all fishing and the closure of commercial striped bass fisheries on the Hudson and Long Island are costing the economy of New York an estimated $40 million each year.

(Sources: EPA, INFORM, *Clearwater Navigator.*)

Tracing the Pollution

In 1985 the New York City-based research organization INFORM published "Tracing a River's Pollution," a comprehensive analysis of toxic discharges into the Hudson River basin. Based on the files recording 185 "point" sources* discharging toxic chemicals into the Hudson River, it was the first inventory of toxic chemicals discharged by identifiable sources to an entire U.S. river basin since the enactment of the 1972 Federal Water Pollution Control Act.[23] It also investigated numerous nonpoint sources of pollution such as farm fields, city streets, mining and construction sites, and forests from which rainfall and melted snow carry heavy metals and other contaminants to the river.

> *"The question we have to keep asking isn't whether the Hudson* has *improved, but whether it's going to* continue *to improve."*
> — John Cronin, Riverkeeper

One of the most remarkable conclusions of the INFORM study is that neither federal nor state regulatory agencies collect data about nonpoint sources of toxic pollution in the Hudson River. Among its other important findings:

- Forty percent of the 555 streams of toxic chemicals identified by INFORM on discharge permit applications are not regulated by federal or state governments, based on the judgments of the permit writers.

* Discharges that can be traced to a specific source such as a plant's drainage pipe.

• Forty-three percent of the streams of known or suspected carcinogens are also unregulated by government.
• The 7 known or suspected carcinogens in INFORM's samples of 26 toxics were released by plants located in 68 cities and towns with a combined population of more than 1.2 million.
• Twenty-seven plants, operated by twenty-two industrial companies, discharged 5 or more of the toxics studied by INFORM, with three companies — General Electric, IBM, and Octagon Process Inc. — each discharging 14 different toxics.

In 1987 INFORM published the second part of its Hudson River report. It found:
• PCBs accounted for nearly 60 percent of the violations of water quality standards over the nine years studied and close to 100 percent of the standards set for the safe consumption of fish. Lead, mercury, and cadmium accounted for most of the other water quality violations.
• Government monitoring of the Hudson was not providing a clear picture of conditions in the river, due mainly to "funding constraints." PCBs, for example, were monitored in only one of the nine years studied and at only one location, forty-five miles upriver. And there were no effective mechanisms to assure that all pollution problems caused by hazardous chemicals had been identified.[24]

INFORM recommended urgent efforts by government agencies and private groups and individuals to find better ways of controlling urban and agricultural runoff, and for New York's Department of Environmental Conservation (DEC) to sample more of the chemicals it regulates and to establish a "mass balance" survey that would report on the amount of chemicals a) entering each plant, b) used in manufacturing at the plant, and c) leaving each plant as waste. Any unaccounted amounts would indicate how much might be going into the Hudson. Hailed by scientists, environmentalists, and government officials as an extremely important source of information for anyone seeking to understand toxic pollution in their rivers, INFORM's recommendations remain unheeded by the state and federal agencies and private companies directly concerned with the health of the Hudson River.

The PCB Legacy

Although the PCB discharges into the Hudson were essentially stopped in 1977 by the DEC, and although 300,000 pounds of this hazardous chemical were dredged from the river, hundreds of thousands of pounds remain in the Hudson's sediments and still present a serious chemical-pollution problem in the river. Despite state health advisories and fishery closures, many people and animals continue to eat PCB-contaminated fish, exposing themselves and their offspring to hazardous levels of the chemical. As striped bass spawned in the Hudson migrate south and north along the Atlantic coast, they carry PCBs with them. Liver tumors in 90 percent of the adult tomcods caught in the Hudson, as well as the disappearance of mink from the river's shoreline, have been linked to the spread of PCBs through the river's food chain. A great horned owl found dead on the bank of the Hudson had PCB levels of

300 parts per million in its brain tissue — 150 times more than the level considered unsafe for humans. According to New York's wildlife pathologist, the cause of death was acute PCB exposure.[25]

PCB Gridlock

In May 1990 the New York State legislature passed a bill barring state officials from spending any money on dredging PCB-contaminated sediment from the Hudson until the federal government reviewed the problem, a process that could take two or more years. Even if an EPA-approved cleanup were to begin, it would take until the end of the decade to complete, according to an official of the DEC.[26] The bill's provision was the result of intensive lobbying by the General Electric Company, which fears it will be stuck with the $280 million for dredging the remaining PCBs from the river. Since 1985, GE, government officials, environmentalists, and local residents have debated blame, cleanup technologies, and financial responsibility. "It's a failure of the body politic to come to terms with a major pollution problem," said John Mylod, executive director of the Hudson River Sloop Clearwater, one of several environmental groups working to protect the river.[27]

The dredging plan proposed by the DEC is based on the use of a technique that loosens the sediments and sucks them up to a waiting barge. The process has been extensively tested and found to be a safe, effective way to remove sediments without spreading a plume of contaminants downstream. Once removed from the river, the sediments must be treated to destroy the PCBs. This can be done chemically, thermally, or biologically at an upriver landfill. Dredging and encapsulation of sediment have in fact been recommended by the EPA for other PCB-polluted waterways including Waukegan Harbor, Illinois; New Bedford Harbor, Massachusetts; the Sheboygan River and Harbor, Wisconsin; and the Saint Lawrence River in New York.[28] General Electric opposes dredging, claiming it would stir up PCBs and send them downstream. In its place the company proposes **bioremediation**, a process that uses bacteria in the river sediment to detoxify the PCBs without removing the sediments from the river bottom. However, according to New York officials, there is no assurance that the technique, which has only been demonstrated in the laboratory, will work in a river as large as the Hudson. Joining GE in opposition to dredging is the town of Fort Edward in Washington County, 190 miles north of New York City. If a PCB landfill is built by the state, it would most likely be located in Fort Edward. "This project is very unwelcome in our town," said Sharon Ruggi of Fort Edward, a member of Citizen Environmentalists Against Sludge Encapsulation.[29]

To Congress: Let's Act

Clearwater and other state and regional conservation groups say that it is time to act on cleaning up the Hudson's PCB problem rather than continuing with further studies. Endorsing the DEC plan, Clearwater urges Congress and the EPA to commit themselves to a permanent cleanup solution for the Hudson River and to pursue General Electric as the responsible

party under the provisions of the Superfund law.* Scenic Hudson, a major New York environmental organization based in Poughkeepsie, also favors the DEC plan. "We still support removing most of the contaminated sediments from the river," said Cara Lee, Scenic Hudson's environmental director. "We are skeptical about the GE approach."[30] In Clearwater's view, the true cost of PCB contamination goes well beyond the cost of cleanup. The lost value of commercial and recreational fisheries since they were closed or restricted due to PCB pollution has been estimated as close to $500 million and will continue to mount until the river is restored, says Bridget Barclay, environmental director for Clearwater, which advocates bringing a natural-resources damage claim against GE.[31]

Polluters Beware!

In addition to Clearwater, two of the most effective Hudson River watchdog organizations are the Hudson River Fishermen's Association and the Riverkeeper Fund, cofounded in 1966 by the Hudson Valley activist and writer Robert H. Boyle. He had heard about English riverkeepers who protect private trout streams from poachers. In the case of the Hudson, Boyle decided, the guardian would pursue polluters, not poachers. In 1983 he hired as riverkeeper John Cronin, a former lobbyist, congressional aide, and commercial fisherman.

> *"Once I sink my teeth into a polluter's ankle, I don't let go."*
> — John Cronin, Riverkeeper

A few month's later Cronin discovered that Exxon Corporation tankers were discharging saltwater ballast into the Hudson and taking on fresh water and exporting it to Aruba. Over a two-year period, 177 Exxon tankers flushed their tanks and removed clean water from the Hudson. The Fishermen's Association filed a complaint with the U.S. Attorney General. Exxon maintained that it had not broken the law, but in April 1984 it agreed to a settlement of $250,000 each to the Fishermen's Association and the Open Space Institute, another environmental group. In a separate settlement, New York State received $1.5 million. Since then the association has taken action against a wide range of violators including the town of Newburgh, Anaconda Wire and Cable, Standard Brands, and the Metro-North commuter railroad. When it comes to discussing the Hudson's improving health, Cronin is wary. "In some ways, I'm fearful that the greedy and selfish will feel more relaxed about plundering it again," he says. "The question we have to keep asking isn't whether the Hudson *has* improved, but whether it's going to *continue* to improve."[32]

Zebra Mussels Arrive in the Hudson

In May 1990 a small, unexpected pest turned up in the Hudson River. The zebra mussel, an inedible shellfish that seldom exceeds 1 1/2 inches in length and that is native to the Black and Caspian seas in Russia, had arrived after a circuitous route by way of the Great Lakes, where it was deposited in ballast

* Superfund was created in 1980 to identify and clean up chemical spills and toxic-waste sites that threaten human health or the environment. See also Jon Naar, *Design for a Livable Planet* (New York: Harper & Row, 1990), pages 235-236.

waters of oceangoing ships traveling the St. Lawrence River from Europe. In only three years the mussel and its fast-swimming larvae worked their passage from Lake Saint Clair near Detroit, west into Lake Michigan, east into Lakes Erie and Ontario, and then via the Erie Canal into the Hudson River. Named for the dark-brown stripes on its light-tan shell and without any known natural enemies in North America to attack it, the zebra mussel reproduces prolifically — a female produces as many as forty thousand eggs a season. The mussels often cluster around water-intake pipes, which provide them with a moving source of food. They are environmentally dangerous because they clog the pipes of power plants and devour microscopic plants at the bottom of the food chain that are normally eaten by native freshwater fish. According to the U.S. Fish and Wildlife Service, the expanding zebra mussel population may add as much as $500 million a year in expenses for public utilities, water companies, factories, and loss of business to marinas, sports-fishing interests, and motels along inland waterways.[33] Although biologists had predicted the eventual infection of the Hudson River by the zebra mussel, its presence in Haverstraw Bay less than 25 miles north of New York City so soon surprised them, said Charles O'Neill, project supervisor of the New York Zebra Mussel Information Clearinghouse at the New York State University at Brockport. "We have a lot of questions on how they got there," he added, noting that the zebra mussel is a freshwater organism not usually found in salt water such as the lower Hudson River.[34]

> *"They can't be cleaned up like an oil spill. In a hundred years, they will be all over America..."*
> — Margaret Dochoda, biologist

In their native habitats zebra mussels are kept under control by diving ducks. Without natural enemies there is no way of controlling their spread other than flushing pipes with water treated with chlorine, copper sulfate, or other polluting chemicals or by laborious hand-cleaning of water intake screens, which is a very expensive operation. Moreover, none of these methods would seriously prevent the mussels from taking away the food from the native fish in the Hudson and other North American waterways. According to Margaret Dochoda, a biologist with the Great Lakes Fishery Commission in Ann Arbor, Michigan, the zebra mussel problem is worse than the Exxon *Valdez* oil spill. "They can't be cleaned up like an oil spill. In a hundred years, they will be all over America, while the oil spill will have been cleaned up."[35]

What You Can Do to Protect the Hudson

1. Urge federal, New York, New Jersey, and local elected officials to get the Hudson River PCB cleanup project under way, with General Electric, not the taxpayers, paying the costs.
2. Join the following organizations that are dedicated to Hudson River preservation, conservation, and restoration:
— Hudson River Sloop Clearwater, 112 Market Street, Poughkeepsie, NY 12601; (914) 454-7673;

— Hudson Riverkeeper Fund, P.O. Box 130, Garrison, NY 10524; (914) 424-4149;
— Scenic Hudson, 9 Vassar Street, Poughkeepsie, NY 12601; (914) 473-4440.

3. Attend town meetings, planning board meetings, and public hearings that relate to the Hudson. Speak up for the creation of waterfront parks and more green spaces along the river.
4. Participate in riverside cleanup days. If necessary, organize them in your own community.
5. Protect the river's shorelines by reporting suspicious dumping or other potentially harmful activities to the New York State Department of Environmental Protection (800) TIPP-DEC.
6. Write to the federal Secretary of Transportation, 400 Seventh Street, SW, Washington, DC 20590, that you don't want the Hudson poisoned by vessel operators who are unfamiliar with the Hudson and that you are concerned about the U.S. Coast Guard's opposition to the mandatory use of river pilots. For more on this, check with the riverkeeper.
7. Urge the mayor of New York City to: a) increase funding for water meter installation, leak detection and repair, and water conservation, because the city uses water from the Hudson during drought emergencies; b) support the creation of a Hudson River Esplanade; and c) oppose ecologically unsound commercial development in the former Westway area.
8. Conserve water in your home and workplace.
9. Find out about the Hudson River Estuarine Research Reserve, c/o Bard College, Field Station, Annandale, NY 12504; (914) 758-5193. It is a program that manages four tidal wetlands on the Hudson River estuary as outdoor laboratories and classrooms.
10. Visit the Museum of the Hudson Highlands in Cornwall, New York; find out about their tidal restoration program (914) 534-7781.
11. Join Shorewalkers, Box 20748, Cathedral Station, New York, NY 10025; (212) 663-2167. They advocate a 330-mile Hudson River trail and preservation.

Sacking the Sac

California's greatest river, the Sacramento runs for 377 miles from the shadow of Mount Shasta in the north of the state through the Central Valley and delta to San Francisco Bay.* Annually the Sac provides recreation for more than 2 million people and $30 million from its salmon fishery.† Its water irrigates hundreds of thousands of acres of farmlands that produce high-value fruits, nuts, and vegetables. The river also generates electricity for more than 1.5 million homes. Its flood-control works have reduced the threat of flooding to the state capital and dozens of Sacramento Valley communities. Yet this progress has been achieved at a costly price. The Sacramento's lush river-

* For more on San Francisco Bay, see chapter 3, pages 91-94.
† Ninety percent of all salmon caught in the ocean between Monterey and San Francisco originate in the Sacramento River.

side vegetation, with wild grape and blackberry vines hanging down into the water from stately cottonwoods, willows, oaks, and sycamores, reminded John Muir of tropical jungles. Today less than 15,000 acres remain of the original 800,000 acres of forest. Sixty-five major dams have been built on the river and its tributaries, reducing its meandering current to what has become a boulder-lined ditch.

As an anti-erosion measure almost 50 miles of its riverbank have been stripped of trees and vegetation and lined with rock by the Army Corps of Engineers in a process known as **riprapping**. Severely disrupting the natural evolution that rejuvenates habitat, this engineering is seen by ecologists as a death sentence for the river's wildlife, which includes winter-run salmon, bank swallows, yellow-billed cuckoos, Swainson's hawks, river otters, muskrats, and beavers. The importance of the Sacramento River as an ecosystem is described by Jim Mayer, water specialist for the *Sacramento Bee*, as follows: "In its natural state the river is the driving force of a complex ecosystem that depends not only on water, but also on the power of a river meandering through the center of a valley. Hundreds of species either rely on or thrive on this process of erosion and deposition, of constant tearing down and building up."[37]

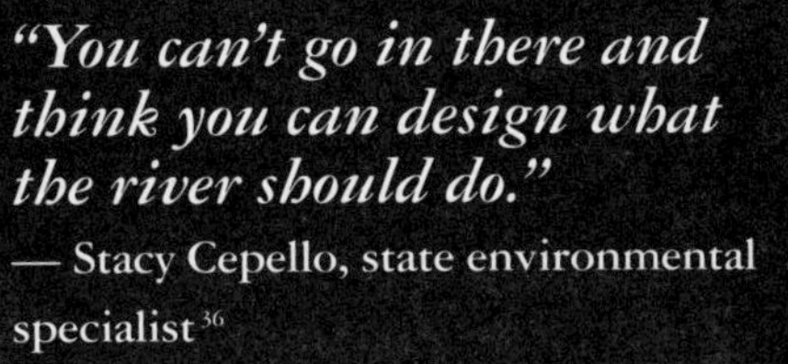
"You can't go in there and think you can design what the river should do."
— Stacy Cepello, state environmental specialist[36]

As more and more water is diverted for agriculture and electricity generation, pollution becomes an ever greater problem. For decades wastes from farms, municipalities, mining, and industry have been dumped into the river. One case is that of the Simpson Paper Company's Shasta Mill, the second largest dioxin polluter in the nation, which released the highly toxic chemical

used in bleaching paper into the river for twenty-four years before it was detected in 1990. Dioxin, like DDT, persists long after its source is controlled; it is picked up by insects, then by fish, and potentially by birds and other larger animals. Even if the dioxin discharge were halted tomorrow, resident trout would be contaminated for life, perpetuating the health hazard, said Harry Rectenwald, a biologist for the California Department of Fish and Game.[38] At the present rate of pollution from such a wide range of sources, ecologists fear that within a few decades the Sac could be a biological wasteland.

The Cantara Spill[39]

On July 14, 1991, a portion of a Southern Pacific train derailed while crossing the Sacramento River on a stretch of track known as the Cantara Loop 45 miles north of Lake Shasta, the state's largest reservoir and the keystone of its federal water system. One of the train's tank cars spilled 19,000 gallons of metam sodium, a powerful soil fumigant, into the river, contaminating a large area of the environment and causing injury to animals, plants, air, and water. For several days poisonous fumes hovering above the river kept biologists from getting a close look at the damage it had wrought. But airplane flights over the scene and videos made by emergency workers in protective suits revealed a virtually lifeless scene unlike any ever witnessed before on a California river.[40] By July 17 a toxic plume from the wreck that had killed hundreds of thousands of fish on its way downriver poured into the lake, still potent enough to kill further large numbers of fish and leaving doubts about how quickly its effects would dissipate. Although it would be many months before the full impact of the spill could be assessed, the effect on wildlife was catastrophic. According to the California Fish and Game Commission, the species confirmed killed included rainbow and brown trout, Sacramento sucker, dragonfly, mayfly, and many other insects. Among the wildlife made ill or threatened by the pesticide are bald eagles, golden eagles, black bears, beavers, river otters, raccoons, and minks.[41] Particularly devastating to anglers was the loss of the native rainbow trout for which the Sacramento is famous. Officials hope that fish surviving in the upper reaches of the river and in Lake Shasta will eventually repopulate the damaged stretch of the waterway. "Nature will do that in time," said Richard May, president of California Trout, "but as impatient fishermen and human beings, we'd probably like to see that be shorter than the ten years it would take nature to do the full job."[42]

Three months after the accident, people living or working along the river found themselves covered with sores or so short of breath that they could no longer work. "We have people who were perfectly healthy marathon runners beforehand and now can barely walk uphill," said Dr. James Cone, assistant professor of medicine at the University of California at San Francisco, who examined thirty victims of the spill.[43] The victims are angry that they were not warned of the pesticide's harmful effects for at least twenty-four hours after the wreck. "We slept all night breathing in that stuff as the plume floated by the house," said Verle Melvin, who had eruptions of scaly red patches on his hands, knees, back, and scalp.[44]

Virtually all of the scientific literature on metam sodium and its more toxic breakdown product, methylisothiocyanate, is based on laboratory experiments on animals. "In a sense," stated Dr. Lynn Goldman, chief of environmental epidemiology for the California Department of Health Services, "[this] is an experiment in which an unsuspecting community was exposed as they slept."[45] According to a damage assessment report by the California Department of Fish and Game in October 1991, "the entire aquatic ecosystem of the Sacramento River was almost entirely eliminated."[46] Based on experience from other river systems and assuming that recovery proceeds "in a predictable, orderly fashion," the overall recovery time for the fish and aquatic resources "may take twenty years." Recovery of vegetation to prespill conditions in the affected areas might take twenty to fifty years, while recovery time for terrestrial wildlife populations, "based on what we currently know," might be five to twenty-five years, the report stated, pointing out that recovery of wildlife species can only occur after their food source and habitat have been restored to suitable levels for survival.[47]

What You Can Do to Protect the Sacramento River

1. Urge your congressional representatives to support annual appropriations for acquisition of critical wildlife habitat and recreation lands along the Sacramento River as well as to support legislation for reforming the federal government's massive Central Valley project.*

2. Help persuade the California legislature, governor, and State Water Resources Control Board to develop an ecologically sound water policy that is based on preserving and restoring riverside forests. Such a policy should be based on the recommendations of a Sacramento River task force authorized by a bill by state senator Jim Nielsen (R-Rohnert Park).

3. Contact the Sacramento River Trust, P.O. Box 5366, Chico, CA 95927; (916) 345-4050, which is dedicated to the preservation, conservation, and restoration of the Sacramento River ecosystem. The Trust has initiated legislation to enhance the river's environment, promoted establishing a "meanderbelt" to allow the river to function in an ecologically sound manner, filed lawsuits to protect winter-run salmon, and pressed for the testing of alternatives to riprapping.

4. Contact the California office of Defenders of Wildlife, 1228 N Street, Sacramento, CA 95814; (916) 442-6386. This group has played a lead role in initiating the proposed Sacramento River National Wildlife Refuge and many other measures affecting the Central Valley region.

5. Support Friends of the River (see the Directory, pages 331-355) and other conservation groups in their class-action suit against the Southern Pacific Transportation Company, claiming that the July 1991 toxic spill was due to gross negligence and cost cutting by the giant hauler.

6. Take part with the Nature Conservancy and the Sacramento River Preservation Trust in restoring wildlife habitat at three sites along the river. For a schedule of events and directions to the restoration sites, call (800) 733-1763.

*For details on pending legislation, see Appendix B, pages 359-364.

Watching Rivers from Vermont to the Rio Grande

The Ottauquechee River flows through the lovely rolling hills of the Vermont countryside. But in the late 1970s it smelled like an open sewer. That was because the lovely old houses along its banks had not-so-lovely old sewage systems that drained right into its waters. Today the river is back. You can canoe on it, fish in it, and swim in it. Getting the Ottauquechee back on a clean course was the work of a local high school teacher who recruited students to help. Setting up thirty sampling stations along the river, they discovered high levels of fecal bacteria and more than four hundred sources of pollution along its 44-mile length. Inspired by their work, Henry T. Bourne, then chairman of the Ottauquechee Regional Planning Commission, and other local residents pushed successfully for the construction of sewage treatment plants. Because some of the homes were too far from the sewer lines, college students were recruited into the program to go door to door urging homeowners to install modern septic systems. One of those students was E. William Stetson, who in 1987 cofounded River Watch Network (RWN), a national organization whose main purpose is to help clean up the nation's rivers through a network of locally-supported grass-roots programs.

Since Ottauquechee, RWN has initiated more than twenty-five programs in the Northeast, the Midwest, and the West that include the Connecticut River (three programs), the Lackawanna, the Mississippi, the Colorado, and the Rio Grande (two programs). It is RWN's policy to enlist students and to provide them with a practical application of their science education by involving them in programs that work cooperatively with local, state, and federal river protection efforts. RWN also serves as a water quality information clearinghouse for all participants working to protect the nation's rivers.

For further information: Contact River Watch Network, 153 State Street, Montpelier, VT 05602; (802) 223-3840.

The Wild and Scenic Rivers Act

In response to the increasing damage inflicted on rivers by dam builders, Congress in 1968 passed the Wild and Scenic Rivers Act to ensure that certain "outstandingly remarkable" rivers remain in a free-flowing state.* The act establishes that designated river sections be either "preserved in their existing state" or "enhanced," and thus mandates a moratorium on their future development, including hydroelectric projects. Eight rivers were designated for immediate incorporation in the National Rivers System with twenty-seven others to be evaluated for eventual inclusion. Depending on their location, proposed river segments are evaluated by a complex process involving the Department of the Interior or the U.S. Forest Service, the White House, the Office of Management and Budget, and, for final decision, Congress. The usual study period is three years, during which no development is permitted on the river.

The river sections are classified into three categories: "wild," "scenic," or "recreational." *Wild* means pristine, untouched by humans, accessible only by

* For further information on this act, see Chapter 9, page 274-275.

foot, horseback, boat, or occasionally by air. Scenic sections are those free of impoundments and most development, but which can in places be reached by road. Recreational river sections may be dammed, show limited development along banks, and be fully accessible by road. In some cases, a river might be divided into all three classifications. If land bordering the river is privately owned, the federal government may acquire a corridor boundary, extending about a quarter of a mile on either riverbank. Once a river is included in the system, it is protected to the extent that no federal agency may approve, license, or help in a project considered harmful to the wild and scenic values. This means no dams, canals, dredging, or use that will degrade the river's water.

Expansion of the Wild and Scenic Rivers System has been slower than its proponents had hoped in 1968. In early-1993 the system consisted of 153 river segments totaling some 10,500 river miles and over 7 million acres of adjacent lands. This represents less than one-third of 1 percent of the nation's rivers. Conservationists say the total should be much higher. American Rivers, for example, wants to double the total in the lower forty-eight states by the year 2000. A major weakness is geographical imbalance of protected river mileage. About 40 percent of its total mileage is in Alaska, where twenty-five wild and scenic rivers were added at one stroke of the pen in 1980 by the Alaska National Interest Lands Conservation Act (ANILCA).* Of the remaining designated rivers, more than half are in California, Washington, Oregon, and Idaho. Many of the system's problems stem from its complex acquisition process and the long-demonstrated ability of a hostile administration to sabotage proposals before they reach Congress.

What You Can Do to Support Wild and Scenic Rivers

1. Take an active part in shaping wild and scenic river proposals in your region.
2. Contact the U.S. Park and Forest services and the BLM to find out which rivers are under review and which agency is responsible for them, and get informed about the rivers themselves.
3. Let the relevant agency and your elected officials know that you support greatly increased wild and scenic designations.
4. Support American Rivers, Friends of the River, American Wildlands, and other conservation groups in their efforts to strengthen the Wild and Scenic Rivers System and to restore or upgrade rivers to "wild" status.

RESOURCES

1. U.S. Government

The main federal agencies dealing with rivers are the U.S. Army Corps of Engineers (Department of Defense), the Bureau of Land Management (Department of the Interior), the Bureau of Reclamation (Department of the Interior), the Environmental Protection Agency, the Farmers Home Administration (Department of Agriculture), the Fish and Wildlife Service (Department of the Interior), the Forest Service (Department of Agriculture), the National

* For more on ANILCA, see chapter 9, pages 284-285.

Oceanic and Atmospheric Administration (Department of Commerce), and the departments of agriculture, energy, and the interior.

For addresses and phone numbers of these and other federal agencies and congressional committees, see the Directory, pages 352-354.

2. Environmental and Conservation Groups

The following are the main groups concerned with protecting and conserving rivers: American Rivers; American Wildlands; Hudson River Sloop Clearwater; Friends of the River; National Audubon Society; River Watch Network; Scenic Hudson; Sierra Club; and Trout Unlimited. Except where provided above, addresses and other information of these groups are given in the Directory.

3. Further Reading

— *The American Rivers Guide to Wild and Scenic River Designation*, Kevin J. Coyle. Washington, D.C.: American Rivers, Inc., 1988.
— *Basin and Range*, John McPhee. New York: Farrar, Straus and Giroux, 1981.
— *Cadillac Desert: The American West and Its Disappearing Water*, Marc Reisner. New York: Penguin Books, 1986.
— "The Colorado: The West's Lifeline Is Now America's Most Endangered River." *Time*, July 22, 1991.
— *Dams and Other Disasters*, Arthur Morgan. Boston: Porter Sargent, 1971.
— *Downriver: A Yellowstone Journey*, Dean Krakel II. San Francisco: Sierra Club Books, 1987.
— *Endangered Rivers and the Conservation Movement*, Tim Palmer. Berkeley: University of California Press, 1986.
— *The Hudson River*, Robert Boyle. New York: W. W. Norton, 1979.
— *Hudson River Estuary Management Plan*, Albany: New York State Department of Environmental Conservation, September 1990.
— *Life on the Mississippi*, Mark Twain. New York: Penguin Books, 1984.
— *The Monkey Wrench Gang*, Edward Abbey. New York: Avon, 1976.
— "Protecting Our Nation's Rivers," Randy Showstack. *American Rivers*, November 19, 1990.
— *A River No More: The Colorado River and the West*, Philip L. Fradkin. Tucson: University of Arizona Press, 1984.
— *Rivers and Lakes*, Laurence Pringle. Alexandria, Va.: Time-Life Books, 1985.
— *Rivers at Risk: The Concerned Citizen's Guide to Hydropower*, John D. Echeverria, Pope Barrow, and Richard Roos-Collins. Washington, D.C.: American Rivers/Island Press, 1989.
— *The Snake River: Window to the West*, Tim Palmer. Washington, D.C.: Island Press, 1991.
— "Water and the Flow of Power," Donald Worster. *The Ecologist*, vol. 13, no. 5 (1983).
— *Water Use on Western Farms*. New York: INFORM, 1982.
— "Wild and Scenic Rivers: Alive But Not Well," Verne Huser. *American Forests*, November 1982.

NOTES

1. Edward Abbey, *The Monkey Wrench Gang* (New York: Avon, 1976), p. 112.
2. The extent to which Lake Mead and other reservoirs are silting up is documented in Marc Reisner, *Cadillac Desert* (New York: Penguin, 1987), pp. 489-94.
3. Linda Kanamine, "Columbia River Tops New List of Waters at Risk," *USA Today*, April 9, 1992.
4. Timothy Egan, "Fight to Save Salmon Starts Fight Over Water," *New York Times*, April 1, 1991.
5. American Fisheries Society report cited in *American Rivers*, Fall 1991, p. 5.
6. Douglas Gantenbein, "Salmon on the Spot," *Sierra*, January/February 1991, p. 31.
7. Sources include: American Rivers; "Tatshenshini Wild," a proposal by American Wildlands, 1991; David Darlington, "Windy Craggy and the River Big as the Ocean," *Wilderness*, Winter, 1990, pp. 50-51; Rick Searle, "Journey to the Ice Age," *Equinox*, January/February 1991; Randy Showstack, "Protecting Our Nation's Rivers," *American Rivers*, November 19, 1990, p. 9.
8. American Wildlands, "Tatshenshini Wild," p. 15.
9. Ibid., p. 16.

10. Paul Gray, "A Fight over Liquid Gold," *Time*, July 22, 1991, p. 22.
11. Reisner, *Cadillac Desert*, p. 127.
12. Reisner, *Cadillac Desert*, pp. 125-150.
13. Ibid., p.128.
14. Source material for this section includes: R. Roy Johnson and Steven W. Carothers, "External Threats: The Dilemma of Resource Management on the Colorado River in Grand Canyon National Park," *Environmental Management*, vol. 11, no. 1, 1987, pp. 99-107; Jim Carrier, "Water and the West: The Colorado River," *National Geographic*, June 1981, pp. 2-33; and Charlene Crabb, "Glen Canyon Controversy Continues," *E Magazine*, May/June 1990, pp. 27-30.
15. Gray, "A Fight over Liquid Gold," p. 26.
16. The main sources for this section are Carl Safina et al., "Threats to Wildlife and the Platte River," National Audubon Society, Environmental Policy Analysis Department Report #33, March 1989, George Laycock, "River of Cranes," *Wildlife Conservation*, January/February 1991, and material supplied by American Rivers.
17. U.S. Fish and Wildlife Service report cited in Safina, p. 6.
18. Laycock, "River of Cranes," p. 61.
19. Based on information from Nell Bogen, "Amid Signs of Change, Platte River Flows On," *Audubon Activist*, March 1991.
20. Cited in John Hay and Peter Farb, *The Atlantic Shore* (Orleans, Mass.: Parnassus Imprints, 1982), p. 41.
21. For a more detailed description of the Hudson River's natural resources, see *Hudson River Significant Tidal Habitats* (Albany: New York State Department of State Division of Coastal Resources and Waterfront Vitalization and the Nature Conservancy), March 1990.
22. *Tracing a River's Toxic Pollution: A Case Study of the Hudson, Phase II* (New York: INFORM, 1987), p. 3.
23. Ibid., p. 1.
24. Ibid., pp. 5-8.
25. Bridget Barclay, "PCBs: G.E.'s Legacy for the Hudson," *UpRiver/DownRiver*, January/February 1991, pp. 16-17.
26. Allan R. Gold, "After 15 Years, Hudson Still Has PCB's," *New York Times*, May 16, 1990.
27. Ibid.
28. Barclay, "PCBs," p. 18.
29. Ibid.
30. Harold Faber, "Experimental Cleaning of Hudson Is Under Way," *New York Times*, August 25, 1991.
31. Barclay, "PCBs," p. 15.
32. Susan Reed and Harriet Shapiro, "Polluters, Beware!," *People*, July 2, 1990, p. 67.
33. Harold Faber, "Aquatic Pest Turns Up Early in Hudson," *New York Times*, November 29, 1990.
34. Ibid.
35. Harold Faber, "Shellfish Is Tiny, but Problems It Could Cause Are Huge," *New York Times*, May 14, 1990.
36. Cited in Jim Mayer, "The Dying River," *Sacramento Bee*, special report, September 17-22, 1989.
37. Ibid.
38. Ibid.
39. This report is based on the report *Natural Resource Damage Assessment Plan, Sacramento River: Cantara Spill*, State of California, Department of Fish and Game, October 15, 1991.
40. Martin Halstuk, "Experts Debate Next Move in Pesticide Spill," *San Francisco Chronicle*, July 19, 1991.
41. Dawn Garcia, "Spill Reaches Shasta," *San Francisco Chronicle*, July 18, 1991.
42. Ibid.
43. Elliot Diringer, "Sacramento Spill Has Many Still Worried," *San Francisco Chronicle*, October 7, 1991.
44. Ibid.
45. Ibid.
46. *Cantara Spill*, p. 108.
47. Ibid.

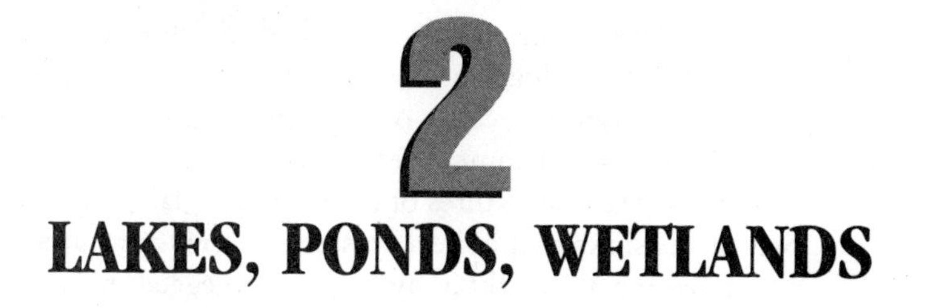

2 LAKES, PONDS, WETLANDS

"A lake is the landscape's most beautiful and expressive feature. It is the earth's eye; looking into which the beholder measures the depth of his own nature. The fluviatile trees next the shore are the slender eyelashes which fringe it, and the wooded hills around are its overhanging brows."

— Henry David Thoreau

Lakes and their smaller versions, ponds, are linked to rivers in that they are all part of the earth's hydrologic cycle (see diagram on page 50). They are also joined to rivers symbiotically, often supplying them with water, while rivers can create lakes and sometimes destroy them. Compared with rivers, which are very long-lasting features of the landscape, lakes have relatively short lives. Serving as sinks that collect wastes from many sources, lakes gradually fill up with sediment brought in by streams or vegetation growing on their banks and eventually become dry land. How long this process might take varies considerably. Some lakes are the result of millions of years of change, while others develop within a very short span of time. The Salton Sea in southern California was formed in 1905 when the Colorado River overflowed the entrance to a canal built to bring irrigation water to the lower end of the California desert. Even today you might see a lake in the making. When a river jumps its channel and cuts off a meandering loop, it gives birth to an oxbow lake, so called because of its shape. You can see a pond arise in a matter of months as the result of beavers spreading a river beyond its banks. And in the desert country of the Southwest heavy rains create temporary lakes by filling basins called *playas*. As we have seen in the preceding chapter, human engineers have now surpassed beavers as the great makers of lakes, especially in the West. Most natural lakes are inland depressions formed by geological activity thousands of years ago, by deposition of silt, driftwood, and other debris in a slow-moving stream. Ponds, which are smaller and shallower than lakes, have geological origins as well as being created by humans and animals.

The Life Cycle of Lakes[1]

Once a lake has been formed, life gradually takes shape in the new water. It begins at the bottom with microscopic plants and moves upward through an intermediate (**limnetic**) zone to the top layer known as the **littoral** or **shore** zone. Characterized by reeds, cattails, water lilies, and deeper down, submerged rooted plants, this zone is populated by numerous larval and adult insects, snails, clams, frogs, and snakes. The limnetic zone extends as far as light penetrates, which in shallow water may be to the bottom. It contains a wide variety of organisms including phytoplanktons, zooplanktons, larger insects, fish, and amphibians. In deeper bodies of water such as Lake Superior the bottom zone is referred to as **profundal.** Its main food source is detritus from the limnetic zone and, depending on the temperature of the lake, lake trout and other cold-water fish, or perch, pickerel, bass, and other warm-water varieties. In lakes whose waters discharge rapidly, many of the microscopic organisms are carried away before they can proliferate. In the relatively shallow Marion Lake in British Columbia, for example, the water exits so fast that the entire volume of water is replaced in fewer than five days. At the other end of the scale, Lake Tahoe in California is so deep that its complete water renewal takes seven hundred years. However, its water is so "pure" and nutrient-poor that its phytoplankton population is relatively sparse. The growth of phytoplankton, crucial to the productivity of a lake, depends on the availability of nitrogen, phosphorous, potassium, calcium, and other minerals brought in by rain and snow, the decay of organisms in the water, and leaching from soil and rocks. Equally important is the supply of solar energy for photosynthesis. For example, less sunlight is generally available in the arctic with its long, dark winters, and correspondingly more is available in temperate and tropical regions. However, in turbid waters sunlight may penetrate only a few feet and photosynthesis is curtailed.

Thermal Stratification

As a heating agent, the sun has an important effect on algal and plant growth. Water is different from other liquids, which are heaviest at the freezing point; it reaches its maximum density at 39 degrees Fahrenheit. This means that the densest water, which is at the bottom of a lake, never freezes, and therefore the organisms living there are not destroyed. The special temperature-density relationship of water causes a condition called thermal stratification or thermocline. In the warmest time of the year most lakes more than 30 feet deep become stratified into a warm, somewhat turbulent upper layer, a middle transitional zone, where temperatures drop sharply, and a bottom layer, where the water is cold and relatively undisturbed. When colder weather comes in autumn, heat is lost to the atmosphere and the temperature drops in the water's upper layer, increasing its density and allowing it to sink and mix with the lower layer. Eventually the lake is completely mixed and stratification is no longer evident.

Stratification plays an important role in a lake's ecology because it inhibits the flow of nutrients across the thermal barriers between the layers. During the warmest months nutrients from decaying organisms sink to the bottom. But in early spring and fall the wind stirs up unstratified waters, allowing nutrients to circulate freely to the sunlit upper layer where generation of food takes place, thus creating seasonal cycles in the bioproductivity of the lake.

Nutrient Input and the Aging Process

There is a close relationship between land and water ecosystems. Through the **hydrologic cycle**, water falling on land from rain or snow runs from the surface or moves through the soil to enter springs, streams, rivers, and lakes. It carries with it silt and nutrients that enrich aquatic ecosystems. In its early stages a lake is described as **oligotrophic**, having little nourishment; its water is clear and rich in dissolved oxygen but poor in essential plant nutrients. The relative lack of fertility inhibits the production of algae and plants, checking the profusion of animal life. In middle age the lake becomes **mesotrophic**, having medium nourishment. At this point nutrients flowing into the lake join those from the lake's own decaying organisms at the bottom. Some of this nourishment is then recycled back into the upper waters, where it is converted by photosynthesis into increased phytoplankton output, which in turn supports larger populations of plants and animals. In the **eutrophic** stage the lake's fertility increases rapidly, generating abundant growth of rooted plants near the shore and floating phytoplankton. "Production of algae becomes dominated by mats of blue-green species that cover the surface like pea soup. The waters... are turbid with

decaying organic matter from the increased populations of both plants and animals."[2] This wholesale decomposition depletes the oxygen at the lake bottom where anaerobic bacteria attack the organic matter, releasing hydrogen sulfide, a gas that smells like rotten eggs. In the eutrophication process increased biological activity is accompanied by marked changes in the types of organisms that inhabit the lake. Bottom-dwelling larvae and nymphs are replaced by midges, which need less oxy-

HYDROLOGIC CYCLE

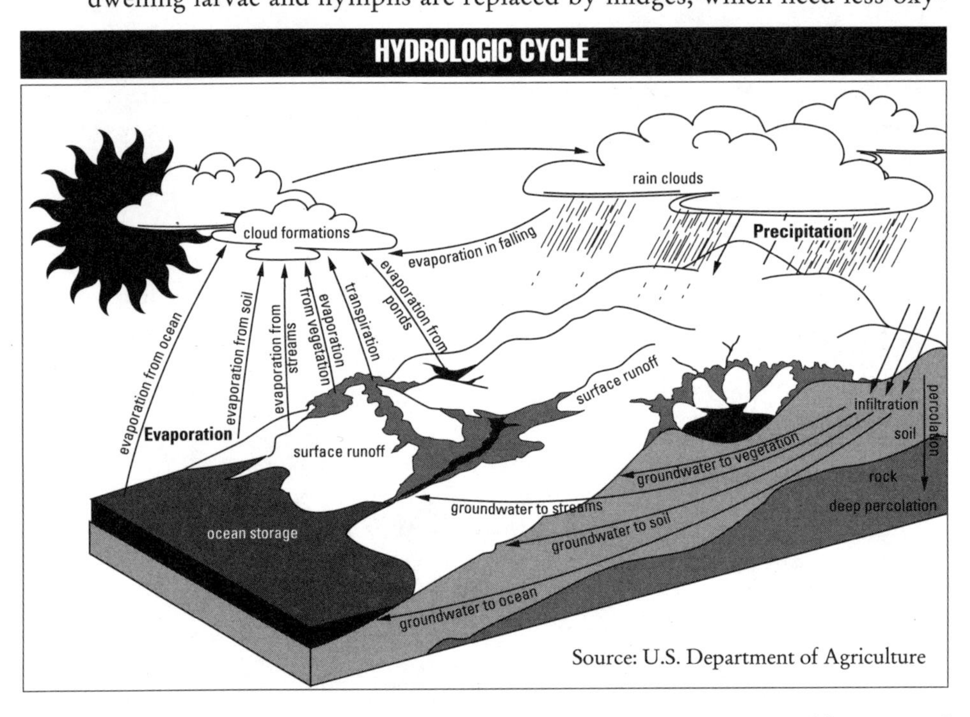

Source: U.S. Department of Agriculture

gen. The number of species declines, although the numbers and biomass of organisms often remain high. As the lake's basin continues to fill with sediment and inorganic substances, the depth of water decreases and the resulting shallowness accelerates the cycling of nutrients and further speeds plant production. The final stage is described by the ecologist Robert Leo Smith: "What develops is a sort of positive feedback mechanism that carries the system to eventual extinction — the filling in of the basin and the development of a marsh, swamp, and ultimately, a terrestrial community."[3]

Cultural Eutrophication[4]

Lakes age at different rates. Some take ten thousand years to accumulate 15 feet of organic sediment. Some with rocky, nutrient-poor watersheds remain youthfully oligotrophic for even longer periods of time. Others, in tropical regions, where heat and humidity speed plant growth and decay, eutrophy more rapidly. And as we shall see below, the process can be hastened by **cultural eutrophication**, premature aging induced by sewage, industrial waste, artificial fertilizers, and other contaminants getting into lakes and ponds.

Lake Washington and Beyond[5]

Lake Washington is an 18-mile-long lake on the eastern edge of Seattle. In 1955 scientists at the University of Washington under the direction of Professor W. T. Edmondson detected the presence in the lake of abundant blue-green algae *(Oscillatoria rubescens)* and plankton associated with the pollution of other clear lakes in populous areas. Because the lake had not yet seriously deteriorated, the scientists were able to get in on the ground floor of a study of the early stages of developing nuisance conditions.[6] In the 1940s sewage and industrial wastes from surrounding enterprises were dumped into Lake Washington, but those responsible for the dumping claimed that the lake was large enough to render the wastes harmless. However, by 1962 the lake was found to be polluted. As a result of studies done by Edmondson and his colleagues, the scientists and their supporters argued that if dumping wastes into the lake was stopped, it would return to its former condition in a short time.

The first stage of diverting the wastes from the lake was completed in 1963, and five years later the dumping into Lake Washington had completely stopped. At an international congress in 1965 Edmondson wagered that within six years the lake would be healthier than it had been in 1950. To win his bet — a bottle of liquor — the transparency of the water* would have to exceed the maximum of 4 meters (13 feet 2 inches) visible at that time. In 1971 at another congress he was able to announce that the transparency was now 4.5 meters (14 feet 10 inches).[7] In 1987, its clearest year to date, the figure had reached 6.4 meters (21 feet). At present Lake Washington is considered to be in excellent condition for a lake in an urban area. As Edmondson reports, it still produces algae, including a small proportion of blue-greens, but the algae are not a nuisance. Well used for all types of recreation, including fishing, the lake continues to be a fascinating subject of scientific research.[8]

Summarizing his conclusions on Lake Washington, Edmondson says it can be a guide for addressing other environmental problems: "The key to the successful solution of a problem is to identify the kind of scientific knowledge required to assess it and the kinds of corrective actions available."[9] One unusual aspect, he explains, was the extent and the character of the scientific background, including studies in 1933 and 1950 showing what the lake was like before pollution, and the chance sighting of the *Oscillatoria* in 1955, which assured that the early stages of deterioration would be studied. Another important feature was the public action by dozens of citizens who volunteered their time and effort to the local educational and planning activities that, combined with strong and intelligent leadership, led to the successful restoration of the lake.†

* Transparency is regarded by scientists as a favorable characteristic for lakes and other bodies of water, except when it is the result of acid precipitation.
† For more on restoration of lakes and other ecosystems, see chapter 10.

Salt Lakes

All lakes contain dissolved minerals, but when they have more than 3 percent* of salt, they are classified as salt lakes. Occurring in arid regions of the world, they occupy almost as much of the earth's surface as the freshwater variety. Most salt lakes start out as bodies of fresh water and get progressively saltier, because their inflowing streams bring high concentrations of minerals or, as is more frequent, because a major geologic movement — occurring when the plates of the earth drifted together or apart — has cut off the lake's outflow channel, preventing the flushing out of salts. In these situations the water in the lake evaporates faster than it can be replenished, leaving behind an ever-increasing content of minerals. The salt lakes of the West are the evaporated remnants of large oceanlike lakes that covered much of the northern basin and range country millions of years ago. Great Salt Lake in Utah, North America's largest salt lake, once covered 20,000 square miles but is now reduced to less than one-tenth of that area, dwindling from 2,400 square miles to 1,600 in the last hundred years. Its salinity averages 22 percent.[10] Saltier even than Great Salt Lake is the Dead Sea in Israel, with a salinity of 28 percent. In such a concentration only a few **halophilic** (salt-loving) organisms can survive. These include a species of green algae and several microorganic species, one of which supplies a purple pigment capable of converting sunlight directly into energy through photosynthesis. The near-lethal salinity of the Dead Sea has given rise to a widespread misconception that salt lakes are biological deserts. However, as we shall see in the following report on Mono Lake, a salt lake can be extremely productive of life.

Mono Lake

For more than a million years the blue expanse of Mono Lake has reflected the jagged peaks of the Sierra Nevada, east of Yosemite National Park. Although melting glaciers during the last Ice Age caused it to overflow, no river has ever succeeded in draining its 338-square-mile basin, which at one time was 900 feet deep. Today the salt-rimmed 60 square miles of milky water you see is "a briny shadow of its ice age self."[11] Mono Lake is in a delicate state of balance between its natural water input — springs, streams, rainfall, and melting snow — and evaporation. Having no natural outlet, the water flowing into the lake eventually evaporates, leaving a concentration of dissolved minerals. Its salinity is about three times that of seawater and its alkalinity is eighty times greater. In an underwater process the resulting carbonates have given rise to towers of porous limestone known as tufa, which are now 6 to 8 feet tall, exposed and stranded as the level of the lake has dropped. Typical of alkaline saline lakes, Mono is highly productive biologically, supporting a very dense population of algae, alkaline flies, and tiny brine shrimp, which in the absence of any other competition — there are no fish in the lake — multiply prodi-

* Compared with 3.5 percent, which is the average salt content of the earth's oceans.

giously. As many as four thousand flies have been counted in a single square foot of shore. The brine shrimp, a species that cannot survive in any other habitat, are so light that it takes about six thousand of them to make a pound. There may be as many as four trillion of them in the lake. Attracted to the enormous supply of algae, flies, and shrimp are hundreds of thousands of migratory birds — Wilson's phalaropes, eared grebes, and the ubiquitous California gulls that nest on islets in the lake.[12]

Drained of Life

Although the lake has always lost great amounts of water from evaporation by the hot desert sun, this loss of some 40 inches a year was made up over the millennia by the inflow from springs and freshwater tributary streams. However, since 1941, when the city of Los Angeles began drawing water for domestic and industrial use from four of Mono Lake's five major feeder streams, the level of the lake has fallen 41 *feet* and the resulting salinity threatens to kill the lake. Mono has already lost half of its water volume while more than doubling its salinity. Partial exposure of the lake bed produces a toxic dust laced with selenium and arsenic.[13] By late 1978 Mono's falling water level exposed enough lake bottom to serve as a land bridge to Negit Island, a nesting site for as many as thirty thousand California gulls, providing access to their eggs and chicks for coyotes and other predators from the mainland. California National Guardsmen were brought in to restore Negit's former isolation. "Twice during the spring of 1979 they tried unsuccessfully to blast the bridge out of existence with explosives. That summer coyotes invaded the island in force... Not a chick survived."[14] To bar the coyotes the state then erected a chain-link fence topped with barbed wire across the 2-mile bridge. But the gulls refused to return, crowding onto smaller islands in the lake, which, as the water recedes will also become joined to the mainland and thus be lost as nesting sanctuaries. If the inflow diversion continues at the present rate, salinity will reach 20 percent soon after the year 2000. In water so salty, ecologists predict that Mono's unique shrimp would die and brine-fly larvae are not likely to survive. The demise of these creatures would in turn mean the loss of a vital feeding station for migratory birds and the transformation of the Mono Lake ecosystem into a virtually lifeless body of water similar to the Dead Sea.

A Big Legal Win

In one of California's most bitterly fought environmental battles, the National Audubon Society, the Mono Lake Committee, and other environmental groups sued the L.A. Department of Water and Power to curtail its diversions of water from the lake. In April 1991 the conservationists chalked up a major victory when the Eldorado County Superior Court ruled that Los Angeles must cease its practice of siphoning water from the Mono basin, effectively barring the city from reestablishing the diversions until the water level rises

> *"Mono Lake is a national environmental, ecological, and scenic treasure [that] should not be experimented with even for a few brief years."*
> — Judge Terence Finney, Eldorado County Superior Court[15]

high enough to protect the lake's wildlife habitat. The ruling extends a previous injunction aimed at maintaining the lake at 6,377 feet above sea level until 1993, when the California Water Resources Control Board was expected to complete a scientific study on the long-term ecological effects of any future diversions. Although the legal victory is encouraging, the fight is not over. Los Angeles has mounted a renewed court challenge to regain access to the water. Nevertheless, Mono Lake activists express hope that L.A.'s combative attitude will change. "We hope the city will wake up and spend its money on conservation, not litigation. Los Angeles can easily conserve enough water to more than make up for the loss from Mono Lake," said Dan Taylor, Audubon's Western regional representative.[16] Ultimately the choice between Mono Lake's survival or death must be based on a wider understanding of the problem by those people who are causing it. As Professor W. T. Edmondson points out, "There is one fundamental contrast between the Mono Lake situation and most environmental problems of the Lake Washington type in which most people observe the deterioration at first hand and see the results of their corrective action. Probably many people in Los Angeles do not know where Mono Lake is and would not care what happens to it if they did."[17]

What You Can Do to Help Save Mono

1. Contact the Mono Lake Committee, executive director Martha Davis, P.O. Box 29, Lee Vining, CA 93541; (619) 647-6595. Ask for a copy of their quarterly *Mono Lake Newsletter*.

2. Contact the National Audubon Society, Western Office, 555 Audubon Place, Sacramento, CA 95825; (916) 481-5332.

3. Petition California State Parks Director Don Murphy, P.O. Box 942896, Sacramento, CA 94296, to not close the Mono Lake Tufa State Reserve because of budget cuts.

The Great Lakes

The five Great Lakes — Superior, Michigan, Huron, Erie, and Ontario — contain 20 percent of all the fresh water on earth. They form a 2,342-mile continuously navigable waterway from the western tip of Lake Superior at Duluth, Minnesota, to the entrance of the Saint Lawrence River at Kingston, Ontario. The entire lake system, shared by Ontario, Minnesota, Wisconsin, Illinois, Indiana, Michigan, Ohio, Pennsylvania, New York, and Quebec, covers some 95,000 square miles, an area greater than the United Kingdom. The Great Lakes ecosystem also includes eighty thousand inland lakes and close to a half million miles of inland streams and rivers. Twenty-seven million United States citizens and eight million Canadians live in the Great Lakes basin region, enjoying its resources for drinking, fishing, navigation, and other purposes. Tragically, the lakes have become seriously polluted over the years. Today toxic chemicals heavily contaminate the water. Fish and wildlife suffer deformity and premature death. Restrictions on drinking water, eating fish, and using beaches are commonplace throughout the Lakes system. The Great Lakes are recipients of human sewage, industrial waste, agricultural runoff, and other chemicals of almost indescribable quantity and toxicity.

An Ecosystem in Trouble

There are many reasons why the damage to the Great Lakes has become so extensive: two centuries of uncontrolled commercial fishing and human tampering with the lakes' tributaries, where fish spawning takes place; building dams for powering water mills, which block the passage of salmon up many rivers; cultivation of fields, leaching soil that silts up salmon spawning grounds; and cutting down forests, which deprives young salmon of the vital shade that cools their water. The disappearance of salmon from Lake Ontario around the turn of the century paved the way for alien marine species — alewives and sea lampreys, in particular — to invade that and other Great Lakes. As it migrated west, the lamprey, meeting few natural enemies, devastated native stocks of trout and whitefish. By 1960 Lake Huron's commercial whitefish catch fell from 3.5 million pounds to 100,000. Lake Michigan's trout harvest dropped from 6.5 million pounds to zero.[18] Since the 1970s efforts to reintroduce native fish and to control invading species have enjoyed some success. Because of tougher pollution controls, the quality of surface water has markedly improved and the number and diversity of many fish have rebounded. "We've made a lot of progress in the lakes," reports Glenda L. Daniel, director of the Great Lakes Federation, a Chicago-based environmental organization. "But along the way we have discovered a lot of problems we didn't know we had."[19]

One of the problems is the role played by airborne pollution and sediment contamination in maintaining a persistent level of toxic contamination in the Great Lakes. In a 1975 study of mud from the bottom of a lake on Isle Royale in Lake Superior scientists found traces of PCBs and dioxin; five years later they discovered toxaphene, a pesticide used on cotton crops in the South. Even though PCBs had not been produced in the region for years, high levels were detected in the air above Lake Superior near Duluth, in the eggs of terns nesting around Lake Michigan's Green Bay, and in over 90 percent of the fish caught in Sheboygan Harbor. In Lake Huron's Saginaw Bay and other areas, nesting double-crested cormorants were found to be hatching chicks with severely crossed bills. That PCBs, DDT, and other toxic chemicals are turning up in the lake water, not just on the muddy bottom, is evidence that the winds are continually depositing these contaminants.[20]

In addition to the pollutants found in lake water, highly contaminated sediment was identified by the International Joint Commission, a U.S.-Canadian agency that oversees management of the Great Lakes, at forty-two areas including Saginaw Bay in Michigan, Waukegan and Sheboygan Harbors in Wisconsin, the Grand Calumet River in Indiana, the Ashtabula River in Ohio, and the Buffalo River in New York.

Bioaccumulation

Although nearly thirty million people get their drinking water from the Great Lakes, they run a greater health risk from eating fish as the result of a process known as **bioaccumulation** or **biomagnification**. Microscopic amounts of

toxic chemicals are absorbed by tiny organisms living in the lakes. These organisms are eaten by small fish that are in turn eaten by larger fish. With each step in the food chain, toxic chemicals keep building up in strength. Also, over their lifetimes, fish absorb large quantities of chemicals through their gills. The resulting concentrations can be millions of times greater in the fish than in the water in which they live. Thus you can ingest as much poison in one meal of Great Lakes fish as in 150 years of drinking the water. In 1989 *National Wildlife* reported that nine out of ten fish tested from Michigan's Great Lakes waters were tainted with toxic chemicals at levels considered unsafe to fish-eating wildlife. In addition, at least sixteen predators throughout the region suffered reproductive problems or population declines linked to toxic chemicals in their food. The report adds a disturbing note: "The wildlife isn't dying of cancer, which is the focus of most environmental regulations designed to protect humans from pollution. Rather, in animals, the harm shows up as birth defects, reproductive problems, and brain, kidney, and liver damage."[22]

> *"If you are a woman who ever intends to have a child, or you are under 16 years of age, don't eat any trout, salmon, or certain other fish from many areas of the Great Lakes. For everyone else, don't eat more than one meal of them a week."*
>
> — U.S. and Canadian public health warning.[21]

Eutrophying Erie

Lake Erie, the smallest and shallowest of the Great Lakes, extends 240 miles from Toledo to Buffalo. Its major source is Lake Huron via a river system that serves industrial Detroit; it also gets water from rivers that drain agricultural land and service Cleveland, Toledo, Buffalo, and Erie. In 1969 one of these rivers, the Cuyahoga, was so contaminated with oil and other flammable wastes that it caught fire. At the same time Lake Erie was pronounced dead, strangled on its own algae. In its earlier existence the lake was part of an extensive network of wetlands, marshes, bays, and rivers that provided abundant shallow-water habitats for fish spawning and for a rich mosaic of plant and animal life. With human settlement came a vast reduction in the wetlands and the introduction of wastes, especially chlorides and sulfates — constituents of human and industrial sewage — and other environmentally harmful chemicals. From the 1930s to the 1970s there was a threefold increase in phosphates from laundry detergents and from fertilizers washed off farmland. However, as a result of state and provincial laws limiting their use in detergents, the amount of phosphates has dropped markedly.

The rapidly growing human population, particularly at the western end of the lake, was accompanied by a corresponding explosion of eutrophying phytoplankton and zooplankton, with shoreline accumulations of the alga *Cladophora* sometimes reaching 50 feet in width and making beaches unfit for recreation. Between 1913 and 1946, when human population tripled, the coliform bacterial count increased at about the same rate. At the same time the organisms at the bottom of the western end of the lake reflected the consequences of pollution and

lower levels of oxygen. The mayfly population, for example, was reduced almost to extinction. Lake herring, blue pike, whitefish, and other commercially significant fish have declined dramatically. Even if pollution were stopped, how long would it take to get rid of the pollutants currently in the Great Lakes? According to the ecologist Eugene P. Kormondy, it would take about twenty years for 90 percent of the wastes to be cleared from Lakes Erie and Ontario and hundreds of years from Lakes Superior and Michigan.[23]

Working Together

In 1985 the IJC urged the two governments to develop remedial action plans for forty-two "areas of concern" requiring immediate attention. Twelve of the areas are in Canada, twenty-five in the U.S., and five extend across the international boundary. The plans were to identify the problems, describe the causes, and work out solutions. In 1989 the Great Lakes Water Quality Board, IJC's principal adviser, reported that the program was progressing well, although it found half of the first plans defining the problems to be inadequate.[24] It will be several years before a final estimate of the program's cost can be made, but one estimator calculates the likely cost for all the areas "to reestablish some semblance of ecosystem integrity" will be in the range of $100 billion to 500 billion.[25] Since 1972 the United States and Canada have spent close to $10 billion to clean the lakes, primarily by building or improving sewage-treatment plants. On Lake Michigan's Green Bay, for example, the twenty-five paper mills and municipal sewage systems that in the 1970s daily spewed some 400,000 pounds of organic waste into the lake now release one-eighth of that amount. Factories, mines, and other point sources have also reduced their pollution. Successful long-term restoration and maintenance of the integrity of water in the Great Lakes ecosystem will depend on the continuing commitment of the United States and Canada.

EPA Program

In 1989 the U.S. EPA's Great Lakes National Program Office (GLNPO) began a five-year program that, like many other EPA efforts, is more concerned with studying problems than with solving them. Its aims include a public education program to protect the lakes from human abuse, finding out what is needed to make fish safe to eat in Lakes Michigan, Ontario, and Erie, restoring "beneficial uses" in all areas of concern, and addressing the problem of bottom sediments to determine if additional controls are needed to restore oxygen levels in Lake Erie. The actual job of pollution cleanup remains complex and expensive. Is it feasible, for example, to bury contaminated sediment with clean sand? In places where shipping channels must be continuously dredged, where can thousands of tons of mud laced with toxic chemicals be put? What are the long-term effects of air and water pollution on the ecosystems of the Great Lakes? These are but a few of the hard questions that must be answered. The crux of the problem was highlighted by William Brah, president of the Center for the Great Lakes: "The battle over conventional pollutants has been essentially won. But we have a long way to go on toxics, which are a more subtle and insidious problem. The things we are finding now are like the canary in the mine, an early warning that toxic chemicals are moving up the food chain."[26]

Zero Discharge Alliance

In 1978 the Great Lakes Water Agreement between Canada and the U.S. to protect the Great Lakes/Saint Lawrence River Basin adopted the "zero discharge" principle — that is, the complete elimination of the production, use, and disposal of persistent toxic chemicals that could get into the Great Lakes ecosystem.

When the IJC opened up public discussion on how to achieve this goal, it became apparent that most people living in the region wanted to move faster than the governmental agencies. In Ontario the New Democratic Party unexpectedly swept to power in the 1990 provincial election on a platform that included the promise to achieve zero discharge by the year 2000. At about the same time, a coalition of environmental and citizens' groups, Zero Discharge Alliance (ZDA), was formed in Windsor, Ontario, and Detroit, Michigan. In addition to Greenpeace, Friends of the Earth, the Sierra Club, National Audubon, and Clean Water Action, the alliance is supported by some fifty grass-roots and activist groups in Canada and the U.S. ZDA presents a strong alternative to the U.S. and Canadian government policy of "one-at-a-time" pollution control. The EPA, for example, establishes how much of a pollutant may be released into the environment by issuing "permits" that allow industry and others (including the government itself) to discharge limited amounts of toxins into the air and water. The agency then tries to police the polluters to force compliance with the permitted limits. The basic flaw in this approach, ZDA points out, lies in the sheer numbers involved: There are more than forty thousand chemicals in use in the U.S. alone and one to two thousand new ones introduced commercially each year. Meanwhile, during its twenty-year effort, the EPA has managed to set "safe" limits for *fewer than one hundred* of these chemicals. Nevertheless, they have issued permits that ignore most chemicals entirely because they have no basis for saying how much is safe.[27]

Persistent Chemicals

Persistent chemicals include mercury, lead, arsenic, cadmium, and other heavy metals which never break down because they are, themselves, elements. Other examples include PCBs, dioxins, chlorine compounds, and DDT. DDT has a half-life of about twenty years in the environment. (A half-life is the time it takes for half of a substance to disappear. After ten half-lives approximately one-thousandth of a substance remains. So, as a rule of thumb, scientists say that chemicals persist in the environment for ten half-lives. Thus DDT, with a half-life of twenty years, takes two hundred years to disappear from the environment. There are thousands of persistent toxic chemicals that are dangerous because they remain available for such a long time to poison plants, animals, and humans.

(Source: Environmental Research Foundation.[28])

ZDA advocates a sweeping ban on the release of persistent toxins by preventing pollution at the source, before it gets into the Great Lakes. For the alliance, the term *discharge* is not limited to a single environmental medium; it applies to "toxic discharges into water, air, landfill, product, etc.... and in all

cases where there is good reason to believe the substance itself is a persistent and/or bioaccumulative toxic or when persistent toxics are generated during its production, use or disposal."[29] The ZDA program urges the International Joint Commission to end the paper industry's use of chlorine for bleaching pulp, ban incinerators in or near the Great Lakes basin, and phase out industrial discharge permits. Every year the chlorine-based bleaching of paper discharges millions of pounds of toxic organochlorines directly into the Great Lakes. The paper industry defends itself by arguing that only "a few" of the compounds in its effluent are environmentally dangerous. Yet in 1989 IJC's Science Advisory Board condemned the bleaching as harmful to ecosystems and human health, recommending that the organochlorines be phased out.[30] Burning of garbage and hazardous waste is the fastest-growing new source of poisons going into the lakes. It releases into the air additional organochlorines as well as mercury, lead, cadmium, and other toxic metals. Toxic incinerator ash goes into landfills, pollutes the groundwater, and eventually ends up in the lakes. The third main element of the Zero Discharge strategy is to get the Canadian and U.S. governments to establish a system of "sunset permits." These would set fixed dates after which industry could no longer produce and discharge toxic substances into the Great Lakes. The three steps, ZDA says, are an absolute minimum for beginning the job of eliminating persistent toxic pollution in the Great Lakes.

What You Can Do to Help Clean Up the Great Lakes

1. If you live in the Great Lakes region, have your city council, church, or other civic organization endorse the Zero Discharge Alliance campaign.
2. Write your elected officials urging zero discharge and ask for supporting legislation at federal, state, provincial, and local levels.
3. Ask managers at local stores to stock nonchlorine-bleached paper. It is becoming increasingly available. Get your local schools, offices, and workplaces to use it.
4. Let the paper companies know that you strongly disapprove of chlorine bleaching of paper and that they should switch to less polluting bleaches such as peroxide. In a major breakthrough in early 1992 *Time* magazine was persuaded by Greenpeace to switch to nonchlorine bleached paper.
5. Urge your local newspapers to use a higher percentage of recycled and nonchlorine bleached paper than they do at present.
6. Get your community to intensify trash recycling instead of overloading landfills or building costly and contaminating garbage incinerators.*
7. For further help contact:
— Greenpeace/Great Lakes Project, 1017 West Jackson Boulevard, Chicago, IL 60607; (312) 666-3305; or 185 Spadina Avenue, #600, Toronto, ONT, M5T 2C5; (416) 345-8408;
— Great Lakes United, State University College, Cassety Hall, 1300 Elmwood Avenue, Buffalo, NY 14222; (716) 886-0142;

* For more on how you can do this, see *Design for a Livable Planet*, pages 26-28.

— Lake Michigan Federation, 1270 Main Street, Green Bay, WI 54302; (414) 432-5253;
— The Michigan Audubon Society, 6011 West Saint Joseph, Suite 403, Lansing, MI 48908; (517) 886-9144;
— The National Wildlife Federation, Great Lakes Natural Resource Center, 802 Monroe, Ann Arbor, MI 48104; (313) 769-3351;
— The Sierra Club, 214 North Henry Street, Suite 203, Madison, WI 53703; (608) 257-4494;
— Zero Discharge Alliance, P.O. Box 32246, Detroit, MI 48232, or P.O. Box 7243, Windsor, ONT N9C 3Z1.
Note: The following national organizations also have programs dealing with lakes, ponds, and wetlands: Conservation Foundation; Environmental Defense Fund; Environmental Law Institute; National Audubon Society; Natural Resources Defense Council; the Nature Conservancy; and the Wilderness Society. For addresses, phone numbers, and details on their activities, see the Directory, pages 331-355.

Walden Revisited

It is ironic that the historic pond near Concord, Massachusetts, intimately associated with Henry David Thoreau should itself become a victim of the attention he so lovingly bestowed on it. "I've been going to Walden for thirty years," says Edmund Schofield, ecologist, president of the Thoreau Society, and member of Walden Forever Wild (WFW). "Now I see a mismanaged landscape and a sick ecosystem."[31] Walden now attracts almost a million visitors a year who are wreaking havoc by stamping out vegetation and causing its shoreline to erode. Most of them swim in the pond, giving it the highest urine content of any freshwater body in the state. In the effort to save Walden, WFW is working for legislation to ban swimming there and to limit car parking in the area. In neighboring Walden Woods, the Thoreau Country Conservation Alliance (TCCA) is fighting construction of a 148,000-square-foot office-park complex slated to go up less than 700 yards from the shore. Its placement, says Thomas Blanding, president of TCCA, "represents the very imbalance that Thoreau warned against."[32] With the support of the Massachusetts Audubon Society, the New England chapter of the Sierra Club, and the Thoreau Society, TCCA pressured the developer to agree to an initial environmental impact study. The company then indicated it would be willing to sell the property for $10 million, more than triple its 1984 purchase price. Meanwhile TCCA is working with government officials to develop a conservation plan for all of Walden Woods. On the opposite side of the pond, also in Walden Woods, the site of a 135-unit proposed condominium project was purchased for $3.6 million in 1990 by the Walden Woods Project, an outgrowth of TCCA. Local zoning codes barring such developments had been waived because the developers said they would provide a minimum percentage of the units for low-income housing. Commenting that many local people and officials were losing touch with Walden's historical identity, Blanding says, "If you can't save the place where the conservation ethic was first asserted, how can you hope to assert the principle elsewhere?"[33]

Contact one or more of the many groups working to preserve this important area:

— Friends of Walden Pond, c/o J. Walter Brain, Concord Road, Lincoln, MA 01773; (617) 259-8823.

— The Massachusetts Audubon Society, South Great Road, Lincoln, MA 01773; (617) 259-9500.

— The Thoreau Country Conservation Alliance, 100 Barretts Mill Road, Concord, MA 01742; (508) 369-3565.

— The Thoreau Society, located in the Thoreau Lyceum, 156 Belknap Street, Concord, MA 01742; (508) 369-5912. For historical information primarily.

— Walden Forever Wild, P.O. Box 275, Concord, MA 01742; (508) 371-2421. Executive director, Mary P. Sherwood.

— Walden Pond Advisory Committee, c/o Walden Pond State Reservation, Concord, MA 01742. President, Kenneth Bassett.

— Walden Pond Historian, Dick O'Connor, 39 Everett Avenue, Watertown, MA 02172.

— Walden Woods Project, 18 Tremont Street, Suite 630, Boston, MA 02108; (800) 543-9911.

Hydrilla Invasion

Hundreds of lakes in Florida and South Carolina are being invaded by hydrilla, an Asian aquatic weed that has found its way across the United States from southern California, hitching rides on boats and trailers, and undisturbed by any known natural enemies. Believed to have originated in India, colonies of the plant appear as tangled green carpets of floating vegetation that spread out like umbrellas near the surface of the water, killing millions of fish and other organisms. In November 1991 some 60,000 acres in half of Florida's public lakes were choked with hydrilla. "It withstands drought, high flows, floods, insects, fish. It has lots and lots of survival methods and is one of the fastest growing plants," warned Jeffrey Schardt, environmental administrator for the Florida Bureau of Aquatic Plant Management.[34] The weed has also spread to South Carolina, fouling half the shoreline and 35,000 acres in Lake Marion and Lake Moultrie, near Rimini in the center of the state. The lakes are considered to be among the best fishing spots in the region. In addition to killing fish, hydrilla's rope-like vines clog up water intake valves of power plants and the propellers of boats. Efforts to control hydrilla in South Carolina now cost close to $2 million a year, but botanists do not believe they have made much headway. Florida reports limited success with herbicides, and the U.S. Department of Agriculture is experimenting with an insect that feeds on hydrilla, but it is too early to tell if this method will be effective, Mr. Schardt said. The botanist who first identified hydrilla in Florida drowned in the early 1970s when he became entangled in the weed underwater while doing research on it.[35]

Freshwater Wetlands

Accounting for 90 percent of U.S. wetlands, freshwater systems include inland marshes, wet meadows, mud flats, ponds, bogs, bottomland, hardwood forests, wooded swamps, and fens. Until relatively recently they were considered to have little or no value. Now, as they disappear, many people are beginning to appreciate them as precious ecological resources nurturing wildlife, purifying waters, checking the destructive power of floods and storms, and helping recharge groundwater supplies and maintain water tables in adjacent ecosystems. They are also used for growing wild rice, cranberries, and other crops, and for many types of recreational activities.

Freshwater wetlands — land flooded all or part of the time with fresh water — cover some 415,000 square miles* in the U.S., one-third in the lower forty-eight states, two-thirds in Alaska. Located on floodplains, alongside rivers, streams, lakes, and ponds, and in isolated depressions upland, they embrace a wide range of ecosystems identifiable by their emergent vegetation: bottomland forests in the Southeast, prairie potholes in the northern Great Plains and Canada, black-spruce bogs and red-maple swamps in the northeastern and north central states, and wet tundra in Alaska. Being formed in basins, these wetlands are naturally surrounded by uplands and are sometimes called island habitats. Providing nutrients from decaying organic matter and minerals that wash in from streams, rivers, and lakes, they attract dense concentrations of organisms that feed on the nutrients and other species that feed on them.

Types of Wetland

Bog: A peat-accumulating wetland with no significant inflow or outflow of water, supporting mosses, especially sphagnum.

Bottomland: Lowlands along rivers and streams, usually alluvial and periodically flooded. They are often forested and sometimes called bottomland hardwood forests.

Fen: A peat-accumulating wetland drained from surrounding soil and generally supporting marshlike vegetation.

Marsh: A wetland that is often or always flooded, characterized by emergent herbaceous vegetation adapted to saturated soil conditions.

Peatland: A wetland that accumulates partly decayed plant matter.

Pothole: A shallow, marshlike pond, found particularly in the prairies of north central U.S. and western Canada.

Slough: A swamp or shallow lake system in northern and midwestern U.S.; a marsh or slowly flowing shallow swamp in the southeastern U.S.

Swamp: A wetland dominated by trees or shrubs.

Wet meadow: A grassland with waterlogged soil near the surface but without standing water for most of the year.

(Sources: National Audubon Society; Milton W. Weller, *Freshwater Marshes*)

* Roughly the size of California (156,000 square miles) and Texas (262,000 square miles) combined.

Inland wetland ecosystems provide habitats for freshwater fish and wildlife including one-third of all the bird species in North America. Birds proliferate in wetlands because most of their natural enemies are inhibited by water. Only a few of their predators — mink and otters, mainly — regularly swim and live in or near water and raise their young there. The other aquatic species either provide food for the birds or, as in the case of beavers, bobcats, or alligators, do not normally bother them.

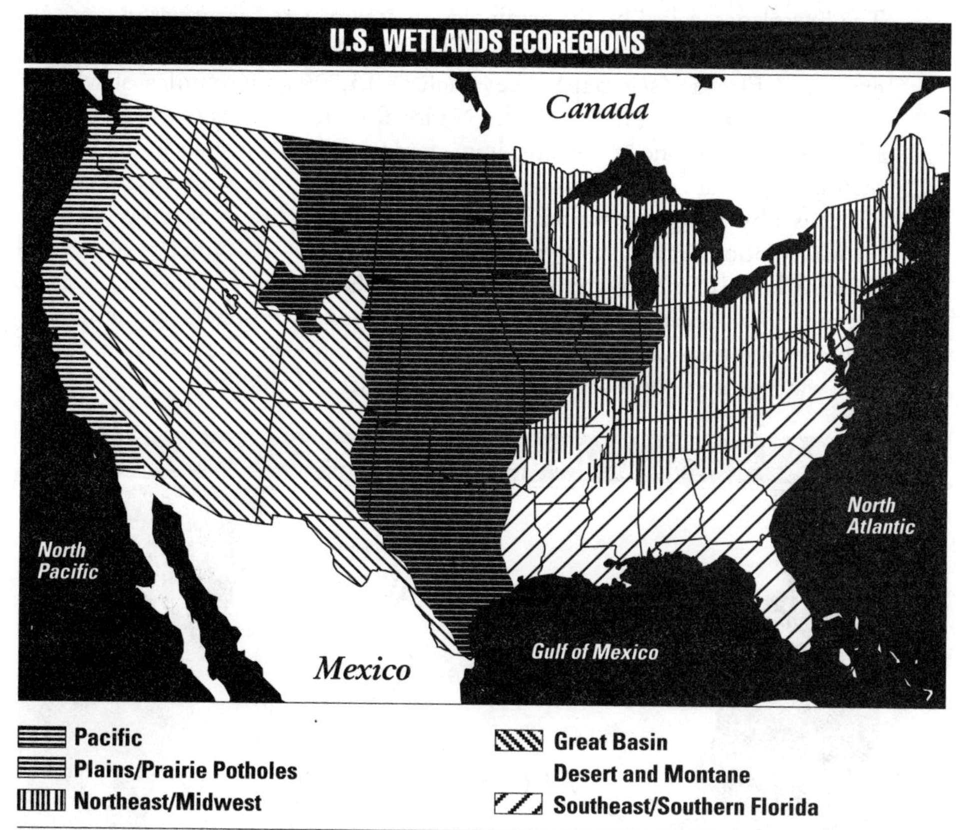

State of the Wetlands

Wetland destruction has been a national tradition for more than two hundred years. In 1764 the Virginia Assembly chartered the Dismal Swamp Company (of which George Washington was a member) to drain 40,000 acres of the Great Dismal Swamp so that its fine lumber could be cut. In the mid-1850s Congress gave 65 million acres of wetlands — an area the size of Arizona — to states, urging them to reclaim and sell them. Between 1940 and 1960 the U.S. Department of Agriculture subsidized the drainage of another 60 million wetland acres for farmlands. Since then the pace of destruction has accelerated. Wetlands have been converted to shopping malls, industrial parks, housing tracts, and many other uses, with more than half of the wetlands in the lower forty-eight states now filled in. Ninety percent of California's and Connecticut's wetlands is gone, as is 95 percent of Iowa's. At the current rate, half of the remaining wetlands in the U.S. will disappear in less than a hundred years.

The Everglades

"There are no other Everglades in the world. They are, have always been one of the unique regions of the earth, remote, never wholly known."

— Marjory Stoneman Douglas, *The Everglades*

The Everglades, called by the Seminoles "grassy waters," is a river of grass 50 miles wide and 100 miles long extending from Lake Okeechobee to the southern tip of Florida (see map). They include 13,000 square miles of watery wilderness, much of it in Everglades National Park, which is twice the size of Rhode Island and one of the most important preserved wetlands in the United States. The terrain is marshy grassland bordered along the western and southern coasts by mangrove swamps. The entire region is low-lying and flat, with naturally poor drainage. Drenched annually by 50 to 60 inches of rain, it remains largely underwater from June until November. In the Everglades National Park green fields of saw grass are dotted with islands or hummocks of hardwood trees and less frequently with thick forests of cypress interlaced with watercourses supporting an incredible variety of wildlife.

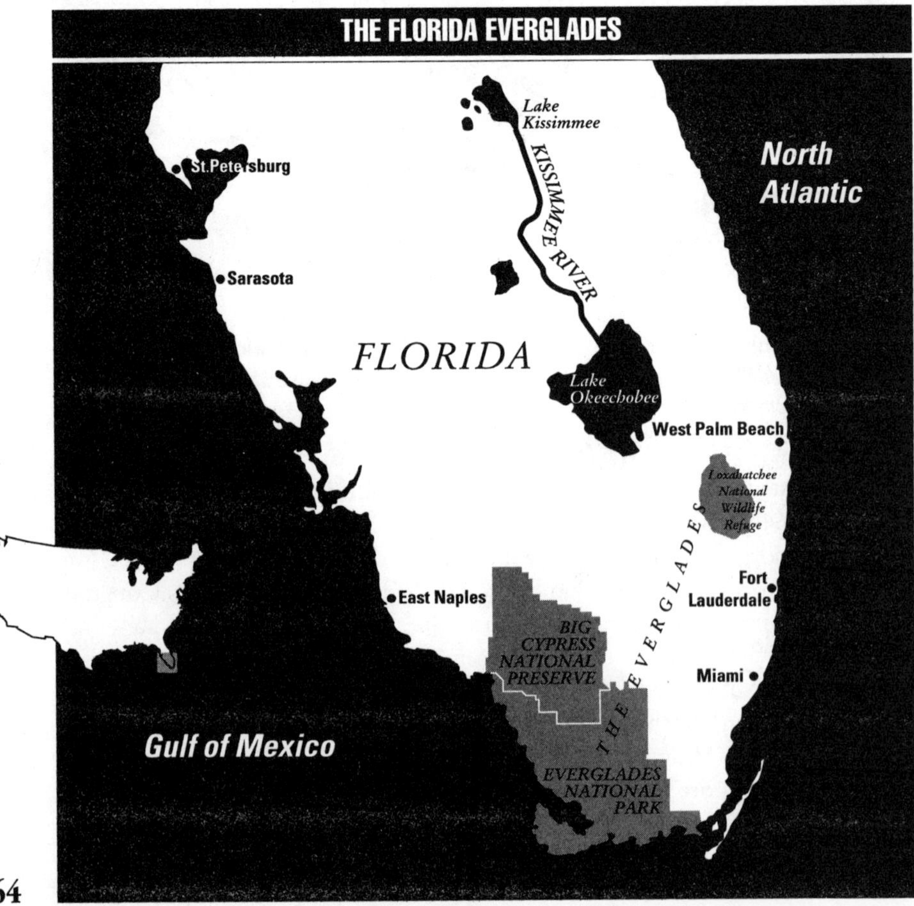

Its proximity to Miami and Palm Beach has drastically changed the Everglades. The relentless campaign to convert Florida's swamps to real estate began in the 1890s. Drained by a half-billion-dollar maze of canals, the Everglades were transformed from a wildlife mecca shared by Native Americans and some ten thousand early settlers into a suburban sprawl of almost five million residents. The Kissimmee River-Lake Okeechobee-Everglades ecosystem that once encompassed 9,000 square miles has been reduced by one-third. "The Everglades were treated as a commodity... [and in the process of draining them] developers reduced a natural work of art to a thing pedestrian and mundane," said Senator Bob Graham, former Governor of Florida, who launched the Save the Everglades Campaign in 1983.[36]

The Survival of the 'Glades

The fragility of the Everglades ecosystem was highlighted in the aftermath of the 1992's Hurricane Andrew, which devastated south Florida and the Gulf of Mexico in Louisiana. In the Everglades National Park Andrew wrecked two visitors' centers, exhibits, employee housing, and important research facilities at an estimated cost of $27 million. Perhaps even greater damage was done as the hurricane left in its wake a 22-mile path of dying vegetation, which may allow seeds from exotic trees and plants, such as Autralian pines and melaleuca trees blown in by the wind, yo take root in the sawgrass marshes and hardwood hummocks. According to Park Service scientists, this could radically change the park's appearance and ecology. Downed vegetation rotting in creeks and bays could produce a tremendous amount of decaying organic matter, robbing the water of dissolved oxygen breathed by fish and other creatures. Significant fish kills have already been noted in three west coast bays of the Everglades. In the longer term, however, if the organic matter is broken down, fish could thrive there again.

In the Louisiana coastal wetlands, rain did more damage than wind, causing some seven million fish deaths in oxygen-depleted waterways. The storm surge with 140-mph winds tore loose "flotant marsh"– peatlike soil and vegetation that floats on fresh water. Without the oil and vegetation, the terrain goes instantly from marsh to water, leaving behind large shallow ponds in which even the slightest wind stirs up the muddy bottom, blocking out sunlight and rendering the ponds ecologically worthless. Andrew also pushed saltwater into many freshwater marshes, setting off "tidal scouring," in which vegetation dies and barren soil erodes and flushes out to sea with the outgoing tide.

Early Assaults

The first attack came from the New York millinery industry whose traders in wild-bird plumes massacred vast populations of herons and egrets on the Everglades' rookeries. Then came Florida's pioneers. In the late 1890s they tried to control flooding in the south of the state by building canals. They drained the wetlands for farming and to build on, causing water tables to drop. In the 1940s, the new farmlands and residential communities flooded, and salt water intruded from the sea, replacing the vital fresh groundwater. Although the newly drained

land was initially very fertile, the peat soil soon dried out and the muck turned to dust, which blew away. In the Everglades Agricultural Area immediately south of Lake Okeechobee, where 700,000 acres of drained wetlands comprise the largest pocket of peat soil in the world, subsidence is occurring at a rate of a foot a decade. In some places only a few inches of soil remain above bedrock.[37] The vast flocks of wading birds, which are most directly dependent on the Everglades' seasonal rhythms of rising and falling water, have all but vanished. Their population, estimated at 2.5 million in the early 1800s, is now thought to be less than 250,000.[38]

Raising Cane

Since World War II, federal programs to divert water for flood prevention, agriculture, and urban development have further upset the natural ecological balance of the Everglades. In 1948 Washington, trying to make the central part of southern Florida economically profitable, sent in the Army Corps of Engineers to build 1,500 miles of canals and levees south of Lake Okeechobee and to drain 700,000 acres of wetland so that they could be used for raising sugarcane and other crops. The network of canals acts as a link or sometimes a barrier between the lake — the Everglades' main source of water — and the Everglades themselves. Now the flow of water depends not on acts of nature but on decisions by engineers. In the 1960s the Kissimmee River, a winding waterway rich in wildlife, was straightened and dredged by the U.S. Army Corps of Engineers for flood control. The resulting ditch supports little wildlife and is heavily polluted with agricultural runoff.

Faced with a record drought in the early 1970s, the state decided to use the new canal system to even out the extremes of wet and dry seasons. What they didn't realize was that the cycle of extremes is vital to the survival of animal and plant life. They also were unaware that the long-term addition of a few inches of water to the Everglades' normal levels would drastically transform the ecosystem from wetland to pond, endangering animal and plant communities in the process. In the late 1970s blooms of toxic algae and other oxygen-sapping plant life caused Lake Okeechobee itself to change dramatically.[39] Water-management officials, concluding the problem was linked to phosphorus from the runoff of cow manure and from the decaying soil of the converted wetlands, diverted the nutrient-rich runoff from Okeechobee and pumped it instead into the Everglades. This they did in the mistaken belief that the area was so large it could absorb the nutrients. According to conservationists, this led to the explosive growth of cattails that are choking the waterways.[40] Growing more than 10 feet tall on their rich diet of phosphorus, cattails tower over the naturally occurring saw grass, crowding it out with their roots and leaves. They also shade out oxygen-producing algae that are the basis of the Everglades' food web. Less desirable species of algae that support fewer organisms proliferate on the nutrients, leaving thick stands of cattails and foul-smelling mud.[41] In 1980-1981 southern Florida suffered a year-long drought during which the Everglades' dried-out muck caught fire, sending clouds of smoke over Miami, Fort Lauderdale, and Palm Beach. At the same time an underground wedge of salt water moved inland, contaminating freshwater supplies.

A Prophet in Her Time

The publication in 1947 of *The Everglades: River of Grass*, Marjory Stoneman Douglas's environmental classic, coincided with the opening of the Everglades National Park, for which she had lobbied for twenty years. It was also one year before the heavy machinery of the Army Corps of Engineers moved in. The book was the first shot in her ceaseless campaign to persuade Florida and Washington to save this unique region. As the *New York Times* wrote in an editorial celebrating Douglas's one hundredth birthday, the book "transformed Florida's perception of Florida's fabled wetlands."[42]

In 1969, in response to a proposal to build a giant jetport in the heart of the Everglades, Douglas founded the citizens' group Friends of the Everglades (FOE), which has achieved victories in the fights to restore the Kissimmee River, clean up Lake Okeechobee, and return the sheet flow of water to the Everglades marshes. For more than twenty years Douglas presided over the annual meeting of FOE, whose work has been influential in former governor Graham's "Save Our Everglades" program for the state of Florida, initiating wildlife preservation laws in Florida, drafting rules governing Florida's aquatic preserves and ecosystem protection for the Everglades. What is perhaps the ultimate testimony to Douglas comes from Roderick Jude of the Sierra Club: "The Everglades wouldn't be there for us to try and continue to save if not for her work through the years."[43]

Growing Threats

In March 1989 game fish in the Everglades suddenly showed mercury levels among the highest ever recorded in the United States. According to the *Florida Naturalist*, burning sugarcane prior to harvest pours 21,500 tons of mercury into the air, land, and water. Additional mercury may come from pesticides and fungicides used by the sugar industry. The Florida Fish and Game Commission estimates that a 75-pound panther, for example, could get a fatal dose of mercury from eating a single raccoon that had dined on six contaminated fish a day for fourteen weeks.[44] In September 1989 alligator hunting in Florida was banned because of high mercury levels in their meat. At the same time the National Audubon Society estimated that the number of wood storks, egrets, and other long-legged wading birds nesting in the Everglades had declined by 90 percent since the 1920s. Washington's response to the threats was inconclusive. In 1988 Congress directed the Army Corps of Engineers to undo the damage it had wrought. However, under pressure from President Reagan, the corps refused to spend the $2.3 million appropriated by Congress for the restoration. At the same time the U.S. Department of Justice initiated a lawsuit seeking to compel Florida to enforce its own environmental laws and to clean up the water flowing into the Everglades National Park and a national wildlife refuge near West Palm Beach. However, sugarcane and vegetable farmers in the Lake Okeechobee area successfully invoked federal rules exempting them from preventing

toxic runoff from their land. In 1991 there were three major proposals for helping restore the Everglades: an Army Corps of Engineers plan to dismantle the 98-mile canal it built in the Kissimmee River and return it back to its original channel at an estimated cost of over $300 million; a corps plan to breach two canals to allow water to flow across a water conservation area and under Route 41, the Tamiami Trail, at an estimated cost of $75 million; and a federal plan to buy 107,000 acres to add to the Everglades National Park at a cost of $40 million. Forming part of a larger ecosystem, the additional land would allow water coming in from the north to spread across the Shark River Slough, a 30-mile-wide basin to the northeast of the park.[45]

Seeking a Settlement

In January 1991 the newly elected governor of Florida, Lawton Chiles, attempted to work out a settlement with the federal government that, it was hoped, could begin the process of saving the Everglades.[46] Seven months later the bitter suit, initiated by the Department of Justice and costing the state $6 million in legal fees, was "settled," but a solution to the basic problem of the Everglades seemed far removed. According to the settlement, which must be ratified by the South Florida Water Management District, the state must clean up polluted water before it flows into Everglades National Park and the Loxahatchee National Wildlife Refuge, two large sections of the Everglades controlled by the federal government.[47] The state is required to build 35,000 acres of marshes south of Lake Okeechobee, the headwaters of the Everglades, so that water contaminated with phosphorus and other nutrients from vegetable and sugarcane fields will be purified by the marshes before it joins the main flow of water that sustains the Everglades. The proposed settlement was opposed by growers and conservationists. The sugarcane and vegetable growers denied responsibility for the levels of phosphorus from agricultural fertilizer runoff, which are ten to twenty times higher in areas immediately south of Lake Okeechobee than in the more remote parts of the Everglades. "There has never been an evidentiary hearing presenting facts that show damage to the Everglades," argued Bill Earl, an attorney for the Florida Sugar Cane League, in an effort to shift the blame elsewhere.[48] Conservationists claimed that the proposal would neither clean up the Everglades nor make the polluters pay their fair share for the damage being done. "We cannot endorse this settlement," said Joe Podgor, FOE's executive director, "because it shifts the responsibility to clean up the pollution from the polluters to the public. We cannot put any faith in the uncertain, and potentially irreversible long-term effects of installing sewage lagoons as the new headwaters for the Everglades."[49]

The SWIM Plan

Friends of the Everglades takes issue also with a proposed storm-water utility/surface-water improvement and management (SWIM) plan included in the Everglades Protection Act passed by the Florida state legislature in response to the federal lawsuit and named (hypocritically) after Marjory Stoneman

Douglas.* The proposed building by the state of sewage ponds to receive the polluted wastewaters from private agribusinesses would provide what FOE calls a "public septic tank without a clean-out contract" that would become a serious environmental and public health threat to the very resources it is supposed to clean up.[50] In July 1991 FOE proposed its own SWIM plan with a five-year timetable, offering regulatory and structural solutions. Year 1 calls for an inventory of and permit applications for all water discharges to public waters and consumptive uses of water within the Everglades Agricultural Area (EAA). For subsequent years it establishes water-quality standards to be met by all discharges from EAA private lands into public waters; three graded steps would allow reasonable time for dischargers to adjust to the new requirements.

The FOE proposal includes creating a public marsh from former farms in the EAA stretching from existing water conservation areas to Lake Okeechobee. This is intended to rehabilitate former farmlands into a "viable functioning link which ultimately reunites the Kissimmeee River/Lake Okeechobee portions of the Everglades Regional Watershed with the Everglades, River of Grass." No discharge would be permitted from the marsh until the water is clean.[51] The plan would also establish a monitoring and research program to evaluate progress of rehabilitation techniques designed to make the marsh a healthy link in the Everglades watershed and ecosystem.

What You Can Do to Help Save the 'Glades

1. Support the ongoing campaigns of the following organizations:
— Friends of the Everglades, 101 Westward Drive #2, Miami Springs, FL 33166;
— Florida Audubon Society, 460 Highway 436, Suite 200, Casselberry, FL 32707;
— Sierra Club, P.O. Box 43071, South Miami, FL 33143;
— The Florida chapter of the Wilderness Society, 4203 Ponce de Leon Boulevard, Coral Gables, FL 33146.
2. Support legislation introduced by Senator Graham and Representative Fascell to expand the Everglades National Park.
3. For background information, contact the Everglades National Park, P.O. Box 279, Homestead, FL 33030; (305) 247-6211.
4. Read Marjory Stoneman Douglas's book.

* Douglas, who was 101 years old at the time and virtually blind, was not given an opportunity to have the text read to her before Governor Lawton Chiles visited her Coconut Grove cottage to sign the bill before the television cameras. "I think they were sort of using me," she said in an interview with Sally Deneen in *South Florida* magazine in August 1991.

Big Cypress National Preserve

Located in southern Florida between Miami and Everglades City, Big Cypress Swamp is the northern extension of the Everglades, from which it is separated only by the Tamiami Trail highway. "Big" refers not to its trees, which are mostly the dwarf pond variety, but to its 2,400 square miles covering an area greater than that of the Everglades. The swamp is actually a mixture of hardwood hummocks (tree islands), wet and dry prairies, marshes, and estuarine mangrove forests. Broad belts of cypresses edge the wet prairies and line the sloughs. The few remaining giant cypresses, some of them seven hundred years old, are survivors of the lumber boom of the 1930s and 1940s when they were used to make pickle barrels, coffins, stadium seats, and the hulls of PT boats.

In an average year 60 inches of rain falls in the lakes of the Kissimmee basin northeast of the swamp and moves slowly southward to Lake Okeechobee and then drains into the Big Cypress and down into the Everglades. Because the land slopes only 2 inches per mile the drainage is slow, extending the wet season by two to three months after the rains taper off in October and providing a steady mix of fresh and salt water in the estuaries along the coast of Everglades National Park. The nutrient-rich combination supports pink shrimp, snapper, snook, and other marine animals important to Florida's fishing industry. The swamp also supplies water for several southwest Florida cities. During the wet season much of the landscape flows with water belly-high to great blue herons, egrets, wood storks, and other wading birds. If you are lucky, you can also see red-cockaded woodpeckers, bald eagles, alligators, mink, deer, and long-legged, long-billed limpkins in Big Cypress.

Florida's first producing oil well was drilled in Big Cypress in 1943. In the 1960s drainage of the swamp was accompanied by land development and speculation schemes. In 1968 plans to build a jetport on the swamp's eastern edge posed a threat to the watershed of the Everglades and led to the creation in 1974 of Big Cypress National Preserve. Run by the National Park Service, national preserves are a new type of federally protected area that allows many uses prohibited in most national parks — hunting, driving off-road vehicles, swamp buggies, and airboats, and existing cattle grazing and airstrips. Oil exploration was permitted until halted by a conservationist lawsuit in 1988. Speeding cars and trucks on the Tamiami Trail and Alligator Alley are the main killers of the endangered Florida panther (less than fifty of which are now left[52]) and thousands of other animals. Only 40 percent of Big Cypress was set aside by Congress for the preserve. Conservationists are generally unhappy with the state of Big Cypress. As Dave Foreman and Howie Wolke say in *The Big Outside*, this preserve "may have the dubious distinction of being the most poorly managed unit in the National Park System."[53]

Corkscrew Swamp

In the northwest corner of Big Cypress is the 11,000-acre Corkscrew Swamp, the most important remaining fragment of virgin full-sized bald-cypress forest in southern Florida. It is also the largest wood-stork rookery in North America. A specimen of an almost lost terrain, Corkscrew — named after the local term for a cypress stand — is now a sanctuary managed by the National Audubon Society.

Kite versus Stork

Locally known as "flintheads" because of the dark-gray unfeathered skin on their head and upper neck, wood storks have been steadily declining in numbers from some twenty thousand pairs in the 1930s to perhaps three thousand pairs today. The main problem is that the loss of wetlands and the change of water levels in the Everglades has severely cut the production of fish that were concentrated in the storks' traditional feeding areas. When water levels rise too high, such fish as there are get dispersed and the storks' catch declines.[54] Ecological dislocations set off by the artificially created water patterns in the Everglades affect storks and other wading birds in ways that are not yet fully understood. Storks and other species are moving their colonies to central and northern Florida and even to South Carolina in search of food supplies. Another bird, the snail kite, a tropical raptor reaching the U.S. only in southern Florida is facing a different kind of fate. One of the first species to be listed as endangered in 1966, it feeds for the most part on the apple snail, which depends on a high water level for its survival. In times of low water, the snail becomes scarce and both adult kites and their chicks starve.[55] The water demands of Florida's human population, combined with drought and changes in the delivery of water to the Everglades, will affect more than wood storks and kites. In the words of Steven Beissinger, a wildlife ecologist and snail-kite expert, "An ecosystem is the sum total of many parts. If you manage it without one of its elements, you are not managing the whole system."

"An ecosystem is the sum total of many parts. If you manage it without one of its elements, you are not managing the whole system."[56]

— Steven Beissinger, wildlife ecologist

Okefenokee Swamp

Known to Native Americans as the land of the trembling earth, Okefenokee is a 412,000-acre, peat-filled swamp on the boundary between Georgia and Florida and one of North America's great unspoiled areas. Covering much of the swamp floor, the peat deposits are so unstable in spots that you can literally cause trees and bushes to tremble by stomping the surface. About four-fifths of the swamp is included in the Okefenokee National Wildlife Refuge, which is administered by the U.S. Fish and Wildlife Service (FWS). Except for some fifty small pine-covered islands totaling 25,000 acres, most of Okefenokee is shallowly flooded and forested with cypress, black gum, bay, and maple. About 60,000 acres are flooded marshland or "prairie," covered with water lilies, pipewort, ferns, maidencane, and a variety of sedges and grasses. Originally forested, these marshes were created during periods of

severe drought when fires burned out vegetation and the top layers of peat. The areas of open water consist mainly of watercourses, prairie lakes and ponds (known as 'gator holes), and the twelve-mile-long Suwannee Canal, dug in 1891. The varied habitat of Okefenokee's swamp forest, prairie, swamp edge, and moist and dry upland supports a correspondingly varied range of reptiles, amphibians, fish, bird, and mammal species that include alligator, armadillo, black bear, roundtail muskrat, sandhill crane, and the endangered wood stork and red-cockaded woodpecker.

Saving the Red-Cockaded Woodpecker

A major long-term FWS program for saving the habitat of this endangered species is the restoration of the longleaf pine/wire-grass community within its natural range in the Okefenokee Refuge. Historically, open stands of longleaf pine and the understory of wire grass and scattered clumps of gallberry and palmetto were maintained by frequent, low-intensity natural fires. However, in the early 1900s the almost total clear-cutting of the longleaf-pine forest disrupted the natural fire cycle, allowing more prolific vegetation to replace the pines and invade the habitat of the red-cockaded woodpecker, the Florida panther, the Eastern indigo snake, and many other species that form part of the longleaf pine/wire-grass community. Many of these species are long-lived and reproductively unprolific. Thus, when forced out of their natural habitats, they are likely to become extinct. Among the restorative measures being taken by FWS are selective thinning in mixed stands of tree to favor longleaf pine and direct seeding of the species on suitable sites. The service is also considering the use of prescribed fires — that is, the intentional warm-season burning in selected longleaf pine stands to replicate the natural fire regime and to encourage the regrowth of longleaf pine and the spread of the wire-grass undercover.[57]

Forested Wetlands

According to the U.S. Department of the Interior, nearly 80 percent of the 25 million acres of periodically flooded bottomland hardwood forests that once existed in the lower Mississippi Valley have been lost to agriculture. What remains still serves as the main wintering grounds for most of the continents's mallards and for virtually all of the wood ducks in the central U.S. It also provides a rich habitat for many varieties of other wildlife and fish.

Prairie Potholes

The North American prairie pothole region covers 300,000 square miles, primarily in western Minnesota, Nebraska, the Dakotas, northern Montana, and south-central Canada. It was formed roughly ten thousand years ago, when Ice Age glaciers receded, leaving a poorly drained terrain pockmarked with millions of small depressions — potholes — most of them less than 10 acres in size. Some are so shallow that they hold water for a few weeks in springtime; others are covered with emergent vegetation and may hold as

much as 4 feet of water for all but the driest seasons; and still others contain large patches of water ringed with marsh plants. Prairie potholes serve as a vast breeding ground for North America's migratory waterfowl. More than half of our duck population is born there. Because these prolific wetlands are scattered among the nation's most fertile agricultural areas, many of them have been drained and turned into cropland. For example, 90 percent of Nebraska's Rainwater Basin wetlands, a focal point in the Central Flyway used by millions of migrating duck, geese, and crane, have been destroyed. However, many of the drained potholes cannot grow crops because the conversion sharply reduces water supply and soil quality of the surrounding farmland. The lack of water increases the danger of fire, especially in the recent times of intense drought. Ironically, some of the drained marshes are now reverting to their natural state.

Washington's Mixed Signals

The first major federal effort to protect wetlands was incorporated in the Clean Water Act of 1972 (CWA).* Under Section 404 of this act, you may not alter or destroy a wetland by discharging dredge or fill material into it without first getting a permit from the U.S. Army Corps of Engineers. For every permit issued, the corps must prepare a statement of finding that explains why the permit was granted. A permit can be vetoed by the Environmental Protection Agency if it determines that the project will cause unacceptable damage to the wetland. Like many laws, this one leaves many loopholes, some large enough to drive a bulldozer through. In the first place, it does not cover agricultural use, which accounts for some 80 percent of all wetland conversions over the past twenty years. It allows for public input in the application process, but the corps does not normally hold public hearings on permits. If you want one, you have to ask for it.† Another problem is that neither the corps nor the EPA has enough staff to monitor compliance with permits. For these and other reasons, including pressure from the White House, the corps has been slow to apply the law. In 1985, pressed by environmental groups and duck hunters, Congress included a "swampbuster" provision in the Food Security Act setting penalties for converting wetlands to fields but allowing farmers to plant perennial crops with impunity.[58]

No Net Loss

In 1987 the Conservation Foundation, a Washington, D.C.-based nonprofit environmental group, was asked by the EPA to organize the National Wetlands Policy Forum. The forum was an alliance of environmentalists, industrialists, government officials, and real-estate developers with a national goal of wetlands protection. Specifically, it called for a policy of

* For more on this act, see chapter 9, page 274.
† For specifics on what you can do to help save wetlands, see pages 75-77.

"no net loss" of wetlands, endorsed by the federal government and by Maryland, Illinois, Massachusetts, Oregon, New York, Washington, New Jersey, and other states. Under this principle, conversion of wetlands is allowed only if the user restores previously converted wetlands or creates new ones. In early 1989 the EPA announced a no-net-loss policy for the nation's remaining wetlands that was thought by the Conservation Foundation, the National Wildlife Federation, and other conservation groups to be a major step forward. Under the new policy, wetlands destroyed by development would have to be replaced in quality and quantity within the same watershed or ecological area. Up until then, federal agencies had sought to prevent wetland loss on a piecemeal basis. This was the first time a federal agency had committed itself to preventing an overall loss of wetlands.

From the outset the new policy was doomed. It didn't halt wetlands loss because little was known about how to restore them and because there was little public money or private financial incentive to fix wetlands degraded by decades of waste dumping, road building, strip-mining, and other damage. Despite widespread acceptance of its principle, early attempts to implement no net loss encountered angry resistance, especially from farmers who see the regulatory efforts as a confiscation of their assets. "I bought my farm and, if the government wants it, they should acquire it the good American way — buy it," said Rick McGown, a Missouri farmer, in his testimony to Congress in March 1990.[59] Then there was the matter of a White House responding to pressure from developers, oil companies, and other antiwetlands interests.

Bushwhacking

In his first election campaign George Bush said he wanted to become "the environmental president." All wetlands, he declared, "no matter how small, should be preserved."[60] Once elected, the President reaffirmed the no-net-loss policy, forming a Domestic Policy Council task force to pursue the matter. In a final gesture of support the President and Congress proclaimed May 1991 American Wetlands Month. So far so good, you might think. Behind the scenes it was another kettle of fish. The new administration pressed Congress to eviscerate the federal wetlands program and to weaken the wetland provisions of the CWA. In response to this maneuvering, the EPA's senior wetlands ecologist resigned from the Domestic Policy Council's technical committee because he felt he was being asked to behave unethically.[61] At the same time an antiwetlands consortium with the misleading name "National Wetlands Coalition" was formed. It was in fact a lobby for real-estate developers, oil and gas industry representatives, farmers, and dam builders that enlisted support from administration and congressional policymakers to open the door to unchecked development of wetlands nationwide. In public hearings blasting the federal regulations as being too tough, the consortium "launched a well-orchestrated grass-roots campaign like we've never seen before," commented Hope Babcock, general counsel for the National Audubon Society.[62] In the summer of 1991 two bills were intro-

duced in the House of Representatives that would classify wetlands according to their supposed value. Those deemed "less valuable" would get little or no protection under Section 404 of the Clean Water Act and thus become open to development. H.R. 1330, introduced by Representative Jimmy Hayes (D-Louisiana), would completely eliminate the EPA's role in the Section 404 process and drastically limit the legal definition of a wetland — to be considered a wetland, an area would have to be inundated by water for twenty-one consecutive days — wiping out with one stroke thousands of critically important seasonal and transitional wetlands. The second bill, H.R. 404 introduced by Representative John Hammerschmidt (R-Arkansas), was considered by environmentalists to be little better than the Hayes version.[63] (For an update on wetlands legislation, see Appendix B, pages 359-360.)

Moving the Goalposts

By August 1991 the Bush strategy was revealed as a cynical ploy to take away 30 percent of the nation's wetlands by redefining them as land that could be used for farming and other purposes.[64] The Bush "revisions" would exempt wet meadows, bogs, floodplains, Alaska moist tundra, and other temporarily flooded wetlands. Removal of these wetlands from the protective umbrella of Section 404 would mean lost habitat for hundreds of wildlife species. It would also leave large chunks of territory exposed to flooding, soil erosion, and water contamination. "It took the agencies that protect wetlands over twelve years to develop a biological definition of wetlands," stated Gerry Paulson, Sierra Club issue chairperson. "It would be unconscionable to allow it to be so easily thrown away."[65] And, as Ed Pembleton, director of Audubon's Water Resources Program, explains, "We simply don't know enough about the functions of wetlands to risk destroying them by excluding whole categories."[66] Immediate reactions from federal scientists indicated that under the new definition almost half of the Everglades might be lost. At the same time conservationists nationwide mobilized in their opposition to perhaps the largest weakening of a regulation in environmental history.

How the Wetlands Can Be Saved

Wetlands protection is beset by a tangle of federal laws and regulations complicated by widely divergent interests at state and local levels. Ultimately, action to save wetlands must be initiated by Congress, the EPA, the Army Corps of Engineers, and other federal agencies, and by an administration truly concerned with the long-term health of this country's ecosystems. In 1982 a Louis Harris national survey showed that 83 percent of the U.S. public believed it "very important" to preserve the nation's remaining wetlands. With this degree of support, saving the wetlands must clearly be a top priority on any conservationist or environmental agenda. Among the environmental organizations taking the lead in wetlands protection are the National

Audubon Society, the Wilderness Society, the Nature Conservancy, Ducks Unlimited, the Conservation Foundation, and the Environmental Law Institute which, with the National Wetlands Technical Council, publishes the *National Wetlands Newsletter*. In addition to participating in the Wetlands Policy Forum, these organizations work at many other levels to preserve and restore wetlands. They are involved in a nationwide effort to influence government policy, educate the public, and help develop specific short- and long-term remedial projects.

The Audubon Campaign

The July 1990 *Audubon* magazine is devoted to a "grand design" for the wetlands. It advocates designating as wetlands all those areas that are components of internationally significant ecosystems and calls upon national governments in the Western Hemisphere to protect wetlands, which provide important breeding and staging points for migrating birds.[67] Preserving wetlands is not enough, states *Audubon* in an effort to convince all federal agencies that have an impact on land use — from the Department of Transportation to the Farm Credit Administration — that *restoration* of wetlands must be an important part of U.S. national policy. *Audubon* also urges the improvement of state wetlands-protection laws and tougher local ordinances. While good laws are a beginning, "wetlands will be protected only if people understand how the laws work and make sure the laws are applied."[68]

Nature Conservancy Project

In 1983 the Nature Conservancy initiated the $75 million Wetland Conservation Project, which included a $25 million grant from the R. K. Mellon Foundation. Since then the Conservancy has protected several of the country's pristine wetlands including the Cache River basin in Arkansas, which draws the nation's largest concentration of winter mallards; the Platte River in Nebraska, which attracts 80 percent of the world's population of sandhill cranes; Crystal Springs, a South Dakota prairie-pothole region of three hundred permanent and seasonal fens and wetlands so critical to waterfowl that it is called "the duck factory"; Cosumnes Valley in California's Central Valley, where only a minute fraction of the original riparian habitat remains; and Ash Meadows in a remote corner of Nevada, the largest oasis in the Mojave Desert, which for its size boasts the highest number of endemic plants, fish, and invertebrates in the continental United States, including more than twenty species found nowhere else on earth. In its 1993 annual report the Conservancy announced that it had teamed up with the California Rice Industry, the California Waterfowl Association, and Ducks Unlimited to flood fallow rice fields in the winter to provide habitat for migratory waterfowl and potentially "creating" thousands of temporary wetlands in the Sacramento Valley.

*What You Can Do to Save the Wetlands**

1. Identify wetlands in your area. Get hold of maps published by the U.S. Fish and Wildlife Service National Wetlands Inventory (9720 Executive Center Drive, Suite 101, Monroe Bldg., St. Petersburg, FL 33702; or phone 800-USA-MAPS), the U.S. Geological Survey, and state and regional agencies.
2. Assess the condition of wetlands. Inventory plants and animals and document them with photos. Describe the wetlands and surrounding land use (e.g., subdivision, highways, agriculture). Identify actual or potential threats such as construction or dumping. Continue to monitor them regularly. If possible, take action to protect wetlands *before* development.
3. Identify owners of wetlands. Find out their plans for the site(s).
4. Develop a strategy. Focus on saving a specific wetland; create a master plan for your community; organize a "wetlands watch" program to monitor existing threats; educate the community; acquire key wetlands.
5. Identify protective mechanisms. Determine zoning for the area. Find out which protective laws and agencies can help. The main federal laws include: Clean Water Act, Section 404 (requiring a permit for altering or destroying a wetland by discharging dredge or fill material into it); Rivers and Harbors Act; Endangered Species Act; Coastal Zone Management Act; Coastal Barriers Resource Act; Food Security Act ("Swampbuster" provision). You should also consult state and local water, coastal zone, and wetlands legislation as well as local planning and zoning regulations.
6. Be proactive. Help set policies, action plans, and legislation before a threat materializes. Attend local planning, zoning, and wetlands meetings and public hearings and make known your concern for wetland protection.

Other Things You Can Do

- Wetlands are peaceful, restorative islands in a hectic world. Take a canoe into a wetland and just listen — to insects, frogs, birds, and the water. Share the experience with friends, especially children, and turn them into wetlands advocates.
- Write to the President about your concerns. Make him move the Wetlands Task Force into positive action.
- Support federal and state legislators and agencies who are working for wetland preservation.
- Educate other legislators to change their votes.
- Join the National Audubon Society, the Nature Conservancy, and other national environmental organizations in their efforts to save the wetlands.
- Work in your community with activist groups such as those described throughout this book.

* Adapted with permission from *Audubon Activist*, March 1991.

RESOURCES

1. International Organizations

— *Canada Center for Inland Waters*, Department of Fisheries and Oceans, P.O. Box 5050, Burlington, ONT L7R 4A6.
— *Canadian Institute for Environmental Law and Policy*, 517 College Street, Suite 400, Toronto, ONT M6G 4A2; (416) 923-3529.
— *Environment Canada*, Great Lakes Environment Office, 25 St. Clair Avenue East, Toronto, ONT M4T 1M2; (416) 973-8632.
— *International Joint Commission*, 100 Oulette Avenue, Windsor, ONT N9A 6T3; (519) 256-7821; or P.O. Box 32869, Detroit, MI 48232; (313) 226-2170.
— *Ontario Ministry of the Environment*, Municipal-Industrial Strategy for Abatement, 135 St. Clair Avenue West, Toronto, ONT M4V 1P5.

2. U.S. Government

The main federal agencies concerned with lakes and wetlands are the Army Corps of Engineers, the Environmental Protection Agency (particularly the Office of Water, Clean Lakes Program; the Great Lakes National Program Office; the Wetlands Hotline: 800-832-7828); and the Fish and Wildlife Service (Department of the Interior). For addresses and phone numbers, see the Directory, pages 331-355.

3. Environmental and Conservation Organizations

Among the leading groups working on lake and wetlands protection are the Environmental Defense Fund, Greenpeace, National Audubon Society, Natural Resources Defense Council, the Nature Conservancy, the Sierra Club, and the Wilderness Society. For details on these and other concerned organizations, see the Directory.

4. Further Reading

Lakes and Ponds

— *The Enduring Great Lakes*, John Rousmaniere. New York: Norton, 1979.
— *The Face of the Great Lakes*, Jonathan Ela. San Francisco: Sierra Club Books, 1977.
— *The Great Lakes: An Environmental Atlas and Resource Book*, U.S. EPA and Environment Canada. Washington, D.C.: Environmental Protection Agency, 1988.
— *Great Lakes, Great Legacy*, Theodora E. Colborn et al. Washington, D.C.: Conservation Foundation, 1990.
— *The Great Lakes Water Quality Agreement: An Evolving Instrument for Ecosystem Management*, National Research Council of the U.S. and the Royal Society of Canada. Washington, D.C.: National Academy Press, 1985.
— *Lakes: Chemistry, Geology, Physics*, Abraham Lerman, ed. New York: Springer Verlag, 1978.
— *The Life of the Pond*, William H. Amos. New York: McGraw-Hill, 1967.
— *The Natural History of Lakes*, M. J. Burgis and P. Morris. New York: Cambridge University Press, 1987.
— *The Pond*, Gerald Thompson and Jennifer Coldrey. Cambridge: MIT Press, 1984.
— *Rivers and Lakes*, Laurence Pringle. Alexandria, Va.: Time-Life Books, 1985.
— *U.S. Progress in Implementing the Great Lakes Water Quality Agreement*, Annual Report to Congress, EPA 905/9-89/005.
— *The Uses of Ecology: Lake Washington and Beyond*, W. T. Edmondson. Seattle: University of Washington Press, 1991.
— *Water for Life: The Tour of the Great Lakes*. Chicago: Greenpeace Great Lakes Office, 1989.
— *Zero Discharge: A Strategy for the Regulation of Toxic Substances in the Great Lakes Ecosystem*, Paul Muldoon and Marcia Valiante. Toronto: Canadian Law Research Foundation, 1988.

Wetlands

— "Audubon in the Glades: Ninety Years of Action," Frank Graham, Jr. *Audubon*, May 1990.
— "Beyond Swampbuster: A Permanent Wetland Reserve," Ralph Heimlich et al. *Journal of Soil Conservation*, September/October 1989.

— *Death in the Marsh*, Tom Harris. Washington, D.C.: Island Press, 1991.
— *Everglades*, Patricia Caulfield. New York: Ballantine Books, 1970. Photography with text by Peter Matthiessen.
— *The Everglades*, Archie Carr. Alexandria, Va.: Time-Life Books, 1973.
— *The Everglades: River of Grass*, Marjory Stoneman Douglas. New York: Rhinehart, 1947. The classic.
— *Freshwater Marshes: Ecology and Wildlife Management*, Milton W. Weller. Minneapolis: University of Minnesota Press, 1981.
— "The Glades," Alice Boone. *Buzzworm*, Autumn 1989.
— "Kite vs. Stork," Frank Graham, Jr. *Audubon*, May 1990.
— "The Last Watering Hole on the Prairie," Gary L. Krapu. *Natural History*, January 1989.
— *Our Nation's Wetlands*, E. L. Horwitz. Washington, D.C.: Council on Environmental Quality, U.S. Government Printing Office, 1978. Interagency survey of wetlands and government policy.
— "Restoring the Everglades," Jeffry Kahn. *Sierra*, September/October 1986.
— *Status Report on Our Nation's Wetlands*, J. S. Feierabend and J. M. Zelazny. Washington, D.C.: National Wildlife Federation, 1987.
— "Sunset on America's Wetlands," Paul Dunphy. *Harrowsmith Country Life*, July/August 1990.
— "Swampbusting in Perspective," Ralph E. Heimlich and Linda L. Langner. *Journal of Soil Conservation*, July/August 1986.
— *Waterlogged Wealth*, Edward Maltby. Washington, D.C.: Earthscan, 1986. Good overview of destruction of wetlands in the U.S.
— "A Wetland...is a wetland...is a wetland," Dale Marsh. *Journal of Soil Conservation*, July/August, 1989.
— *Wetlands*, W. J. Mitsch and J. G. Gosselink. New York: Van Nostrand Reinhold, 1986.
— *Wetlands*, William A. Niering. New York: Alfred A. Knopf, 1988.
— "Wetlands at the Crossroads," Jacquelyn L. Tuxill. *E Magazine*, May/June 1990.
— *Wetlands: Losses in the United States, 1780s to 1980s*, Thomas E. Dahl. Washington, D.C.: U.S. Department of the Interior, Fish and Wildlife Service, 1990. Excellent and graphic summary of the wetland losses.
— *Wetlands: Their Use and Regulation*, Office of Technology Assessment. Washington, D.C.: U.S. Congress, OTA-0-026, 1984.

NOTES

1. Basic source material for this section includes Edward J. Kormondy, *Concepts of Ecology* (Englewood Cliffs, N.J.: Prentice-Hall, 1984), pp. 248-58; and Laurence Pringle, *Rivers and Lakes* (Alexandria, Va.: Time-Life Books, 1985), pp. 113-33.
2. Pringle, *Rivers and Lakes*, p. 122.
3. Robert Leo Smith, *Ecology and Field Biology* (New York: Harper & Row, 1980), p. 214.
4. Based on "Lake Aging and Eutrophication," *The Lake and Reservoir Restoration Manual* (Washington, D.C.: EPA Office of Water, 1990), pp. 28-31.
5. This is the subtitle of W. T. Edmondson's *The Uses of Ecology* (Seattle: University of Washington Press, 1991), on which this section is based. The author is internationally recognized for his pioneering research on aquatic ecology and limnology.
6. Edmondson, *The Uses of Ecology*, p. 14.
7. Ibid., pp. 35-36.
8. Ibid., p. 52.
9. Ibid., p. 53.
10. Ibid., p. 127.
11. Susan Benner, "California's Ancient Inland Sea," *New York Times*, March 24, 1991.
12. Pringle, *Rivers and Lakes*, p. 128; see also Edmondson, *The Uses of Ecology*, pp. 159-60.
13. Robert Budert, "New Plan for Mono Lake," *Nature*, October 12, 1989, p. 478.
14. Pringle, *Rivers and Lakes*, p. 132.
15. Robert Service, "Judge's Ruling Good News for Mono Lake and Its Wildlife," *Audubon Activist*, July/August 1991, p. 8.

16. Ibid.
17. Edmondson, *The Uses of Ecology*, p. 163.
18. Pringle, *Rivers and Lakes*, p. 154.
19. Wayne A. Schmidt, "Are Great Lakes Fish Safe to Eat?" *National Wildlife*, August/September 1989.
20. Theodora E. Colborn et al., *Great Lakes Great Legacy?* (Washington, D.C.: Conservation Foundation, 1990), pp. 113-130.
21. Schmidt, "Are Great Lakes Fish Safe," p. 17.
22. Ibid.
23. Kormondy, *Concepts of Ecology*, p. 256.
24. Ibid., p. 204.
25. Ibid., p. 205.
26. William E. Schmidt, "In the Great Lakes, Some Pollution Defies the Cleanup," *New York Times*, July 2, 1989.
27. Ibid.
28. *Rachel's Hazardous Waste News*, #225, March 20, 1991.
29. Zero Discharge Alliance, *Statement of Principles*, Detroit, Michigan, and Windsor, Ontario, undated.
30. Great Lakes Science Advisory Board, Report to the IJC, 1989 Report Highlights, pp. 3-19.
31. John H. Houvouras, "On Once-Wild Walden Pond," *Sierra*, May /June 1989, p. 94.
32. Ibid.
33. Ibid.
34. "Alien Weed Is Choking Lakes in Southern U.S.," *New York Times*, November 17, 1991.
35. Ibid.
36. Jeffry Kahn, "Restoring the Everglades," *Sierra*, September/October 1986, p. 41.
37. Ibid.
38. Ibid.
39. Charles E. Little, "Rural Clean Water: The Okeechobee Story," *Journal of Soil and Water Conservation*, September/October 1988, pp. 386-90.
40. Jeffrey Schmalz, "Pollution Poses Growing Threat to Everglades," *New York Times*, September 14, 1989.
41. Nicole Duplaix, "South Florida Water: Paying the Price," *National Geographic*, July 1990, pp. 104-5.
42. Richard E. Mooney, "Marjory Douglas's Everglades," Editorial Notebook, *New York Times*, April 8, 1990.
43. Cited in James LeMoyne, "Everglades Sentinel on Watch at 100," *New York Times*, April 8, 1990.
44. *Earth Island Journal*, Fall 1991, p. 3.; see also "Mercury in Everglades Fish Worries Experts," AP report, *New York Times*, March 14, 1989.
45. Keith Schneider, "Returning Part of the Everglades to Nature for $700 Million," *New York Times*, March 11, 1991.
46. AP report, "U.S.-Florida Deal Raises Everglades Cleanup Hope," *New York Times*, February 22, 1991; see also Schneider, "Returning Part of Everglades to Nature."
47. "Florida, Federal Government Settle Everglades Lawsuit," *Ecology USA*, July 15, 1991, p. 135.
48. Anthony DePalma, "U.S. and Florida Settle Suit on Everglades Water," *New York Times*, July 12, 1991.
49. Joe Podgor, statement to the South Florida Water Management District, July 26, 1991.
50. FOE position paper, May 20, 1991, p. 2.
51. Friends of the Everglades' *Proposed Alternative for Surface Water Improvement and Management (SWIM): A Plan for the Repair and Revitalization of the Everglades*, Miami, Fla., July 1, 1991, p. 4.
52. For details on the panther-recovery program see Jon R. Luoma, "Born to Be Wild," *Audubon*, January/February 1992, pp. 50-58.
53. Dave Foreman and Howie Wolke, *The Big Outside* (Tucson, Ariz.: Ned Ludd Books, 1989), p. 371.
54. Frank Graham, Jr., "Kite vs. Stork," *Audubon*, May 1990, p. 109.
55. Ibid., p. 108.
56. Ibid., p. 110.
57. Based on material supplied by the Okefenokee National Wildlife Refuge, August 1991.

58. Ralph E. Heimlich and Linda L. Langner, "Swampbusting in Perspective," *Journal of Soil and Water Conservation*, July/August 1986, pp. 219-24; see also Heimlich et al.,"Beyond Swampbuster," *Journal of Soil and Water Conservation*, September/October 1989, pp. 445-50.

59. Quoted in William K. Stevens, "Efforts to Halt Wetlands Loss Are Shifting to Inland Areas," *New York Times*, March 13, 1990.

60. Cited in Donald Bennett, "The Peril to the Wetlands and Why We Must Care," letter published in *New York Times*, May 4, 1991.

61. Ibid.

62. Allen St. John, "Wetlands Now Imperiled by Industry-backed Bills," *Audubon Activist*, May 1991, p. 1.

63. Ibid., p. 4.

64. Tim Searchinger and Douglas Rader, "Bush's Cynical Attack on the Wetlands," *New York Times*, August 19, 1991.

65. Cited in "Deregulating the Wetlands," *Sierra Club National News Report*, April 12, 1991, p.2.

66. St. John, "Wetlands Now Imperiled," p. 4.

67. Peter A. A. Berle, "A Grand Design for Wetlands," *Audubon*, July 1990, p. 6.

68. Ibid.

3
THE EDGE OF THE SEA

"For no two successive days is the shore line precisely the same... Today a little more land may belong to the sea, tomorrow a little less. Always the edge of the sea remains an elusive and indefinable boundary."

— Rachel Carson

The coastal and estuarine region is where two major environments of the earth — land and sea — come together and affect each other. Although this zone represents less than 10 percent of total ocean area, its shallow waters are so rich in nutrients that they produce 90 percent of all known ocean plant and animal life. They nurture most of our commercial marine fisheries while receiving an extraordinary amount of pollution from human sewage, industry, transportation, and agriculture. If you have ever lived near the coast or spent vacations at the shore, you have your own idea of what a coast should look like. It could be the windswept dunes of Cape Cod, the coral reefs of Florida, the fog-shrouded rocky beaches of Oregon, or a thousand other places from California to the New York island, from the Redwood Forest to the Gulf Stream waters.

The basic configuration of coasts is the result of age-old geologic processes set in motion by the difference in temperature between the earth's inner core and outer mantle, the impact of glacier movement, and other elemental forces. The Atlantic and Gulf coasts are indented with inlets, bays, and estuaries and protected by a chain of barrier islands from Texas to Maine, while the Pacific coast is relatively straight, with broad beaches and steep cliffs. The condition of the coastline changes with the rise and fall of sea levels. Having almost reached its present level about three thousand years ago, the sea continues to rise some 4 inches a century. As the sea rises, coastal areas become flooded and the shoreline moves back toward the land. Rivers and remnants of glaciers, which helped shape the landmass of the continent as we know it today, continue imperceptibly to carve valleys that are flooded by the rise in sea level, forming bays and estuaries. The sediment they carry is eventually dumped into the sea, creating glacial deposits and deltas. Then, reworked by currents and tides, these sediments supply much of the sand to the beaches. The shoreline, the boundary between land and sea, is the most dynamic part

Coastal Facts

- Ninety percent of the world's human population lives within 200 miles of a coast.
- Of the thirty-two largest cities in the world, twenty-two are located on estuaries.
- The tidal shorelines of the conterminous U.S. extend approximately 38,000 miles.
- According to the Army Corps of Engineers, more than 40 percent of the U.S. coastline, excluding Alaska, is eroding at an average rate of about 1.3 feet a year, with the highest rate (about 6 feet a year) along the Gulf Coast.[1]
- Public ownership accounts for approximately half of the total shoreline of the Atlantic and Pacific Oceans and the Gulf of Mexico, but a much smaller share of the erodible shores of the Great Lakes.
- Developers have destroyed or built over between 20 and 50 percent of U.S. coastal habitat areas.

(Sources: David. A. Ross, *Introduction to Oceanography*, and *Environment* magazine[2])

of the coastal zone, subject to tides, winds, waves, changing sea levels, and, because more than half the U.S. population lives within 50 miles of the coast, to a great deal of human impact.[3] Here you find rocky shores, replete with algae, barnacles, starfish, and similar organisms living in intertidal pools, sandy beaches where sand dollars, ghost crabs, and other creatures adapt by burrowing into or adhering to the sand, coastal wetlands teeming with algae, worms, crabs, and crustacea, barrier beaches that defend against the incoming sea, and in the warm coastal zones of tropical and subtropical seas, coral reefs, made up mainly of calcium carbonate secreted by algae and small coral animals, supporting abundant marine life.

The plants and animals that inhabit the coastal regions must adapt to continuous stress from tidal currents, waves, wind, sand movement, changes in temperature, water depth, salinity, and other factors. Turtles develop tough outer shells; other species shelter in rock crevices or under seaweed. Kelp, rockweed, barnacles, mussels, and other immobile organisms survive the fierce pounding of the surf by attaching themselves to rocks and other hard surfaces.

Estuaries

Estuaries are semiclosed parts of the coastal zone where the action of tides mixes fresh water from rivers with salt water from the oceans. Taking many forms according to the geology of the region and other factors, they include marine embayments, bays, sloughs, lagoons, and inlets.

Estuarine Environments

- U.S. estuaries cover an area one-tenth as great as the nation's total land.
- They provide a rich environment that serves as the breeding or feeding ground for roughly three quarters of the seafood catch in U.S. waters, valued at more than $5 billion.
- Half of the country's ten most valuable commercial seafood varieties — Gulf shrimp, sockeye salmon, oyster, South Atlantic shrimp, and blue crab — depend on estuaries for their survival.
- Estuaries and near-coastal waters generate $8 billion annually through the sportfishing industry.

(Source: National Oceanic and Atmospheric Administration)

Until recently the importance of estuaries was little appreciated, largely because much of their life consists of microscopic algae and minuscule creatures that live in the water or are hidden in the mud or sand. Yet estuaries are highly productive ecosystems, supporting much more plant and animal growth per acre than the most fertile farmland. According to the National Audubon Society, the mean annual production of an estuary is eight times as great as that of a cornfield.[4] Estuaries provide food and habitats for many land and marine animals for at least part of their lives, including untold numbers of birds and waterfowl, saltwater fish, clams, oysters, shrimp, and lobsters that are harvested commercially. Despite their rich fertility, these ecosystems are easily com-

promised because their productivity is based on only a relatively few species of plants and animals that can adapt to the wide variations of salinity, temperature, or sediment. Estuaries preserve the quality of water for wildlife and humans by diluting and filtering huge quantities of pol-

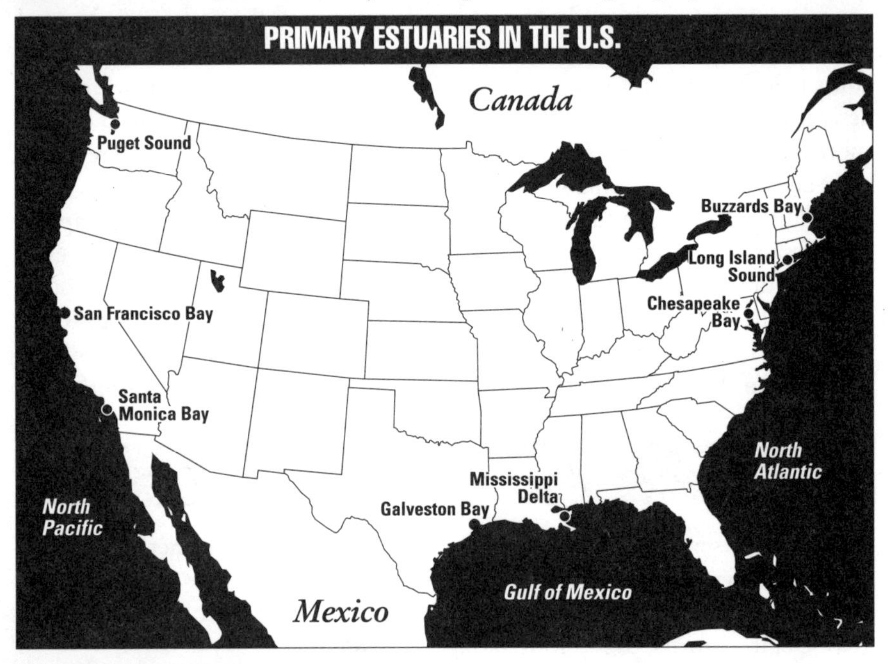

> *"Only over the last couple of years have people begun to perceive [estuaries] as a national resource base that is every bit as important as prime agricultural land in the Midwest or national parks throughout the country."*
> — *Bud Ehler, director, National Oceanic and Atmospheric Administration's Oceanography and Marine Assessment Office*

lutants dumped into the rivers and oceans. In conjunction with barrier islands and coral reefs, they help protect low-lying coastal regions from flooding and other damaging effects of storms and hurricanes. Their sheltered waters make excellent harbors and offer navigable routes to inland commerce, industry, and agriculture. It is no accident that many of North America's — and the world's — great ports are sited on estuaries. They include New York, San Francisco, Seattle, Vancouver, Halifax, Miami, and New Orleans.

There are two basic types of estuary: ocean-dominated, where the river's contribution is relatively small (e.g., Chesapeake and Delaware bays) and river-dominated, where the river's influence is felt far out to sea (e.g., the Mississippi). In general, sea-dominated estuaries are richer in food production. In the following pages we focus on six estuarine ecosystems — Chesapeake Bay, San Francisco Bay, Puget Sound, Long Island Sound, Buzzards Bay, and the Mississippi Delta — highlighting the problems of water pollution, loss of wetlands, decline of marine life, overfishing, and massive erosion from channelization that threaten these and many other estuaries throughout North America.

Chesapeake: Polluting the Heritage

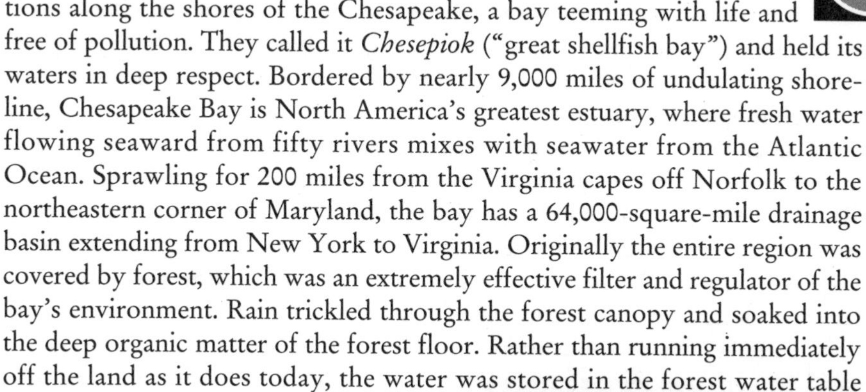

The Algonquin and other Native Americans lived for many generations along the shores of the Chesapeake, a bay teeming with life and free of pollution. They called it *Chesepiok* ("great shellfish bay") and held its waters in deep respect. Bordered by nearly 9,000 miles of undulating shoreline, Chesapeake Bay is North America's greatest estuary, where fresh water flowing seaward from fifty rivers mixes with seawater from the Atlantic Ocean. Sprawling for 200 miles from the Virginia capes off Norfolk to the northeastern corner of Maryland, the bay has a 64,000-square-mile drainage basin extending from New York to Virginia. Originally the entire region was covered by forest, which was an extremely effective filter and regulator of the bay's environment. Rain trickled through the forest canopy and soaked into the deep organic matter of the forest floor. Rather than running immediately off the land as it does today, the water was stored in the forest water table and fed into the rivers by steady underground seepage. The original watershed included 5 million acres of wetlands that filtered pollutants running off the land and buffered the bay against the extremes of drought and flood. The wetlands also nourished an extremely rich wildlife. Since the seventeenth century the Chesapeake has become a veritable food factory, providing successive generations of fishermen with a cornucopia of oysters, steamer clams, blue crabs, striped bass, and other fish. As much as 70 percent of the fish in the coastal waters of North America is spawned or spends a critical part of their lives in the Chesapeake. It supports several thousand full-time commercial seafood harvesters and hosts well over two million people who fish and hunt there for sport every year.[5]

> *"One hundred years ago there were so many oysters in the bay that they filtered the water every four or five days. Today it takes about a year to accomplish the same task."*
>
> — William K. Reilly, administrator, *EPA*

The abundance of marine life in the estuary makes it easy to forget how vulnerable it is to human and natural intrusions. However, in the last twenty-five years the bay has lost to pollution over 60 percent of its underwater vegetation, which provides a critical habitat for millions of birds and fish and the means by which the estuary cleans itself from sediment and other contaminants. As a result of Hurricane Agnes in 1972 an estimated two million barrels of oysters and almost all of the clams were killed by a sharp drop in salinity that occurred when the Susquehanna River flooded fifteen times more fresh water than is normal into the bay. Even more significant for the future of the Chesapeake and other marine ecosystems are long-term changes affecting grasses that form a vital link in the food chain, because so many organisms — including other plants, crustaceans, mollusks, and fish — depend on them for survival. Excess nutrients — mainly nitrogen and phosphate fertilizers drained from nearby farmlands — cause surface algae to multiply so rapidly that they block sunlight and kill bottom grasses. Also, when they die, algae generate bacteria, which consume the oxygen that would otherwise be available for plants and fish.

The Land-Sea Connection

In 1983 the EPA reported that the intensity of human population growth, urban development, and agricultural activity had caused unprecedented changes in the biological communities of Chesapeake Bay, including dramatic declines in major fisheries.[6] Underlying the Chesapeake's vulnerability is the relation between its land and its water. Though long and broad, the bay is extremely shallow — its average depth is less than 22 feet — providing insufficient water to flush out the volume of pollutants it now receives. Deterioration of the bay prompted the EPA, Maryland, Virginia, Pennsylvania, and the District of Columbia to initiate a joint program cutting across political boundaries to protect marine systems, especially underwater vegetation. This required limiting pollution from **nonpoint** sources — farms, parking lots, lawns, and other areas from which water runs off — and **point** sources, which include more than 2,750 wastewater treatment plants.

To curb nonpoint-source pollution, the concerned states agreed to promote "best management practices" (BMPs), which include a seed planting method using dead plant material as a mulch; this can reduce rain runoff by 50 percent and significantly reduce loss of sediment, nitrogen, and phosphorus. In Fairfax County, Virginia, another BMP known as "wet ponds" removes up to 90 percent of the silt and 80 percent of the phosphorus runoff from roads.

To combat the phosphorus problem, Maryland in 1985, Washington, D.C., in 1986, and Virginia in 1987 enacted phosphate detergent bans aimed at reducing the amount of phosphorus entering the sewer systems and ultimately the bay. After the Maryland ban went into effect, the state found that it could save its customers more than $2 million a year in sludge treatment and disposal costs. During the first year of the ban Maryland municipalities saved about $4.5 million on chemicals alone. Meanwhile the amount of phosphorus being discharged into the bay from sewage plants *not* having to remove phosphorus was reduced by 20 percent. In Virginia the amount of phosphorus entering sewage treatment plants has been cut by more than 30 percent.[7]

The State of the Chesapeake

> *"The bay we would restore is somewhat comparable to a person whose lungs are half clogged, whose liver and kidneys are shutting down… {it} is also an amputee."*
>
> —Tom Horton, the Chesapeake Bay Foundation[8]

The health of the bay, Horton says, depends on what we put into it (pollutants), what we take out (fish), and what we do to maintain the natural features of the bay's ecosystem. During the 1970s the three bay states — Maryland, Virginia, and Pennsylvania — enacted laws designed to control sediment from developing lands. Among the techniques employed were "filter fences" of straw bales and black plastic cloth, and the building of settling or "wet" ponds to catch and filter water draining from development sites. Such controls cannot completely eliminate pollution, but they can reduce it by as much as 90 percent in weight. However, the finer, lighter particles of sediment escape the controls and stay suspended in the

water.[9] Despite some progress from these and other efforts, the quality of water and aquatic life is far from satisfactory: a few successes on a few rivers; some preliminary evidence that a couple of species may rebound; the cold satisfaction that things would have been worse had nothing been done. But there is no clear trend toward any systemwide rebound. Moreover, there is increased impact on the environment caused by human population that continues to grow across the bay region without any talk of limits. As Horton and other concerned citizens recommend, "To restore the bay, we must reduce our impact now and then reduce it further to offset another 2.6 million people coming into the watershed by the year 2020."[10]

Turning the Tide?

The Chesapeake Bay Commission, the multistate panel monitoring the efforts to save the estuary, believes that controlling and guiding growth is its first priority. Maryland has introduced stringent safeguards for protecting and restoring wetlands, recognizing their important role in filtering runoff and stabilizing coastlines by subduing waves and slowing erosion. The state has established a 1,000-foot buffer zone along its bay shore in which development, especially housing subdivisions, is severely limited. At the same time a plan to develop a 400-acre island to house a sports complex in the James River off Newport News, Virginia, was dropped after a study by the Virginia Institute of Marine Studies (VIMS) indicated that the island would interfere with oyster larvae. Another VIMS study warned against the U.S. Navy's proposed testing of electronic weapons to withstand the electromagnetic pulse generated by nuclear explosions, because the pulse could alter magnetite, a substance in turtles' brains that is vital to their navigation.

Signs of progress in the Chesapeake include regrowth of underwater grasses and the swimmability of the Potomac River, a major tributary that was dangerously toxic before the building of modern sewage plants. Baltimore Harbor, once thought killed by industrial pollution, now boasts newfound beds of clams and the healthy stirrings of mudworms. Striped bass, shad, the famous Chesapeake oyster, bald eagles, and peregrine falcons are beginning to make a comeback. In 1990 there were enough striped bass (called rockfish by local residents) to permit the first fishing season for this popular species in eight years. However, after just a few days the season had to be closed because of too much angling. "That is a sign of just how fragile our progress is in cleaning up the Chesapeake," said William C. Baker, president of the Chesapeake Bay Foundation, a citizens' group that monitors the bay and promotes its conservation. "Our realistic hope is that we can return to a bay in which species aren't endangered and pollution and human pressure are manageable."[11]

The fragility of the ecosystem was underscored by an oyster catch in 1989 a mere tenth of what it was a hundred years ago and by the absence of three-quarters of the formerly plentiful grass. There are no simple answers, Baker said. For example, just when experts thought they had figured out how to handle the problem of excess nitrogen from sewage, they discovered that acid rain was bringing it into the estuary. According to an Environmental Defense Fund study, a quarter of all nitrogen contributed by human activity to Chesapeake

Bay originates in acid rain falling directly on the bay or its watershed. Only fertilizer runoff surpasses this, contributing one-third. The atmospheric nitrate deposits arise from nitrogen oxide emissions due to fossil-fuel combustion primarily from motor vehicles and power plants. An additional 14 percent of the nitrogen reaching the bay comes from atmospheric ammonium, largely from ammonia emissions related to agriculture, the report said, emphasizing the importance of the atmosphere as a medium for transfer of nitrogen in Chesapeake Bay and other East Coast estuary watersheds.[12]

The key recommendations of the EDF report were: The federal government should regulate motor vehicles, power plants, and other stationary sources that result in a reduction of 40 percent in nitrogen oxide emissions, consistent with reductions planned for other nitrogen sources to the bay; states in the Chesapeake Bay watershed should implement stringent nitrogen-oxide emissions limitations for new and existing stationary sources, while improving motor vehicle inspection and maintenance programs; federal and state governments should vigorously reduce nitrogen from sewage-treatment plants, fertilizer and other nonpoint-source runoff. Unless such measures are implemented, the EDF scientists warn, "the quality of East Coast waters will continue to deteriorate."[13]

What You Can Do to Protect Chesapeake Bay

1. Be inspired by the action taken by the Chesapeake Bay Foundation and the Natural Resources Defense Council. Frustrated by Maryland's refusal to get tough with the Bethlehem Steel Company's discharging 500 million gallons of heavily polluted wastewater a day into the bay, they filed suit in federal court under the citizens'-suit provisions of the Clean Water Act. Relying on the company's own discharge reports, the court ruled that Bethlehem was liable for over two hundred excess discharges during the five years before the suit was brought. In 1987 Bethlehem settled out of court for $1.5 million which contributed to local cleanup and monitoring efforts.*

2. Support the Chesapeake Bay Foundation (for address and phone number, see the Directory, pages 331-355). Their aim is to restore the bay through legal and scientific advocacy, land conservation, management and planning, and to provide on-the-water environmental education for students, teachers, and the general public.

3. Contact the Alliance for the Chesapeake Bay, 6600 York Road, Baltimore, MD 21212; (301) 377-6270; get their comprehensive "Chesapeake Citizen Directory," a sixty-page listing of citizen groups, government organizations, advisory committees, and trade associations working to protect the bay.

4. Read *Turning the Tide* by Tom Horton and William M. Eichbaum (Washington, D.C.: Island Press, 1991), an excellent report on Chesapeake Bay based on an ecosystem approach, with comprehensive recommendations on how it can be restored.

5. Understand the connection between your lifestyle and the environment.†

* For more on this case, see *Design for a Livable Planet*, page 249.

† See chapter 10, pages 331-355, and *Design for a Livable Planet*, especially pages 313-17.

San Francisco Bay: Too Much Salt[14]

The largest estuary on the West Coast, San Francisco Bay is a complex system of several bays and tributaries. The central part, usually known as the Bay, extends from the Golden Gate Bridge south to San Jose and north to Richmond and San Rafael, opening into San Pablo Bay, of which the Petaluma River, Sonoma Creek, and Napa River are tributaries. Extending to the east are the Carquinez Strait and the great expanse of waterways, marsh, and wetlands called Suisun Bay, into which the Sacramento and San Joaquin rivers drain the entire great Central Valley of California, converging into the San Francisco Bay system. The estuary contains the most extensive wetland habitat in the western United States, on which many endangered or rare species are dependent, including the California clapper rail, the black rail, and the salt marsh sparrow. It supports more than 150 species of fish and provides an important wintering habitat for migratory birds, a paradise for boaters, and home for more than six million residents of the Bay Area — the fourth-largest metropolitan area in the United States. It also has the dubious distinction of being the most altered by human hand of all United States estuaries. San Francisco Harbor, one of the most beautiful in the world, is severely polluted with sewage, oil spills, oil-refinery waste, petrochemical effluent, and chemical runoff from agriculture.

> *"Water is the lifeblood of California."*
>
> — Harry Seraydarian, director, Water Management Division, EPA Region 9

The damage began in 1848 with the gold rush. Hydraulic sluicing of the ore washed millions of tons of dirt and rock downstream, choking creeks and rivers throughout the drainage basin, destroying oyster beds and salmon spawning grounds. Tidal marshes and freshwater marshlands were drained for farmland, saltworks, and urban development. Of the original 850 square miles of marshland less than 50 remain intact today. The demand for irrigation water for agriculture south and north of the bay led to the diversion of the Sacramento and San Joaquin rivers. The latter flows through the highly irrigated Central Valley, picking up heavy runoff of fertilizers, herbicides, and pesticides as well as salt leached from the farmland. In 1978 a channel was built to drain the valley's wastewater into the Kesterson Reservoir, a marshy national wildlife refuge 50 miles south of San Francisco. Selenium and other chemicals leached from farmland soil so devastated wildlife that Kesterson, now officially closed, is known as "the Three Mile Island of irrigated agriculture." In February 1991 selenium discharges from six giant oil refineries that line San Pablo and Suisun bays in the inner estuary were found to be contaminating endangered species as well as popular game fish and ducks.[15] Compounding the problem is a prolonged drought, which is depriving the bay of its natural cleansing by river water. Selenium levels in the eggs of the endangered California clapper rail living in a marsh near the Chevron refinery were five times higher than in unpolluted areas and more than twice as high as at other locations in the bay. Mussels tested near another refinery had levels ten times higher than elsewhere in the bay and the highest ever recorded by the California Mussel Watch.[16]

Diversionary Tactics

Unfortunately for California, its water must travel a long way from source to use. Seventy percent originates north of Sacramento and 80 percent is consumed south of the capital. To convey water from north to south, the state built the world's largest artificial water system with its major supplier, the San Francisco Bay-Delta Estuary, supplying drinking water for 40 percent of the state and irrigation water for much of its agricultural lands. Today nearly 40 percent of the estuary's natural freshwater flow is removed for local consumption upstream and within the delta, and another 24 percent is diverted from the delta through state and federal water projects for agricultural and municipal use in central and southern California. The resulting low flows interfere with the spawning and migration of chinook salmon, striped bass, steelhead trout, and American shad, which live some or all of their adult life in salt water but move upstream into fresher water to spawn. The low flows of fresh water also weaken the food chain on which shrimp, clams, waterfowl, and other creatures depend by pushing their sources of nutrients upstream. At the south end of the delta, millions of young fish are sucked into diversion pumps because of a reverse river flow created by the pumps during the dry season. The diversion allows the influx of too much salt water from the ocean, upsetting the estuary's complex community of interdependent plants and animals, which rely on a consistent supply of fresh water. As the salt moves upstream, it affects the quality of water for human consumption and crop irrigation. Freshwater flow into the estuary, already reduced by 60 percent, is expected to be cut another 15 percent by the end of the decade. The net result is drastic loss of tidal wetlands and wildlife habitat, intensified land-use pressure, and increased pollution.

WHAT IS BEING DONE TO SAVE SAN FRANCISCO BAY

At the federal level: Overall authority for safeguarding San Francisco Bay and other national estuaries lies with Congress via the Clean Water Act (CWA), the Rivers and Harbors Act, and other laws.* Prime responsibility for administering this legislation is vested in the EPA. However, in California, as in many other states, CWA enforcement is delegated to a state agency, the State Water Resources Control Board, which oversees nine regional water quality control boards that are the actual permitting authorities. On a day-to-day basis the Army Corps of Engineers holds life-and-death power over the bay because it handles all applications for wetlands development. As we saw in the preceding chapter, the corps has been historically prodevelopment, destroying countless ecosystems in its wholesale building of gigantic dams, reservoirs, and water projects across the country.† Although its experience in preparing and managing regional plans for local estuaries is very limited, it could, given a positive lead and support from Washington, focus on restoring, not destroying, ecosystems.

* See chapter 9, pages 274.
† See also pages 66-68 and 105-106.

At the state and regional level: The leading state and regional agencies addressing San Francisco Estuary problems are the California Department of Fish and Game (preservation and development of bay wetlands), the Regional Water Quality Control Board (monitoring discharges into the bay), and the State Water Resources Control Board (responsible for balancing the needs of the environment, agriculture, and urban users). The board is under conflicting pressure from the environmental community and from central- and southern-California water contractors who are jealously protective of their water rights. One major problem is that farmers get federally subsidized water at less than one-tenth the price paid by urban Californians, who pay close to the actual cost. With no incentive to save water, farmers waste enormous amounts, often using it to raise cotton and other crops that require much more water than the region naturally provides and that can be more economically produced in other parts of the country where there is adequate rainfall. The state-established San Francisco Bay Conservation and Development Commission (BCDC) is a regional agency with permit jurisdiction over open bay waters and all land surrounding the bay within a 100-foot shoreline band. Although BCDC has stringently controlled development within this area, it lacks authority over most diked former baylands, which constitute a large part of the area. BCDC also has a delicate balancing act, supporting maritime development while encouraging marsh restoration, continued use of salt ponds, and preserving the 85,000-acre San Suisun Marsh for agricultural use, duck-hunting clubs, and wildlife refuges.

The San Francisco Estuary Project

This is a public and intergovernmental agency entrusted with coordinating the programs of the EPA, the state, and the Association of Bay Area Governments, a planning agency to which all Bay Area local governments belong. Its decision-making management conference represents some two hundred environmental organizations, fish and wildlife advocacy groups, boating and marina associations, and manufacturing and commercial interests.

The project's goal is to develop a biologic inventory of the bay, evaluate existing laws, policies, and programs, and develop by 1993 a comprehensive conservation management plan to restore the estuary. Established in 1988 and funded for only five years of research and planning, it has no acquisition or regulatory powers. Implementation is left to local governments and the public.

What You Can Do to Help Save San Francisco Bay

With a long tradition of environmental activism, the Bay Area is home to a multitude of groups concerned with protecting its natural resources. Among national organizations headquartered there are the Sierra Club, Earth Island Institute, the Trust for Public Land, and the Rainforest Action Network. Other national organizations such as the National Audubon Society, Greenpeace, Nature Conservancy, NRDC, and

EDF maintain very active offices in the area.* In addition you can find many other regional and local action-oriented groups in the San Francisco/ Oakland/Berkeley area that focus primarily on issues related to the estuary. They include:

The Bay Institute of San Francisco, 10 Liberty Ship Way #120, Sausalito, CA 94965; (415) 331-2303. This is a public research group whose work includes "government watching" such as monitoring the state's Bay-Delta Hearings. It publishes an impressive list of books, research reports, and a quarterly newspaper, *Bay on Trial.*

The Bay Keeper, Building A, Fort Mason, San Francisco, CA 94123; (415) 567-4401. An on-the-water pollution-prevention project founded by Michael Herz, it logs pollution incidents and other illegal activities in the bay (over 250 recorded in the first eighteen months), reviews dredging and discharge records of the Coast Guard, the Army Corps of Engineers, and the Regional Quality Control Board, and runs a "rogue's gallery" of bay polluters and pollutants. *The Bay Keeper* educates boat owners, marina operators, and the general public on pollution prevention, helps initiate legal action against offenders, and works with government regulatory agencies.

Citizens for a Better Environment, 501 Second Street, Suite 305, San Francisco, CA 94107; (415) 243-8373. A 30,000 member statewide environmental research and advocacy group fighting toxic hazards, air pollution, and other environmental problems. In the estuary, CBE focuses on industrial wastes released through municipal sewer systems, reduction of freshwater flows, and dredge disposal. It uses research, public advocacy, litigation, and negotiation to get large-scale violators to reduce their pollution. It publishes reports, fact sheets, and the quarterly *CBE Environmental Review.*

Friends of the River, Building C, Fort Mason, San Francisco, CA 94123; (415) 771-0400. Formed originally to save the Stanilaus River from the New Melones Dam, FOR is now involved in protecting many of the rivers and streams that are tributaries to the estuary. It holds an annual Rivers Conference and publishes a bimonthly newsletter, *Headwaters.*

Save San Francisco Bay Association, 1736 Franklin Street, Oakland, CA; (415) 452-9261. Formed to stop the filling of the bay, it monitors and works with government agencies to preserve and restore wetlands, protect water quality, and create new shoreline parks, rails, and other means of public access. It publishes an excellent quarterly newsletter, *The Bay Watcher,* and well-documented fact sheets.

Puget Sound

From the Olympic Peninsula northward the Pacific coast is deeply eroded by the impact of glaciers, forming a series of narrow, steep-sided fjords and estuarine systems. In addition to Alaska's Prince William Sound of Exxon *Valdez* fame,† two of the most spectacular of these estuaries are located in the north-

* For addresses and telephone numbers, see the Directory, pages 331-355.

† For more on the Exxon *Valdez* spill, see pages 97 and 115.

western corner of Washington. Nestled between the Cascade and Olympic mountains, the Puget Sound watershed encompasses more than 16,000 square miles. At the heart of the basin lie the resources of Puget Sound, an intricate network of waterways, inlets, and bays extending 90 miles inland from the straits of Juan de Fuca and Georgia. The surface area of the watershed is roughly 80 percent land and 20 percent water — fresh, estuarine, and marine. The common outlet, known as Greater Puget Sound, includes the straits south of the Canadian border.

Puget Sound has a shoreline of 2,400 miles. Its waters cover former beaches of sand and gravel, and its extensive area encompasses salt marshes, mud flats, and protected bays of mixed sand, mud, and cobblestone; it is also marked by islands, rocky reefs, headlands, and sandy and gravelly beaches and by numerous streams and rivers that flow into it. On the surface Puget Sound looks pristine. Its deep, cold depths help keep the water clear, while daily tides move pollution toward the Pacific Ocean. Upon closer investigation, however, the situation is quite different. The sound contains a number of **sills** (shallow areas), which prevent most of the water and the pollutants it carries from reaching the ocean. Instead, they tend to sink to the bottom, creating toxic "hot spots" along the floor of the sound's inner bays. This has led to liver tumors and reproductive failures in the bottom-dwelling fish there.[17] Another problem is bacterial pollution from defective septic systems and farm animals, which forced the closing of nine commercial shellfish beds between 1986 and 1990. Moreover, the region's rapid increase in population and development is causing a continuing loss of wetlands and other fish and wildlife habitats.[18]

Seeking Sound Solutions

In 1986 the state Puget Sound Water Quality Authority initiated a program for cleaning up the heavily polluted estuary. It is in part funded by an eight cents-a-pack surtax on cigarettes, which brings in about $50 million a year to support local water quality projects in the form of grants. The state pays 50 percent of the cost for water-quality facilities such as sewage treatment plants and 75 percent of the cost for planning, education, and other operations. The discharge of waste from all sources is closely monitored by the authority and other state agencies working with private industry to reduce the flow of effluent into the sound. The project also includes limiting runoff from farmland, stringent construction zoning in the critical watershed area, and an areawide educational pro-

gram that, in the words of *Time* magazine, teaches everything from the history of the sound to what not to put down the kitchen sink.[19]

The control of pollution is promoted as everybody's task. High-school students take water samples and local citizens are trained to spot oil spills and other types of water pollution. According to Thomas Hubbard, a water-quality planner for Seattle, bridge tenders are great at calling in violations. "They are up high, and when they see a black scum or a little slick, they let us know about it."[20] Although the cigarette tax is a good beginning, it brings in only a small part of what is needed to finance the cleanup. According to the Water Quality Authority, it will cost about $600 million to upgrade the remaining primary sewage treatment plants to the desired secondary level. Reducing pollution from storm-water runoff may cost $50 million to $160 million a year; and reduction of nonpoint-source pollution is expected to add another $12 million a year.

Funding from local governments and private efforts has had some success, especially in obtaining $53 million from the legislature in 1990 for the acquisition of wetlands and other habitats. Some federal funding has been received also for research and educational purposes. But all of the funding efforts fall short of what is needed to complete the task. New sources being considered by the state Puget Sound Finance Committee include a tax on commercial marine fuels (currently exempt), a fee charged to motor vehicle manufacturers for new cars or trucks registered in the state, increases to the fish and shellfish tax, and an excise tax on the leasing of public lands. All of these proposals will undoubtedly be opposed by a number of special interests. Most conservationists agree that public education is an important key to gaining support for the long-term cleanup of Puget Sound and other similarly polluted areas. As residents and businesses learn more about how they can contribute to the cleanup process, they often become more willing to contribute to the solutions.

What You Can Do to Protect Puget Sound

1. Work through your elected representatives for a strong commitment by federal, state, and local governments to the protection of coastal waters.

2. Support the ongoing educational and advocacy programs of the Puget Sound Alliance, 4516 University Way, NE, Seattle, WA 98105; (206) 548-9343. PSA watchdogs all agencies concerned with Puget Sound, taking legal action where necessary to enforce environmental laws. In addition to taking part in government hearings and planning processes, the alliance organizes numerous volunteer programs that include wetland field trips, kayaking, and other boating trips on the sound. It also supervises the Puget Soundkeeper, who can be contacted at (800) 42-PUGET or, in the Seattle area, at (206) 548-9343.

3. Contact the Puget Sound Water Authority, Abbot Raphael Hall, Mail Stop PV-15, Olympia, WA 98504; (206) 464-7320 or (800) 54-SOUND. Ask for a copy of their latest "State of the Sound Report."

4. Get information from the Washington State Department of Natural Resources, Public Lands Building, Olympia, WA 98504; (206) 753-5327; and the State Department of Ecology, Olympia, WA 98504; complaint number (206) 649-7000.

Valdez Spill Revisited

Three years after the Exxon tanker *Valdez* spilled 11 million gallons of crude oil into Prince William Sound, Alaska, the harm to fish, marine, and other resources was found to be much more severe than many marine biologists had originally believed. In March 1992 the Exxon *Valdez* Oil Spill Trustee Council, made up of six federal and state officials overseeing the spill restoration effort, reported extensive harm to sea otters, orcas (killer whales), harbor seals, seabirds, and fish. Contradicting a national advertising campaign by Exxon, which contended that the sound's ecological vitality had been restored to the pristine conditions prevailing before the spill, the report found:

- Between 3,500 and 5,500 sea otters died of acute petroleum poisoning. In 1990 and 1991 otter carcasses were found on beaches, and recent surveys show a continuing decline in the otter population.
- The social structure of orcas appears to be breaking down, with some mothers abandoning their calves.
- Hydrocarbons have accumulated in bile and blood of river otters, which eat mussels that are still contaminated with oil in many parts of the sound.
- Thousands of bald eagles, sea ducks, loons, cormorants and other bird species have died, and oil continues to disrupt bird nesting sites.
- The spill affected migrations of salmon fry, major migrations of birds, and the primary reproductive period for most species of birds, mammals, fish, and invertebrate species.
- Thirty-five archaeological sites, including burial and home sites, were damaged, mostly on federal and state land.[21]

Long Island Sound: Slowly Suffocating

> *"Action must be taken immediately."*
> —Julie Belaga, director, EPA Region 1.[22]

Often called the Urban Sea, Long Island Sound (LIS) is a 1,300-square-mile estuary stretching 110 miles as the sea gull flies or six times that distance around the coastline from New York City to the eastern tip of Long Island; it is 21 miles wide at its widest point, near the Connecticut River. Its north shore takes in the entire southern boundary of Connecticut and Westchester County, New York, while its south shore is the northern coast of Long Island. Distinct from other estuaries by being open at both ends, the Sound is an extraordinary natural and economic resource that for generations has been used as a waste dump by municipalities, industry, the military, boaters, farmers, and others.

Since the 1950s, coinciding with postwar population growth and suburban sprawl, the Sound has been beset by a barrage of ailments, especially **hypoxia** — low levels of oxygen in the water triggered primarily by excessive

inputs of nitrogen that throw the Sound's ecosystem out of balance. Of the more than 90,000 tons of nitrogen that enter it every year, 55 percent is generated by human activity. Of this latter amount, well over half comes from municipal sewage treatment plants. In New York City, for example, fourteen plants struggle to keep up with their daily load, most of them working over capacity. "This means the wear and tear of treat-

ing the sewage results in the plants breaking down sooner than they should," said Nina Sankovitch, a lawyer with NRDC in New York.[23] Many other cities face a similar crisis. The pollution of Boston Harbor, cleverly manipulated by George Bush in his 1988 election campaign, underlines a basic problem. The $6.8 billion Massachusetts state budget does not have enough funds to fully treat sewage *before* it goes into the pipe. Thus the building of longer pipes merely puts the pollution somewhere else, in this case Cape Cod Bay, at least until 1999, when a new treatment plant is due to go on-line. As President, Bush ignored the harbor that served him so well as a TV commercial, offering no additional federal funds for a real cleanup. In February 1992 a ruptured undersea pipe spilled millions of gallons of untreated garbage from San Diego and fifteen other towns into the Pacific Ocean, demonstrating that the problem is not confined to communities built in earlier centuries. The San Diego system was built in 1963 but designed to serve a population of about 250,000; today it has to cope with a population of close to two million in San Diego County.[24]

Why Hypoxia Is a Problem

The amount of oxygen dissolved in water is critical to the health of marine ecosystems. Too little can lead to stress or death of the organisms that depend on its presence. In estuaries (where the flow of water is more restricted than in a river or the open sea), particularly those impacted by pollution, oxygen levels may fall from normal to zero, a condition known as **anoxia** — absence of oxygen. Saturation, the largest amount of dissolved oxygen (DO) that water can hold at equilibrium, varies with temperature and salinity but is about 7.5 milligrams per liter of water when the water is 72 degrees Fahrenheit, a typical summer temperature. Biologists generally consider three

LIS Stats

- More than five million people live within 15 miles of its shores.
- Fifteen million people inhabit its watershed, which reaches into five states and part of Canada.
- One-tenth of the U.S. population lives within a one-hour drive of its shores.
- Every year its commercial and sports fisheries generate $175 million in revenues; six million people flock to its public beaches; 750,000 recreational anglers fish its waters; and 200,000 boats — one of the largest recreational fleets in the world — are registered at 20,000 slips along its coastline .

parts per million (ppm) to be the minimum DO for sustained health of marine life. When DO falls below this level, hypoxia exists. At this point, stressed marine organisms may become ill, die, or move to oxygen-richer waters. Hypoxia can occur naturally in the deeper areas of coastal waters such as Long Island Sound during periods of warm, stable weather. The surface waters heat up and form a layer over the bottom waters, which are denser because of greater salinity and cooler temperatures. The result is a **pycnocline**, a marked density gradient that restricts mixing of water between the two layers. Oxygen added to the surface water by wave mixing and photosynthesis of marine plants is thus prevented from penetrating into the depths, where it is needed to replace oxygen consumed by marine organisms and the decomposition of organic material. Although natural conditions may lead to extended periods of hypoxia — during a long, hot spell, for example, greatly increased human activity near the Sound has correspondingly increased nutrient and carbon loads, extending the length of time and the area of the hypoxia well beyond what had occurred naturally.

Sound Study

In 1985 the federally funded Long Island Sound Study (LISS) was initiated as part of the National Estuary Program to identify and research the most pressing environmental problems of the Sound and to produce a plan for their solution. In addition to hypoxia, which was found to be the most critical, the problems include pathogens from sewage plants, which have forced beach closings; toxic contaminants affecting tens of thousands of acres of shellfish beds and other marine resources; floating and beached garbage; and other wastes.[25] Every summer since 1986, when water sampling began, the western Sound has experienced hypoxia. In 1987 the condition extended from the Throgs Neck Bridge in New York City eastward to the Bridgeport-Port Jefferson line (see map), with DO levels dropping to nearly zero. At Hempstead Harbor on Long Island's north shore no oxygen was found near the bottom and very little at the surface. As a result there were no fish, while only a fifth of the normal population of crabs, starfish, and other bottom dwellers were alive. Although DO levels have not been quite so low in subsequent years, the total area affected was larger, extending to the New Haven-Shoreham line.[26] Paradoxically, while the Sound is cleaner in some respects than it was twenty years ago because less untreated sewage and industrial

chemicals are now dumped into it, thanks to improvements brought about by the Clean Water Act and other legislation, its cumulative problems add up to "alarming signs of a deteriorating overburdened resource."[27]

The major causes of Long Island Sound's precarious health are population growth and suburban sprawl in the region, which have increased the number of houses, schools, offices, and shopping malls and put a great strain on the infrastructure, especially on outmoded wastewater-treatment plants, which often overflow in rainy weather, pouring raw sewage into the Sound. Suburbanization also means more private septic systems (which seep pollutants into the groundwater that eventually find their way into the Sound), more cars (and auto emissions), more roads (and runoff of auto pollutants), more lawns (running off fertilizers and herbicides), and more seeping landfills, all contributing to the breakdown of the Sound's ecosystems. Another result of suburban development has been the filling in of tidal wetlands (for shopping malls, office buildings, and industrial parks, for example), depriving the Sound of the wetlands' natural filtering action, in addition to losing nurseries and feeding areas for the Sound's marine life, as well as fisheries (including ocean fisheries). The loss of wetlands is estimated at 60 to 80 percent on the Sound overall, and even greater in the western part.

Cleaning Up Long Island Sound

In December 1990 the federal government, New York, and Connecticut agreed to begin cleaning up the Sound by curbing discharges of pollutants into it. The states were required to upgrade nineteen coastal sewage-treatment plants in New York and twenty-four in Connecticut to remove the pollutants before they reach the water. However, no provision was made to underwrite the cost of the project. New York State and City were asked to bear most of the costs because their plants are larger than Connecticut's. The city disagreed on the grounds that its declining population has produced lower nitrogen levels from its sewage-treatment plants and that accordingly it was producing less pollution than it was being charged for. Even if the city paid its share of the estimated $6 billion cost of cleaning up the Sound and built the nitrogen-removal plants, the continued growth of pollution from sources other than treatment plants in Long Island and Connecticut would largely eliminate any gains, said Albert Appleton, New York City's commissioner of environmental protection."[28]

Because of the urgency of the hypoxia problem, New York, Connecticut, and the EPA agreed as a first step to put a cap on nitrogen output allowed to designated sewage plants without waiting for completion of the LIS study due in February 1993. They further agreed to work together for a proposed "no net increase" in nitrogen from nonpoint sources. Also placed on the agenda after prodding by the LISS Citizen Advisory Committee were recognition of the need to protect and enhance tidal and inland wetlands and riparian zones as key ecological components of the LIS basin and agreement to look at land-use issues, which are basic to the LIS problem. Despite the priority given to hypoxia, the problems of toxics and

pathogens are extremely serious. Toxic concentrations in harbor sediments, for example, are described by the LIS Watershed Alliance as a "ticking time bomb."[29] To oversee completion of the study and to coordinate implementation of the plan, the federal government in early 1992 set up a Long Island Sound office in Stamford, Connecticut, with a second office at the State University of New York at Stonybrook.

LIS Watershed Alliance

The alliance, which was created in January 1991 at the concluding conference of the National Audubon Society's yearlong Listen to the Sound citizens' hearings campaign, cosponsored by more than 175 organizations, is a network of groups united by a concern for Long Island Sound and by a common vision for its future. Its basic goals are: restoration of the Sound's water quality to a standard that will support swimming and fishing and the greatest possible biodiversity of the Sound's marine life; education to cultivate a sense of stewardship and lifestyle changes necessary to support a healthy Sound; development kept in balance with the needs of the LIS ecosystem; and close coordination of government programs. In late January 1992 the Watershed Alliance and the National Audubon Society Northeast Regional Office organized an LIS Citizens Summit Conference, which endorsed an action program for the Sound based on incorporating an LIS Restoration Act (H.R. 3660) into the Clean Water Act. Cosponsored by all members of the congressional LIS caucus, this act would increase federal funding to repair and upgrade the sewer and sewage treatment infrastructure, create strong controls on polluted runoff and ground seepage and provide strong wetlands protection. Also it would authorize $250 million, with a state and local match of 30 percent to be spent over five years, to fund demonstration water quality improvement projects in six harbors around the Sound. Dealing with nonpoint as well as point pollution, H.R. 3660 would be the first major step toward improving water quality in Long Island Sound and serve as a prototype for other estuaries around the country.

What You Can Do to Help Save the Sound

1. Become informed. Begin by following LIS reports in the newspapers and other media. This will help you identify organizations, programs, and officials who share your concerns. Among the groups to contact for further information are:
— *LISS*, NY Sea Grant Extension Program, 125 Nassau Hall, SUNY, Stony Brook, NY 11794; (516) 632-8737, contact Melissa Beristain; and
— *Sea Grant Marine Advisory Program*, University of Connecticut Cooperative Extension, 43 Marine Street, Hamden, CT 06514; (203) 789-7865, contact Chester L. Arnold, Jr.
2. Join a group. To find out which one suits your needs best, contact one of the following:
— *Long Island Sound Watershed Alliance,* representing a broad spectrum of organizations and members including environmentalists, anglers, garden

clubs, civic organizations, and conservation commissions. It is chaired and coordinated by the National Audubon Society. For information on groups in your area, contact Jane-Kerin Moffat, P.O. Box 313, Cos Cob, CT 06807-0313; (203) 629-1248.

— *Clean Sound Inc*, 20 Ojibwa Road, Shelton, CT 06454; (203) 629-6195. John Toth, president. An all-volunteer group that organizes beach and shoreline cleanups soundwide.

— *LIS Taskforce*, 85 Magee Avenue, Stamford, CT 06902; (203) 327-9786. Robert Teeters, president. Concerned with education and advocacy issues. *LIS Regional Information Service*, at the same address, provides detailed information for grade school through postgraduate level.

— *Long Island Soundkeeper Fund*, P.O. Box 4058, Norwalk, CT 06855; (203) 854-5330. Terry Backer, Soundkeeper. The fund supports monitoring, legal action, and intervention in the regulatory process to stop pollution and wetland destruction.

— *The Sounds Conservancy*, P.O. Box 826, Essex, CT 06426; (203) 767-1933. Christopher Percy, president. Dedicated to protecting and restoring the natural resources of six sounds of southern New England, including LIS, Fisher Island, Block Island, Martha's Vineyard, and Nantucket.

— *SoundWatch*, P.O. Box 104, City Island, NY 10464; (212) 885-9566. Susan Bellinson, president. An advocacy and education group focusing on water quality in the western Sound.

3. Contact elected officials. Do this directly or through the *Bistate LIS Marine Resources Committee*, which is composed of state legislators and environmental officials from New York and Connecticut dedicated to identifying and coordinating bistate actions affecting LIS. You can get a list of members by telephoning either (203) 240-0480 or (516) 669-9200. *LIS Caucus* is a bipartisan coalition of ten congressional representatives whose districts abut the Sound. You can reach them through their district offices or in Washington.

4. *If you are one of the fifteen million people who live in the LIS watershed (or if you live in any other watershed)*: As the LIS Taskforce reminds us, everything you put down your sink drain ends up untreated in the Sound. Sewage treatment plants do not treat chemicals. All those containers in your bathroom or kitchen whose labels warn "Caution — Harmful or fatal if swallowed" are also harmful to marine life. Instead of harmful chemicals use environmentally friendly alternatives such as baking soda or white vinegar, salt, and water as a cleanser.*

Respect the water and earth generally by conserving water, electricity, and fuel. Recycle. And use only biodegradable products.

Buzzards Bay: Model for the Nation

Buzzards Bay is a semienclosed bayment located along the southeastern coast of Massachusetts. It is 28 miles long, has a mean width of 9 miles, covers 235 square miles, and has a mean depth of 36 feet. The groundwater and surface drainage

* For the minihandbook "Common Household Products and Safer Alternatives," see *Design for a Livable Planet*, pages 298-305.

basin surrounding the Bay covers approximately 400 square miles. The drainage transports pollutants associated with land uses and human activities of some 250,000 people who live in the region and many more who vacation along its shores. Recognized by Congress in 1984 as one of four estuaries selected for intensive study, the Buzzards Bay Project began in 1985 with the goal of assessing sources of pollution and recommending how to protect coastal water quality and the health of living resources in the Bay. The project focused on pathogen contamination of shellfish beds and swimming areas, nutrient enrichment and eutrophication of coastal waters, and contamination of shellfish and finfish by toxic chemicals. Three years later the Bay was designated an estuary of national significance and included in the EPA's national program. Receiving 75 percent of its funding from the federal government and 25 percent from the state, the project is administered by the Massachusetts Office of Environmental Affairs and the EPA. It is directed by a management committee consisting of federal, state, regional, and community representatives, who are advised by local town officials, a panel of scientists and technical specialists, and the Coalition for Buzzards Bay, an independent citizens group. In the summer of 1991 the Buzzards Bay conservation and management plan — one of the first national estuary programs to be completed — was hailed at a congressional hearing as a "blueprint for how to get government at all levels working with local citizens to protect their coasts from pollution."[30]

According to Joseph Costa, Buzzards Bay project manager, the critical ingredient for successful implementation of the plan is adequate funding for the variety of actions that must be taken to restore and protect the Bay. Although Congress had authorized $4 million to carry out the plan's initial recommendations, additional funds are required to implement the necessary longer-term pollution-cleanup operations, Costa said. At issue is the amount of support that Congress will devote to the preservation of Buzzards Bay and other estuaries within the National Estuary Program (NEP).* Currently the EPA funds only the development plans, but not the implementation of their recommendations. When this provision was written into the Clean Water Act, it was hoped that individual states would be able to finance the remediation activities. However, according to Jeff Benoit, director of the Massachusetts Coastal Zone Management, "Given the current fiscal situation of the Commonwealth and other NEP states, it is unrealistic to expect any new state expenditures for implementation."[31]

The Mississippi Delta: Vanishing Coast

The coastal zone of Louisiana is one of the United States' premier natural resources, containing 40 percent of the country's coastal wetlands. Created by the nation's largest river, the Mississippi, which drains 40 percent of the lower forty-eight states and significant areas of two Canadian provinces, this zone is the largest and most active delta in North America. The range, diversity, and productivity of plants and animals that inhabit the region are

* For more on this program, see pages 107-109.

extraordinary. The wetlands support the largest coastal fin and shell fishery in the U.S., producing 30 percent of the country's annual commercial harvest. They are the wintering ground of two-thirds of the ducks and geese that migrate in the Mississippi Flyway. The delta also provides a natural means of water treatment and flood protection. Louisiana's on- and offshore coastal habitats produce a fifth of the U.S. production of oil and almost a third of its natural gas. The Mississippi River ranks as the country's most important inland navigational waterway.[32]

For some ten thousand years the Mississippi River has been "delta switching" every thousand years or so, causing some land areas to build while others deteriorate. The river builds a delta out into shallow shelf areas until its course becomes long, sinuous, and inefficient. It then changes its course, following a shorter, more efficient route to the Gulf of Mexico, thereby switching the delta's location. With a new delta always building, net coastal land gain was historically between 1 and 2 square miles a year. Since the turn of the century, however, the land has been eroding and sinking at an accelerating rate. It is now responsible for 80 percent of the nation's annual loss of coastal wetlands. According to estimates by the U.S. Army Corps of Engineers, the continued erosion of coastal Louisiana within thirty to fifty years could result in a 30 percent decline in fisheries, a huge loss in real estate values, a need to enlarge hurricane levees, and relocation of hundreds of miles of highways, railroads, utility and telephone lines, and more than 1,500 miles of oil and gas pipelines. "The relocation of tens of thousands of families who, for generations, have lived and prospered in the coastal zone is an appalling possibility."[34]

> *"If erosion continues unchecked, scientists say that four counties will have all but disappeared in 50 years and the waves will be lapping over the New Orleans suburbs."*
>
> — Newsweek[33]

Upsetting the Balance

The growth of the river-dominated Mississippi Delta depends on a delicate balance between the amount and type of sediment that the river brings, how much this sediment is compacted, the rise and fall of sea level, the underlying geology, and the effect of tidal and wave forces where the river meets the sea.[35] In the case of Big Muddy there is continual subsidence of sediment because the river carries mostly mud, which is composed of clay and water. As the mud settles, water is squeezed out by the pressure created by new sediment deposited on top, causing the land of the delta to sink. In the natural course of events river-borne sediments are added to the delta, and the extra layers offset the settling of earlier deposits. However, the balance was altered when people built embankments or levees in an effort to control the annual spring flooding and excavated channels to improve navigation. As a result less mud settled on the delta, while the land continued to sink.[36]

In the nineteenth century the riverbanks were weakened by the removal of plants and trees to aid steamboat navigation and agriculture. Since

the 1930s petroleum and gas companies have dredged more than 10,000 miles of canals to get to the reserves located in the region. To facilitate navigation and prevent the river from overflowing, the Army Corps of Engineers built hundreds of miles of levees and artificial waterways. This prevented the buildup of sediments that would otherwise offset natural settling and maintain the delta. Instead of nurturing the marshes, the river sediments now flow straight to sea, where they are lost in the deeper waters of the Gulf. Further upsetting the delta ecosystem, dredging of navigation channels allows salt water to creep inland and eat at the vegetation. "What those canals have done to the marsh is the same as the insidious, cumulative effect of hundreds of small cuts to the body," says Eugene Turner, a coastal ecologist at Louisiana State University.[37]

The Mississippi Delta has an enormous freshwater flow and a tidal rise and fall of only about a foot. Less dense than salt water, the fresh water floats on top and spreads out in a thin layer that gets progressively wider as it moves seaward. Underneath, a wedge of salt water slowly pushes upstream, extending 150 miles from the river's mouth. In this part of the estuary you can catch bottom-dwelling marine fish while boating on the surface of water that is (theoretically) fit to drink. The edges of the estuary are also essentially fresh water, home to bulrushes, water striders, whirligig beetles, and other freshwater flora and fauna.

Saving Coastal Louisiana

In 1986 a citizens' group, the Coalition to Restore Coastal Louisiana, was formed to bring together more than a hundred concerned groups, including the Louisiana Wildlife Federation, the Terrebone Parish Government, and the League of Women Voters. It creates major policy incentives, treating the delta as a dynamic ecosystem rather than on a piecemeal basis. The coalition has set up a citizens' coast watch to monitor wetlands projects that include plugging old canals against saltwater intrusion and restoring sand dunes by using recycled Christmas trees as silttrapping brush fences. Freshwater flows are being returned to some marshlands, according to a report from Rob Gorman, a Catholic Church social worker and one of the coalition's cofounders. "We want to ensure that funds are spent wisely," he said, but warned that tough decisions lie ahead. "Gates installed in levees can restore flows of fresh water and silt to dying marshlands. However, such projects can be opposed by delta residents who don't want to be relocated from revived floodways."[38]

A task force was created in the Louisiana governor's office to plan and coordinate wetland protection. In 1989 voters created a wetland restoration fund, supported by gas and oil revenues, that in 1990 came to $26 million. In 1990 the U.S. Congress passed a bill sponsored by Louisiana senator John Breaux, providing $36 million a year for more wetland projects. Several different ways to save the Louisiana wetlands are being tried. One is marsh management, which involves the building of levees, weirs, and other structures to limit saltwater intrusion into the marshes. Another approach relies on putting the river back into the marshland. The plan, designed by the Army Corps of Engineers with an estimated cost of $150 million, is to build enormous structures that will channel a flood of fresh water to push back

the salt water and prevent it from poisoning the plants whose roots hold the marsh together. A third approach, which many experts think the state and the corps are neglecting, is the strengthening of barrier islands at the edge of the Gulf. In the view of Dr. Shea Penland, associate director of the Louisiana Geological Survey, ignoring these islands is particularly shortsighted because they can be strengthened quite cheaply. One such project — building a 3,200-foot dike on one of the Isles Dernieres to repair severe hurricane damage — was successfully carried out for less than $1 million and has survived more than five years of Gulf weather.[39] While scientists, engineers, and public officials debate the merits of different initiatives, there is general agreement on one conclusion: Without a major coordinated state/federal effort, the Louisiana wetlands will not stand much chance of survival.[40]

What You Can Do to Help Save the Louisiana Wetlands

1. Support the *Coalition to Restore Coastal Louisiana*, 8841 Highland Road, Suite C, Baton Rouge, LA 70808; (504) 766-0195. Get a copy of their seventy-page *Coastal Louisiana: Here Today and Gone Tomorrow* and their quarterly newsletter, *CoastWise*.
2. Contact the *Louisiana Wildlife Federation*, 337 South Acadian Throughway, Baton Rouge, LA 70806; (504) 344-6707. They are one of the most active members of the coalition.
3. For more technical information, get in touch with the *Center for Wetlands Research*, Louisiana State University, Baton Rouge, LA 70803; (504) 388-1558.

The Gulf of Mexico

Sometimes known as the American Sea, the Gulf of Mexico is enclosed on three of its sides by the United States, with a 1,631-mile coastline* longer than the entire Pacific coast of California, Oregon, and Washington, and equivalent to the distance from Providence, Rhode Island, to Miami along the Atlantic shore. The Gulf is an exceptionally productive sea, annually yielding over 2.5 billion pounds of fish and shellfish, representing about 40 percent of U.S. commercial fishery landings. Its coastal estuaries, wetlands, and barrier islands provide critical habitat for very large populations of wildlife, including shorebirds, colonial nesting seabirds, and 75 percent of the migratory waterfowl crossing the United States. To a great extent, these resources are coupled with the extensive coastal wetlands of the Gulf, which make up about half of the national total.

The Gulf plays an important role in the current energy supply of the U.S., providing more than 70 percent of the oil and 97 percent of the gas produced offshore. While human population has historically been sparse compared with the northeastern and southwestern coasts, the demographics are changing rapidly. Today some forty-five million people live in Gulf coastal

* Florida (Gulf), Alabama, Mississippi, Louisiana, and Texas.

states, and their numbers are increased by a large seasonal influx of tourists and part-time residents, especially in wintertime.

Ecological Integrity at Risk

In addition to the environmental impact from the expanding population, the Gulf of Mexico is affected by human activities from many other regions: Two-thirds of the continental U.S. drains into it. Among the problems highlighted in a 1991 report by the EPA-funded Gulf of Mexico Program are: extremely low levels of oxygen, attributed to excess nutrients in coastal waters and estuaries — 3,000 square miles of bottom waters known as the "dead zone" have been identified off the Louisiana and Texas coasts; extensive losses of marshes, mangroves, and sea-grass beds; serious coastal erosion in all five Gulf states; freshwater diversion (from flood control, navigation, recreation, agriculture, and growing human populations) leading to saltwater intrusion in coastal wetlands and causing serious damage to ecosystems and drinking water supplies.[41] There are also major problems caused by pesticides, toxic chemicals from the petrochemical industries centered around Corpus Christi Bay, Galveston Bay, the lower Mississippi River, and Pensacola Bay, and urban and residential development. The program, which was created in 1988 to develop a management strategy for protecting, restoring, and maintaining the health and productivity of the Gulf, has achieved little more than convening a Gulf symposium in 1990, drafting reports on habitat loss, freshwater inflow, and nutrient enrichment, and formulating "an action-planning process." As we shall see in the case of the National Estuary Program described below, the EPA's emphasis has focused more on planning than on action.

> *One-sixth of the U.S. population lives in Gulf coastal states.*
> — U.S. Bureau of the Census

Federal Programs

The three main federal efforts to address coastal waters issues are the National Estuary Program (NEP), the National Estuarine Research Reserve System (NERRS), and the National Marine Sanctuary Program (NMSP). In 1987 Congress established NEP to identify significant estuaries, protect and improve their water quality, and enhance their living resources. If, after being nominated by a state or states, an estuary is selected, its program is administered by a management conference consisting of the EPA and other interested federal agencies, state, interstate or regional agencies, local government, industry, public and private institutions, and the general public. The conference's task is to prepare a plan of action defining pollution controls, resource management priorities, and other objectives. In general, the EPA provides the guidelines while state and local agencies carry out the recommended strategies. In 1990 Congress allocated a modest $1 million for each of the seventeen estuaries in the program, the

bulk of which went for research. By contrast, Chesapeake Bay alone gets $20 million from the federal government under a separate budget.

Beyond Planning

Bringing together federal, state, and local governments, the estuary program has moved slowly — a result, critics say, of its timidity. "It is not really a legitimate response to our nation's coastal crisis," says Chuck Fox, legislative director of Friends of the Earth. "I can't name a single success of this program."[42] A similar criticism is made by Doug Rader of the North Carolina Environmental Defense Fund. Even EPA officials are worried about the emphasis on planning rather than action. "One of the challenges we've been faced with. . . is not to let this program do nothing but generate reports," said Mary Lou Sochia, an EPA administrator for the program in Washington, D.C.[43] One major problem, most parties agree, is the need to generate funding for more than research, planning, and public hearings. The Clean Water Council, a coalition of engineering and consulting firms, estimates it will cost $118 billion to build new wastewater-treatment plants or repair old plants to meet higher water-quality standards. Another worry, expressed by David Miller of the National Audubon Society and cochair of the Long Island Sound Study's Citizens Advisory Committee, is that the program fails to impose controls on new shoreline development and farming.

The National Estuarine Research Reserve System (NERRS)

NERRS was set up as part of the Coastal Zone Management Act of 1972 to address threats to the nation's estuaries. Managed by the National Oceanic Atmospheric Administration (NOAA), it incorporates eighteen reserves in fourteen states and Puerto Rico. Collectively they protect some 300,000 acres of different coastal regions and estuarine types that exist in the U.S. and its territories. They serve as field laboratories where studies are carried out on natural and human processes occurring within the seventeen estuaries covered by the federal estuary program. Included in the reserve system are 4,000 acres of the most biologically productive wetlands along 100 miles of the tidal Hudson River; four of North Carolina's estuaries; 193,758 acres of the Apalachicola Reserve in the panhandle of Florida; Elkhorn Slough, a tidal estuarine embayment on the central California coast; Rookery Bay in southwest Florida, one of the few relatively pristine mangrove estuaries left in North America; Sapelo Island off the Georgia coast; South Slough in Coos Bay, Oregon; Padilla Bay in Washington; and the Tijuana River estuary in California.[44] In its educational programs NERRS creates a direct link between scientists, educators, and the public, with classes, workshops, and guided tours for schools, youth groups, conservation and community organizations, families, and individuals.

National Marine Sanctuary Program

Managed by NOAA's Office of Ocean and Coastal Resource Management, this program is designed to protect and manage significant marine areas in the U.S. coastal waters for the long-term benefit of the public and particularly to

preserve or restore their conservation, recreational, ecological, and esthetic values. Unlike NERRS, which can acquire public lands, this program applies exclusively to underwater areas.

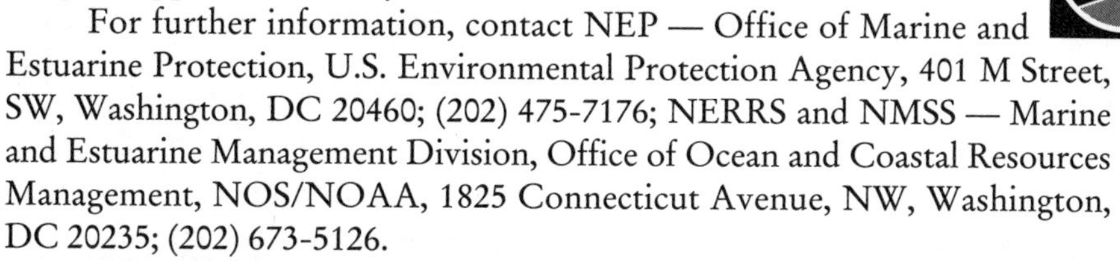

For further information, contact NEP — Office of Marine and Estuarine Protection, U.S. Environmental Protection Agency, 401 M Street, SW, Washington, DC 20460; (202) 475-7176; NERRS and NMSS — Marine and Estuarine Management Division, Office of Ocean and Coastal Resources Management, NOS/NOAA, 1825 Connecticut Avenue, NW, Washington, DC 20235; (202) 673-5126.

What You Can Do to Save Estuaries

The ultimate success or failure of estuary programs will depend largely on the political leadership available in different locations. As we noted earlier, the Chesapeake Bay cleanup is succeeding because it got strong support from the governors of Virginia and Maryland, who pushed it along from the outset.

1. Encourage Congress to increase its efforts to protect our coasts, estuaries, and other waterways.

2. Support the work of national environmental and conservation groups that focus on estuary programs.

3. Increase your effectiveness by joining with the citizens' action groups and various organizations working on the issues described in this chapter.

4. Become more aware of how your own lifestyle may be affecting the degradation of estuarine ecosystems. For more details, see "What You Can Do to Save Wetlands and the Coast," page 119.

Coastal Wetlands

Coastal wetlands, which include salt marshes, salt meadows, mud flats, and mangrove swamps, usually form integral parts of an estuary system's bays, inlets, or lagoons. Often the dominant feature of a coastal region, they are transitional zones between the sea and uplands.[45] Often consisting of low-lying meadows flooded by tidewater from the sea or saturated by floodwater draining from the uplands, they can be described as nutrient-pumping stations, taking in essential salts and minerals from the sea as the tide comes in and releasing organic material into the estuarine system as the tide goes out. Such a process is greatly dependent upon specialized plants — grasses, sedges, rushes, and similar vegetation — adapted for growth in semiaquatic, salty conditions that would kill most land plants. Perhaps more importantly, coastal wetlands transform nutrients arriving via rivers, groundwater, and the atmosphere to organic matter and other, more biologically available forms such as ammonia.

Salt marshes consist generally of upper (drier) and lower (wetter) areas. The lower marsh, usually flooded by two tides a day and dominated by a straight-standing saltwater cordgrass (*Spartina alterniflora*), gets a steady supply of nutrients from the sea, from surface waters and/or groundwater. The higher marsh, covered by salt water only in high spring floods, produces

a salt meadow grass (*Spartina patens*), which lies down in thick mats on the marsh surface, covering hosts of insects and other small creatures. In the coastal wetland environment an organism must be able to adapt to extreme changes in salinity and to survive out of water or underwater for limited times depending on the level of the tides. At the bottom of the food chain throughout marshes and on mud flats is an abundance of algae on which many other creatures depend for their basic food source. Twice a day the tide washes bits of algae and other debris (mainly dead salt-marsh grass) onto mud flats and creeks. There they are decomposed by bacteria and fungi into a rich soup that nourishes marine worms, clams, mussels, and other suspension feeders. These feeders then pass huge amounts of water through their systems, filtering out the algae, bacteria, and other nutrients and recycling nutrients into the salt-marsh ecosystem by producing waste materials that provide food for "higher" organisms along the food chain.

Predation in coastal wetlands is also attuned to the tidal cycles, with fish and birds playing complementary roles. At high tide fish are the major predators roaming over the flats and digging in the soft bottom. Horseshoe and blue crabs, flounder, and other large fish prey on burrowing worms and small clams. At low tide the birds take over, with herons, egrets, and other shorebirds voraciously consuming huge numbers of snails and other small creatures on the exposed surface of mud flats. Because of their abundant productivity, coastal wetlands are a prime source of supply for a commercial and recreational fishing industry whose annual value is now estimated at $15 billion. The precarious ecosystems of the wetlands are subjected to other human disturbances, especially from developers who prize their valuable water frontage. Even the mosquito, universally hated by humans, contributes importantly to the wetlands ecology. Its eggs and larvae provide the main food for many of the insects, fish, and birds that inhabit the mud flats, creeks, and salt pans. Before the passing of the federal Wetlands Acts of 1986 and 1989, hundreds of thousands of acres of salt marsh were destroyed in an effort to eradicate mosquitoes and make the land more habitable for humans. However, according to many ecologists, the ditching of salt marshes has not appreciably reduced the mosquito population.

Pocosin Wetlands

On peat soil built from centuries of slow-decaying plants, the Pocosin wetlands along the coasts of North Carolina and Virginia are shaped by periodic fires from natural causes. Covered with evergreen shrubs and vines, they regulate the flow of fresh water to nearby coastal estuaries and support many species of fish, birds, and other wildlife. They are an essential component of the Atlantic Flyway. The ecological balance of the Pocosins is threatened by a combination of human activities that include draining the marshes, mining them for peat fuel, and a penchant for controlling fires.

Mangrove Swamps

The name *mangrove* covers a species of tree that can live partly submerged in the relatively salty environment of coastal swamps, some species being more salt-tolerant than others. Mangrove swamps are located in the world's warmer regions, where air temperatures average 75 degrees Fahrenheit or more. They grow on flat or gently sloping coasts and are the tropical or subtropical counterparts of salt marshes. Their ability to tolerate salt water allows them to survive. Some mangroves excrete the salt from special glands, others isolate it in inactive tissues within the plant. In marginal coastal areas, mangroves intercept nutrients running off the land and stem the flow of estuarine and tidal waters, trapping silt and contributing to the buildup of mud banks. As these banks become free of water, the mangrove species are replaced by other vegetation. Mangroves create a natural buffer against hurricanes and other storms and, over long periods of time, act as an important force in reshaping coastlines.

More than 525,000 acres of mangrove swamps are located in coastal Louisiana, Texas, and southern Florida. They are dominated by several tree varieties, particularly the red mangrove, which forms dense wildernesses, largely impenetrable except by flat-bottomed boat. These swamps are nurseries for shrimp, sea trout, and other fish. The mangrove swamps of the Florida Everglades are home for the cougar, crocodile, bald eagle, and other rare species as well as many other more common animals.*

Barrier Islands

Composed mainly of sand and shells, about three hundred of these islands stretch in a broken chain along the Atlantic and Gulf coasts from Maine to Texas. As their name indicates, they act as shock absorbers to blunt the force of approaching waves and storms, helping protect the shore and coastal wetlands from flooding and erosion. Some barrier islands, including Cape Cod, Massachusetts, Coney Island, New York, Atlantic City, New Jersey, and Miami Beach, Florida, are heavily developed with roads, houses, shopping malls, and other structures that will sooner or later be damaged or destroyed by major storms and the steady pressure of tides and currents moving the islands toward the mainland.

Cape Hatteras National Seashore

North Carolina's Outer Banks, situated about midway along the Atlantic seaboard, are a thin, broken strand of islands curving out into the ocean and then back again in the sheltering embrace of the mainland coast. These barrier islands have survived the onslaught of sea and wind for thousands of years. Today their long stretches of beaches, sand dunes, marshes, and woodlands are set aside as Cape Hatteras National Seashore under the protection of the National Park Service. Although Hatteras is at the ocean's edge, no clearly defined boundary marks where land ends and sea begins. As with all barrier islands, the sea is dominant, determining what grows on them. Oddly shaped, dwarfed trees are severely pruned by salt-laden winds. Some shorebirds catch small fish or crabs carried by the waves; others probe the sand or dig under shells for insects, worms, or

* For a detailed report on the Everglades, see chapter 2, pages 64-71.

clams. Salt marshes, where twice a day tides exchange and replenish nutrients, are a source of food for birds and other creatures year-round.

In fall and winter, storms batter the islands with fierce winds and waves. Over the years you can bear witness to the retreat of the shoreline from these violent attacks and from the erosions of time and tide.

Virginia Coast Reserve

Twenty miles northeast of Norfolk, Virginia, on the state's eastern shore is the Virginia Coast Reserve, composed of a chain of fourteen barrier islands stretching 55 miles from the Virginia-Maryland border to the mouth of Chesapeake Bay. It includes 8,000 acres on the mainland of the shore and adjoining bays. The reserve's system of unspoiled beaches, maritime vegetation, forests, and salt marshes makes it one of North America's richest natural treasures. The coastal bays between the mainland and the islands constitute the finest coastal aquatic system on the East Coast and are a rich spawning ground for fish and shellfish. The islands provide shelter for more than 250 species of raptors, songbirds, shorebirds, and wading birds. In the spring, pods of bottlenose dolphin migrate along the Atlantic waters off the islands, and in summer juvenile loggerhead sea turtles reside in the shallow coastal bays.

Once home to Algonquin Indians, these barrier islands were taken over by white farmers and livestock raisers in the seventeenth century. By the late 1800s lavish hunt clubs and bathing and lifesaving stations flourished until they were destroyed by storms in the early twentieth century. Today the reserve is the region's leading producer of vegetables, grains, and soybeans. Its waters support a multimillion-dollar seafood industry, which along with sportfishing and tourism is a mainstay of the local economy.

Protective Planning

In partnership with the eastern shore's local community and state and federal agencies, the Nature Conservancy has begun a program for economic growth sensitive to the ecological needs of the area. Through conservation easements,* the program is designed to protect mainland farms as a critical buffer to the core of the preserve. Development of protection planning to promote balanced zoning and land-use laws is another important plan of action. Ongoing conservation and research work continues at the reserve's Long-Term Ecological Research Station, which was established through a partnership between the Nature Conservancy, the University of Virginia, and the National Science Foundation.

The State of the Coasts

In January 1989 a congressional report warned that the coastal waters of the United States were under assault from many directions: "From the contaminated sediment of New Bedford Harbor to the closed beaches of Long Island, from the declining shellfish harvests of Chesapeake Bay to the rapidly

* Rights granted to use the land of another person for a limited purpose.

disappearing wetlands of Louisiana, from the heavily polluted waters of San Francisco Bay to the Superfund sites in Puget Sound, the signs of damage and loss are pervasive."[46] Continuing damage to beaches, estuaries, bays, and other areas from pollution, development, and natural forces presents a far greater threat than the dangers of pollution in the open ocean, the report stated. The basic cause of the troubles is the increasing concentration of people living along the coasts. Today more than half the U.S. population lives within 50 miles of the shore.

> ***"We have an absolutely horrendous problem on our hands."***
>
> — Representative Gerry E. Studds, chairman, Fisheries and Wildlife subcommittee of the House Committee on Merchant Marine and Fisheries

Report Recommendations: The most pressing need is for more money for marine and coastal research, monitoring and regulation, coastal zone management, and water quality programs. Protecting coastal waters should be given higher priority by the EPA and NOAA as well as by individual states to whom federal aid might be suspended if they fail to enforce water standards. The Coastal Zone Management Act should be used to require land-use planning to protect coastal waters, and coastal protection laws should be employed more aggressively to crack down on polluters.

Some of What Goes into Our Coastal Waters Every Year

- 5 trillion gallons of wastewater from 1,300 factories;
- 2.3 trillion gallons of sewage;
- Massive amounts of gasoline, pesticides, and other poisonous substances as runoff from city streets and farms, every time it rains or snows;
- Acid precipitation from air pollution;
- Oil spills from tankers and freighters;
- Pollution from offshore drilling;
- Garbage and oil from commercial and recreational boats.

(Sources: Natural Resources Defense Council, Cousteau Society)

"Ebb Tide for Pollution"

In 1989 the Natural Resources Defense Council (NRDC) published an action plan for cleaning up coastal waters. Few regions have escaped the effects of coastal pollution, the council reported. Nearly one-third of the productive shellfish areas in the U.S. were contaminated. Three sites within Puget Sound in Washington were so polluted by heavy metals as to be declared Superfund cleanup sites. New York banned the sale of striped bass caught in its state waters after the levels of PCBs surpassed FDA allowable limits. In 1989 the Florida Department of Health advised consumers to avoid eating largemouth bass and warmouth fish after these game fish were found to have the highest levels of methyl mercury ever recorded in the U.S.* Swimming in California's

* See also chapter 2, page 67.

Santa Monica Bay was declared unsafe because of extensive and chronic untreated sewage dumping.[47] The NRDC plan advocates imposing strict new nationwide pollution controls for industry, requiring companies to find pollution-free ways of doing business, instituting a national program for safe household waste collection and disposal, and enforcing current laws to control toxic pollution. The plan calls for elimination of raw sewage, upgrading coastal sewage plants, prohibition of new or increased coastal discharges, strengthening of pretreatment programs, and the ending of ocean dumping of sewage sludge, medical waste, plastics, and dredged spoils. Coastal development would be limited by requiring land-use plans, zoning ordinances, and coastal management plans to protect water quality. The plan recommends creating filter strips and buffer zones between new developments and coastal water bodies, including estuaries and rivers. It would also replace subsidies and tax incentives that presently encourage coastal development with incentives to preserve coastal habitat.

> *"By building our largest cities along our coasts and continuing to crowd into environmentally sensitive coastal areas, we have placed an immense burden on the very ecosystem on which we depend for so much."*
> — *Ebb Tide for Pollution*, NRDC

In addition to instituting national programs to protect coastal waters from oil spills and toxic runoff, NRDC urges long-term protection of sensitive marine areas as ocean sanctuaries, adoption of strict new water quality standards to protect coastal habitats, fish, shellfish, and wildlife, and the designation of currently clean waters for special protection to make sure they stay clean.[48] Important as these proposals are, it should be pointed out that implementing them would not be easy, requiring the investment of considerable funding that could run into hundreds of millions of dollars.

Closing the Beaches

In 1989 and 1990 there were more than 2,400 beach closures and pollution advisories in ten of the most populous states on both coasts of the U.S.,* NRDC reported.[49] The most important cause was high levels of bacteria attributable primarily to sewage and storm water contamination. Documenting the reasons for the closings, NRDC found almost all of them were the result of antiquated sewer systems. The NRDC findings were confirmed by the EPA, which said that the household sewage pipes in 1,200 communities in the East, Northwest, and Midwest also carry water draining from cities and suburbs. After heavy rains, sewers designed to carry household sewage cannot handle the additional volume of storm water, which overflows into waters used for swimming and fishing, combining large amounts of raw sewage and other waste with the storm water. The overflow picks up bacteria and viruses from dog, bird, and other animal droppings; lawn chemicals; and

* California, Connecticut, Delaware, Florida, Maine, Maryland, Massachusetts, New Jersey, New York, and Rhode Island, home to 34 percent of the total U.S. population.

street debris, pouring filthy water into rivers, bays, and the oceans and contaminating beaches and seafood.[50]

The worst contamination is in the coastal waters of New York, New Jersey, and Connecticut, the NRDC report said. New York City alone has more than five hundred storm water outlets that regularly overflow during storms, pouring hundreds of millions of gallons of untreated water into the harbor and Long Island Sound. Storm water runoff and raw sewage spilling into coastal waters accounted for shutting down 750 beaches in the three states in 1990.

A major problem identified by NRDC is the absence of a federal mandate setting uniform (or even minimum) beach closure standards. "Current federal EPA *recommended* guidelines allow beaches that meet levels considered 'safe' to contaminate 19 out of every 1,000 swimmers with gastroenteritis* (19,000 out of a million)."[51] Beach-closing and warning standards vary widely from state to state, with Maine the most stringent and New York the most lax. A bacterial concentration requiring closure in New Jersey, for example, would leave the beach open in New York. Although the EPA has authority to set minimum water quality requirements for states, it has never exercised this right for bacterial standards.[52] According to the EPA, the technology for cleaning up coastal water pollution is readily available, but the cost would be enormous. Rebuilding sewers and adding water treatment plants to eliminate the problem could cost $60 billion, the agency said.[53]

Oil Spills and Offshore Drilling

No type of coastal pollution is more dramatic than a catastrophe such as the Exxon *Valdez* spilling 11 million gallons of crude oil into Alaska's Prince William Sound, the full ecological damage to which has not yet been determined. Yet there are literally thousands of other oil spills occurring annually that barely make the headlines in our national media but which are creating massive damage to coastal ecosystems. America's increasing dependency on imported oil has prompted the administration to press for the opening of vast undersea tracts off Alaska, California, the East Coast, and the Gulf of Mexico for the drilling of oil and of natural gas, one quarter of which comes from this source. Offshore drilling is associated with severe contamination of the marine environment, particularly in light of the 3.25-million-gallon Chevron well blowout in Santa Barbara, California, in 1969, which devastated beaches and birds, and the 200-million-gallon spill from the Campeche accident in the Gulf of Mexico ten years later. Today, with more than 3,500 offshore drilling platforms already operating around the United States, the threat is not only from discharges into the sea of oil and gas but also from untold quantities of drilling solution, lubricants, pulverized rock, mud, human waste, and even contaminated rainwater that drips off the equipment.

In February 1991 the EPA proposed tightening its regulations for discharges from offshore platforms in federal waters, which in most states begin three miles from the shore. However, in response to administration pressure,

* Inflammation of the stomach and intestines.

federal requirements were kept less stringent than those of several states where increased drilling is projected. Alabama, for example, has a "zero-discharge" rule, which environmentalists say is a model of ecological safety that should be adopted nationwide for drilling close to the shore.[54] None of the oil companies wanting to tap the huge natural gas reserves beneath Mobile Bay agreed willingly to abide by Alabama's requirements. In the early 1980s, after Mobil Oil was discovered pouring untreated sludge and toxic drilling mud into the bay, the company paid the state a $2 million penalty and at least $500,000 to dredge up the chemical-soaked bay bottom. Suspicion about the oil industry's willingness to comply with state environmental laws remains strong among anglers, shoreline homeowners, and many others who believe that offshore drilling should not be allowed to proceed without strict controls.[55]

Coral Reefs

If you have ever gone snorkeling in southern Florida, the Caribbean, Hawaii, or other subtropical or tropical waters, you know the incredible beauty of coral reefs. But did you know that they are living structures composed mostly of calcium carbonate secreted by photosynthesizing red and green algae and small coral animals?

Although coral reefs are massive, they are built up by an intricate association of countless millions of plants and animals, living and dead. At their foundation is the tiny coral polyp, which secretes lime (i.e., calcifies) to form a rocklike cup about itself. The proliferation of polyps into colonies creates the photogenic patterns we associate with the coral skeleton. The formation of the colony is aided by the polyps' relationship with minute algae that live in their tissue, in which oxygen and carbohydrates from the plants are exchanged for carbon dioxide and nutrients given off from the polyps and seawater. This relationship appears to be essential to calcification, which builds the reef. This biological partnership, combined with the many ways in which reefs are able to capture nutrients from the air and the sea, make them highly productive ecosystems.

Coral reefs provide ideal habitats for fish, crabs, lobsters, and thousands of other invertebrate species. They are also frequented by hawksbill turtles and dolphins. Dugong, conch, rays, and other creatures exist in surrounding sand flats and sea-grass meadows. The white sands surrounding many reefs are created by waves and certain fish and other organisms that abrade the limestone produced by corals. Like rain forests, coral reefs provide great biological diversity, including a number of chemical compounds that show promise of being effective against viral diseases, arthritis, and other ailments. Certain of these chemicals are produced in self-defense by sponges and other organisms that cannot move around.

In addition to the direct benefits of tourism, fishing, and medical research, reefs provide more subtle economic benefits by avoiding costs that would otherwise be incurred. For example, they act as self-repairing breakwaters, playing a similar role to that of barrier islands in helping protect beaches and shorelines from damage by storms and erosion.

Reefs under Siege

Commercial fishing, mining of limestone for building materials, water pollution, motor and fishing boats, tourism, and perhaps global warming are among the increasing threats to coral reef ecosystems. In southern Florida scientists and lawyers from the Environmental Defense Fund (EDF) are working with other environmental groups and local organizations to stem the flow of excess nutrients and pollution from farming and urban development into the Everglades and the Florida Reef Tract, the third largest coral-reef system in the world.[56]

Three times since 1979 coral reefs in many parts of the world have turned white, a process that occurs when the symbiosis between coral polyps and algae breaks down, greatly reducing coral growth and threatening the framework of the reef itself. Bleaching is caused by local factors such as siltation and oil pollution; it can also be caused by slight elevations of the already-high temperatures experienced by the corals. Many mass-bleaching events have occurred when temperatures were slightly above normal. Therefore, many scientists think that the bleaching may intensify as global warming proceeds. EDF staff and other environmentalists are trying to reduce the risk of further mass events by ensuring that greenhouse gas emissions are reduced sufficiently to protect sensitive ecosystems from global warming.

Damage Control

As EDF marine biologist Dr. Rodney Fujita points out, many dangers to coral reefs are preventable and reversible.[57] Better management of fisheries could ease commercial fishing threats. Using wastewater nutrients to grow seaweed could reduce too many nutrients going into coral reefs, while producing valuable byproducts. New industrial and farming techniques could cut down or eliminate runoff pollution at the source. We could also use alternative building materials instead of destroying reefs to get them, while boaters and snorkelers should be made more respectful of these beautiful and extremely valuable ecosystems.

Mysterious Malady

In August 1991 a highly infectious disease was found to be killing off the common black-spined sea urchin (*Diadema antillarum*) in the Florida Keys and the Caribbean and in the process endangering the region's delicate coral reefs. "We may be witnessing the largest extinction of a marine animal ever recorded," said Robert Bullis, director of the Laboratory for Marine Animal Health at Woods Hole, Massachusetts.[58] The demise of the sea urchins could have a devastating effect on the coral. The spiny creatures are considered by marine biologists to be "regulators," because they graze on algae that, if allowed to get out of control, can smother corals.

In 1983 and 1984, a crash in the sea-urchin population in the western North Atlantic and Caribbean seas marked the largest die-off ever witnessed for a marine invertebrate: On many of the reefs 95 percent of the urchins perished.[59] Another die-off took place in 1990, and more began to die in the spring of 1991. "There are so few urchins left that it's hard to see if new mor-

talities are occurring," said Ernest Williams, a researcher at the University of Puerto Rico.[60] Scientists suspect that a virus, bacterium, or protozoan, perhaps transported from the Pacific Ocean to the Caribbean in ship bilge water carried through the Panama Canal, may be responsible. They also note that pollution and general degradation of inshore waters could be contributing factors.

What Is Killing the Palms?

Florida's state tree, the sabal palm, is dying by the thousands along the coast, and researchers suspect that the towering trees may be among the early victims of global warming.

In early 1991 a landowner near Yankeetown on Florida's west coast asked state foresters to look at his dying palm trees. They were surprised to find a die-off centered along a 40-mile stretch of Gulf of Mexico coastline from Cedar Key to south of Homosassa Springs, as well as pockets of dying trees on the east coast in Jacksonville, and at the entrance of Tampa Bay. The problem is concentrated in low-lying areas, where scientists say that salt water is destroying the trees' root systems.[61]

Researchers suspect that increased burning of fossil fuels has led to a rise in the world's oceans. Although fluctuating ocean levels are a normal phenomenon, evidence now suggests the current rise is a good deal faster than normal. For example, as part of a 1985 study of Florida's northern Gulf coastline, University of South Florida marine scientists and geologists studied tide gate measurements that had been recorded at Cedar Key for nearly 70 years. The found an overall water level rise of more than 3 inches, "a rate over three times that of the geologic record."[62]

Who Is Minding the Shore?

In the United States authority for the management of the coastal zone is divided among different federal, state, and local agencies, many of which have overlapping and sometimes conflicting interests and responsibilities. Under the National Coastal Zone Management Acts of 1972 and 1980, the federal government provides assistance to the thirty-seven coastal and Great Lakes states and territories to help them develop voluntary programs for safeguarding those coastlines *not* under federal protection — about half the nation's shoreline, excluding Alaska. With such a large percentage in private hands, it is difficult for federal or state agencies to provide the kind of enforcement authority needed for an ecologically sound program of coastal zone protection. Of individual states, California, North Carolina, and Alabama are considered to have the strongest programs, but even these come under constant threat from developers, oil companies, and other interests.

Federal response to coastal erosion has been applied mainly to urban shorelines on an emergency basis. Hard shoreline protection is generally not favored by Washington because of its economic cost and because it means loss of recreation, habitat, and aesthetic values. Soft protection — replenishing beach sand, sometimes called beach nourishment — is more common. Federally sponsored projects at Virginia Beach, Virginia, Ocean City,

Maryland, and other beach resorts have achieved marginal successes, despite heavy sand losses during winter storms.[63] The authority vested in the Army Corps of Engineers under the Clean Water Act to regulate dredge and fill operations in U.S. waters[64] has been applied mainly to estuarine and wetland areas, but not to beaches, sand dunes, or uplands facing open water.

What You Can Do to Save Wetlands and the Coast

1. Conserve water whenever possible.
2. Use biodegradable or phosphate-free detergents and shampoos.
3. Instead of pouring household cleansers, paints, and other chemical wastes (including water "fresheners") down drains or toilets, recycle them at a hazardous waste center. If you don't have one locally, encourage your local department of health to organize a household toxics collection day when such items will be taken to a specialized dump site.
4. Buy organically grown fruit, grains, and vegetables. This prevents nitrates — the main component of artificial fertilizers — from seeping into the groundwater and running off into rivers, estuaries, and coastal waters.
5. Conserve energy at home, at work, and especially in your car. The less oil we use, the less demand there will be to drill for oil in ecologically sensitive coastal and wilderness regions. It will also lessen the need to transport oil by tanker, reducing the frequency of oil spills.
6. Avoid using antifouling marine paints containing chemicals harmful to marine life.
7. Boycott persistent polluters. Inform them of your decision and encourage others to do the same.
8. Join a civic association, coastal-advocacy group, or other action-oriented group involved in efforts to preserve and restore the coast. If none exists in your area, talk to like-minded people and form your own group.
9. Get to know the quality of the bodies of water — streams, rivers, estuaries, and coasts — in the area where you live.
10. Contact the agency or agencies responsible for maintaining water quality. Start with your local water-control board, department of health or natural resources, or your local government office. Check also with the EPA. Push for tighter controls.
11. Support strong land-use planning. Lobby your local government and state coastal management agencies to adopt policies that reduce the extent of development along streams, wetlands, and the coast. Support measures that leave mature trees standing when excavating for new developments and that limit construction activities that disturb topsoil and produce runoff of increased sediment in coastal streams and waters.
12. Let your elected officials know how you feel about quality of life in coastal regions. Encourage them to take action against industrial, agricultural, and governmental indifference to coastal water pollution. Let them know that you are concerned with coastal and wetlands protection legislation pending in Congress.

RESOURCES

1. U.S. Government

— The main congressional committees dealing with coastal and marine issues are: the House Merchant Marine and Fisheries Committee; the House Public Works and Transportation Committee; the House Interior and Insular Affairs Committee; the Senate Environment and Public Works Committee; and the Senate Commerce, Science, and Transportation Committee. For further information, see the Directory, pages 352-354.

— The main government agencies are:

Department of the Army, U.S. Army Corps of Engineers

Department of Commerce, National Oceanic and Atmospheric Administration; the National Marine Fisheries Service; the National Ocean Service, Marine Pollution Programs Office, Ocean and Coastal Resource Management

Department of Defense, U.S. Coast Guard

Department of Energy, Sub Seabed Disposal Program, Geologic Repositories Office

Department of State, Oceans and International Environmental and Scientific Affairs Bureau; the Oceans and Fisheries Affairs Bureau; and the Office of Ocean Law and Policy

Department of Transportation, U.S. Coast Guard

Environmental Protection Agency, Office of Marine and Estuarine Protection; Office of Marine Mammal Commission; and the Gulf of Mexico Program Office

2. Environmental and Conservation Organizations

Virtually all of the mainstream environmental organizations are concerned directly or indirectly with coastal issues. Those with a more particular interest include the American Littoral Society; American Society of Limnology and Oceanography; Clean Ocean Action; Clean Water Coalition; Coastal Conservation Association; Coastal Society; Center for Marine Conservation; Cousteau Society; Greenpeace; Oceanic Society; Sea Shepherd; and the Surfrider Foundation. For further information, see the Directory, pages 331-355.

3. Further Reading

— *America's Coastal Resources*. Oversight report of the Committee on Merchant Marine and Fisheries, December 1988, Serial no. 100-E. Washington, D.C.: U.S. Government Printing Office, 1989.

— *And Two If by Sea: A Citizen's Guide to the Coastal Zone Management Act and Other Coastal Laws*, Beth Millemann. Washington, D.C.: Coast Alliance, 1986.

— *At the Sea's Edge*, William T. Fox. New York: Prentice-Hall, 1983.

— *Atlantic & Gulf Coasts*, William H. Amos and Stephen H. Amos. Audubon Society Nature Guides. New York: Knopf, 1989.

— *The Atlantic Shore*, John Hay and Peter Farb. Orleans, Mass.: Parnassus Imprints, 1982.

— *Barrier Island: Proceedings of Forum and Workshop*, edited by Barbara S. Mayo and Lester B. Smith, Jr. National Park Service, North Atlantic Region, 1980.

— "Beaches and Barrier Islands," Robert Dolan and Harry Lins. *Scientific American*, July 1987.

— *Chesapeake Bay: Nature of the Estuary*, Christopher White. Tidewater, Va.: Tidewater, 1989.

— *Coastal Alert*, Dwight Holing. Washington, D.C.: Island Press, 1990.

— *Coastal Marshes: Ecology and Wildlife Management*, Robert H. Chabreck. Minneapolis: University of Minnesota Press, 1988.

— "Coral Reef Protection: Briefing Paper for the Intergovernmental Negotiating Committee, U.N. Conference on Environment and Development, Nairobi, Kenya, September 9, 1991," R. M. Fujita, K. Gjerde, T. J. Goreau, and M. Epstein. Available from EDF, New York.

— *Ebb Tide for Pollution*. New York: Natural Resources Defense Council, 1989.

— *The Edge of the Sea*, Rachel Carson. New York: Signet, 1955.

— *Edge of the Sea*, Russell Sackett. Alexandria, Va.: Time-Life Books, 1983.

— *A Field Guide to the Atlantic Seashore*, Kenneth L. Gosner. Boston: Houghton Mifflin, 1978.

— *Land's Edge: A Natural History of Barrier Beaches from Maine to North Carolina*, Michael L. Hoel. Newbury, Mass.: Little Book Publishing, 1986.

— *Life and Death of a Salt Marsh*, John and Mildred Teal. New York: Ballantine Books, 1977.
— *Life at the Edge*, edited by James L. Gould and Carol Grant Gould. New York: W. H. Freeman & Co., 1989.
— *Listen to the Sound: A Citizens' Agenda for Long Island Sound*, David J. Miller and Jane-Kerin Moffat. New York: National Audubon Society, January 1991.
— *Looking at the Bottom*, Sarah Clark. New York: Environmental Defense Fund, 1990.
— *Pacific Coast*, Bayard H. McConnaughey and Evelyn McConnaughey. Audubon Society Nature Guides Series. New York: Knopf, 1988.
— *Polluted Coastal Waters*, Diane Fisher et al. New York: Environmental Defense Fund, 1988.
— *Portrait of an Island*, Mildred and John Teal. Athens, Ga.: Brown Thrasher Books, 1981.
— "Saving the Nation's Great Water Bodies," *EPA Journal*, November/December 1990. Whole issue.
— *Seashores*, Herbert S. Zim and Lester Ingle. New York: Golden Press, 1955.
— *Testing the Waters: A Study of Beach Closings in Ten Coastal States*, Jennifer Kassalow and Diane Cameron. New York: Natural Resources Defense Council, August 1991.
— *The Thin Edge: Coast and Man in Crisis*, Anne W. Simon. New York: Harper & Row, 1978.
— *Turning the Tide: Saving the Chesapeake Bay*, Tom Horton and William M. Eichbaum. Washington, D.C.: Chesapeake Bay Foundation and Island Press, 1991.
— *Waves and Beaches*, Willard Bascom. New York: Anchor/Doubleday, 1980.

NOTES

1. U.S. Army Corps of Engineers, "National Shoreline Study," cited in David A. Ross, *Introduction to Oceanography* (Englewood Cliffs, N.J.: Prentice-Hall, 1988), p. 330.
2. Rutherford Platt et al., "The Folly at Folly Beach," *Environment*, November 1991.
3. U.S. Environmental Protection Agency, Office of Water, Office of Marine and Estuarine Protection.
4. *Audubon Society Nature Guide: Pacific Coast* (New York: Knopf, 1988), p. 76.
5. *Chesapeake Bay: Its Beauty and Bounty* (Annapolis, Md.: U.S. Fish and Wildlife Service, n.d.), unpaginated.
6. U.S. Environmental Protection Agency, *Chesapeake Bay: A profile of environmental change* (Philadelphia, 1983).
7. *Detergents, Phosphorus, and the Bay*, Chesapeake Bay Foundation (1989).
8. "The State of the Bay," *Mid-Atlantic Country*, October 1990.
9. William C. Baker and Tom Horton, "Runoff and the Chesapeake Bay," *EPA Journal*, November/December 1990, p. 13.
10. Ibid.
11. Quoted in B. Drummond Ayres, Jr., "Stirrings of Hope in Redeeming Chesapeake Bay," *New York Times*, December 2, 1990.
12. Diane Fisher et al., *Polluted Coastal Waters: The Role of Acid Rain* (New York: Environmental Defense Fund, 1988), pp. 46-66.
13. Ibid., p. 5.
14. Much of the background information in this section comes from Andre Cohen's *An Introduction to the Ecology of the San Francisco Estuary*, produced by the Save San Francisco Bay Association for the San Francisco Estuary Project, 1990.
15. Jane Kay, "Selenium in Bay Threatens Wildlife," *San Francisco Chronicle*, February 17, 1991.
16. Ibid.; see also Greg Karras, "Pollution Prevention: The Chevron Story," *Environment*, October 1989.
17. Annette Frahm, "Financing the Cleanup of Puget Sound," *EPA Journal*, November/December 1990, p. 29.
18. Ibid.
19. "Our Filthy Seas," *Time*, August 1, 1988, p. 50.
20. Ibid.
21. "Valdez Spill Toll Is Now Called Far Worse," Associated Press report, *New York Times*, April 18, 1992.

22. Michael W. E. Didriksen, "Long Island Sound Cleanup Begins," *New England Environmental Network News*, Winter 1991, p. 14.
23. Personal communication, March 9, 1992.
24. Robert Reinhold, "Break in Pipe Spews Sewage Near San Diego's Shore," *New York Times*, February 6, 1992.
25. See Donna D. Turgeon et al., "Toxic Contaminants in Long Island Sound," National Oceanic and Atmospheric Administration, December 1989; and "Toxic Contamination in Long Island Sound," LISS Fact Sheet #10, June 1990.
26. Information based on EPA reports and especially *LISS Study: Status Report and Interim Actions for Hypoxia Management*, December 1990.
27. Allan R. Gold, "L.I. Sound Is So Polluted It Faces Long-Term Damage, Scientists Say," *New York Times*, July 6, 1990.
28. Allan R. Gold, "L.I. Cleanup Is Official's Goal," *New York Times*, December 1, 1990.
29. Personal communication with Jane-Kerin Moffat, coordinator, LIS Watershed Alliance, January 27, 1992.
30. Representative Gerry Studds, chair, House Subcommittee on Fisheries and Wildlife and the Oceanography Subcommittee to examine Buzzards Bay, cited in *Bay Watch*, June/July 1991, p. 1; for details of the plan, see *Buzzards Bay: Comprehensive Conservation and Management Plan, Volume 1, Management Recommendations and Action Plans*, U.S. Environmental Protection Agency, Massachusetts Executive Office of Environmental Affairs, August 1991.
31. Buzzards Bay Project, *Bay Watch*, Fall 1991, p. 2.
32. Background information in this section is based on *Coastal Louisiana: Here Today Gone Tomorrow?*, Coalition to Restore Coastal Louisiana (Baton Rouge, 1989).
33. "Louisiana's Bayou Blues," *Newsweek*, June 22, 1987, p. 54.
34. Ibid., p. 7.
35. For a more detailed report on this phenomenon, see Jeff Hecht, "The Incredible Shrinking Mississippi Delta," *The New Scientist*, April 14, 1990, pp. 36-40.
36. Ibid.
37. "Louisiana's Bayou Blues," ibid.
38. Wesley Marx, "Citizens and the Gulf of Mexico," *EPA Journal*, November/December 1990, p. 10.
39. Cory Dean, "In Louisiana, a Critical Wetland Habitat Founders," *New York Times*, November 20, 1990.
40. Peggy Rooney, "Louisiana's Wetlands Calamity," *EPA Journal*, September/October 1989.
41. *The Gulf of Mexico Program* (Stennis, Miss.: Environmental Protection Agency, 1991), p. 2.
42. Elissa Wolfson, ed., "Sound Policies," *E Magazine*, July/August 1991, p. 17.
43. Ibid., p. 18.
44. For further information on the national estuarine programs, see *Current: The Journal of Marine Education*, whose volume 10, no. 1 1990 is devoted to estuaries.
45. Material in this section is based primarily on the following sources: William T. Fox, *At the Sea's Edge*, (New York: Prentice-Hall, 1983); Robert Chabreck, *Coastal Marshes* (Minneapolis: University of Minnesota Press, 1988); and Michael L. Hoel, *Land's Edge* (Newbury, Mass.: Little Book Publishing, 1988).
46. Philip Shabecoff, "Citing Decline in Coastal Waters, House Report Urges New Policy," *New York Times*, January 24, 1989.
47. *Ebb Tide for Pollution* (New York: National Resources Defense Council, 1989), pp. 3-9.
48. Ibid, pp. 36-37.
49. *Testing the Waters: A Study of Beach Closings in Ten Coastal States* (New York: Natural Resources Defense Council, August 1991), pp. 7-14.
50. Ibid.; see also Keith Schneider, "Old Sewers Tied to Closing of Nearly 2,400 Beaches," *New York Times*, August 15, 1991.
51. *Testing the Waters*, p. 3.
52. Ibid., p. 25.
53. Schneider, "Old Sewers."
54. Keith Schneider, "Alabama Leads Push for Cleaner Offshore Drilling," *New York Times*, March 10, 1991.
55. Ibid.
56. Rodney M. Fujita, "Coral Reefs: Undersea Battle for Survival," *EDF Letter*, January 1991.

57. Ibid.

58. "Mysterious Sea Urchin Malady Puts Caribbean Reefs at Risk," *Ecology USA*, August 12, 1991, p. 155.

59. Ibid.

60. Ibid.

61. "Environmental Change, Not Disease, Killing Florida Palms," *Ecology USA*, April 6, 1992, p. 63.

62. Ibid.

63. Platt, "Folly at Folly Beach," p. 9.

64. 33 U.S.C. 1344, Section 404.

4 OCEANS

"Every spoonful of ocean water contains life on the order of 10^2 to 10^6 bacterial cells per cubic centimeter, plus assorted microscopic plants and animals, including larvae of organisms ranging from sponges and corals to starfish and clams and much, much more."

— Sylvia Earle, chief scientist of the National Oceanic and Atmospheric Administration (NOAA)[1]

The ocean is the earth's dominant feature, covering 70 percent of its surface and containing more than 90 percent of its life. Yet despite its size and abundance, scientists know relatively little about it, even its shallow inshore waters. Only one excursion has ever been made to the deepest part of the ocean (7 miles down), and that was for a mere twenty minutes in 1960. As Earle points out, even if you include samples obtained in nets and remotely deployed instruments, "less than 10 percent of the ocean has been sampled, and much of it has not been more than superficially mapped."[2] What we do know, however, is that life originated in the sea and that the world's oceans nurtured broad categories of plants and animals millions of years before flowers, trees, insects, birds, mammals, and other land organisms evolved. We know also that the ocean exerts an enormous influence on the atmosphere, the climate, and on the landmasses it surrounds.

Sea Stats

- The sea is the principal reservoir for water on earth. Its 317 million cubic miles (1.33 billion cubic kilometers) contain 97.2 percent of the world's entire supply of water.
- The deep-ocean floor occupies almost 30 percent of the earth's surface. It is covered by sediment hundreds of yards thick, much of which has accumulated particle by particle over millions of years.
- Ocean features persist much longer than land features because there is no erosion analogous to that by glaciers, wind, or rivers on land.
- Sunlight can penetrate to a depth of 350 to 600 feet, depending on the amount of sediment, debris, and pollution in the water.
- The Pacific Ocean occupies more than one-third of the earth's surface and contains more than half of the earth's free (unfrozen) water. With an average depth of 13,200 feet (4,000 meters), it is the deepest, coldest, and least salty of the oceans.
- The Atlantic is shallower, with an average depth of 10,800 feet (3,310 meters), the result of a continental shelf, shallow marginal seas, and the Mid-Atlantic Ridge.
- The Arctic Ocean is a landlocked arm of the Atlantic including all water north of Eurasia and North America. During much of the year it is covered by sea ice in thick chunks that freeze, break apart, and refreeze to form a solid surface to a depth of 10 feet or more.
- The Gulf Stream transports fifty times more water than all the world's rivers.

(Sources: *Cousteau Almanac*; M. Grant Gross, *Oceanography*)

The Oceans' Role

Oceans store the sun's energy as heat, which they distribute through their currents and by evaporation as part of the hydrologic cycle that moves and recycles water throughout the biosphere.* However, even without major ocean currents, the coastal zones of any continent enjoy milder climates than

* See diagram on page 50.

its interior because water loses heat more slowly than land does in winter and absorbs it more slowly in summer.[3] The sea also serves as a vast storehouse of dissolved oxygen and carbon dioxide, which help regulate the composition and temperature of the atmosphere. In addition to supporting an enormous amount of marine life, oceans supply large quantities of phosphates, iron, oil, natural gas, and other resources (including food) consumed by humans. The water in the oceans acts as a climatic buffer, moderating air temperatures and keeping ocean temperatures in the general range of 32 to 98.6 degrees Fahrenheit. Neither the amount of liquid water at the earth's surface nor the overall salt content of the oceans have changed drastically since the retreat of the continental glaciers from North America and Europe five thousand years ago. Although 4 billion tons of dissolved salts are brought each year by rivers to the oceans, they are removed at about the same rate through chemical reactions with sediments and particulate matter. Although sea salt has a relatively constant composition — it is mostly sodium and chlorine — the salinity* of the ocean varies throughout, being influenced by precipitation, evaporation, and movement of water masses. Salinity also affects the ocean by lowering the temperature of initial freezing. In contrast with fresh water, which begins to freeze at 32 degrees Fahrenheit and retains that temperature until only ice remains, seawater, as it approaches 32 degrees, begins to exclude the salt, which causes the brine to become still more salty. This pushes the freezing point for the remaining liquid still lower, requiring the temperature of the ocean to drop before additional ice is formed. Thus, in the arctic the initial freezing temperatures in the sea are lower than in freshwater ecosystems. This has the effect of keeping the temperature of the ocean bottom generally above the freezing point even in arctic and antarctic regions.[4]

Circulating Currents

Driven by the energy of the sun and the moon, by wind, and by the earth's rotation, the sea is continually moving. Ocean currents are caused by the slow exchange of polar and equatorial waters brought about by the sun's unequal heating of the surface of the sea and by winds blowing diagonally toward the equator from the northeast and southeast. Flowing through the ocean like rivers, powerful surface currents carry huge volumes of water from the poles to the tropics and back, holding its original temperature through thousands of miles of its journey. Beneath the surface other currents can create deep-sea cataracts three times higher than the tallest waterfall on land. Surface ocean currents are primarily wind driven. In both equatorial hemispheres the trade winds generally cause the water to flow westward. In the region of light and variable winds between the hemispheres (known as the **doldrums**) there is an equatorial countercurrent that flows east. In addition to the equatorial currents, there are major currents flowing eastward — the **West Wind Drift** in the Southern Hemisphere, the **North Atlantic Current** (which is a continuation of the **Gulf Stream**), and the **North**

* Salinity is the measure of the amount of dissolved salts in a given volume of water.

Pacific Current (an extension of the **Kuriosho**). These currents move large volumes of water in an east to west direction and generally remain in the same climatic zone.

When the water flows reach the continents, they are split into **boundary currents** that flow north to south. In addition to transporting heat from the tropics to the poles, these currents form boundaries to the circulation of coastal waters. Because of the earth's rotation on its axis, the currents

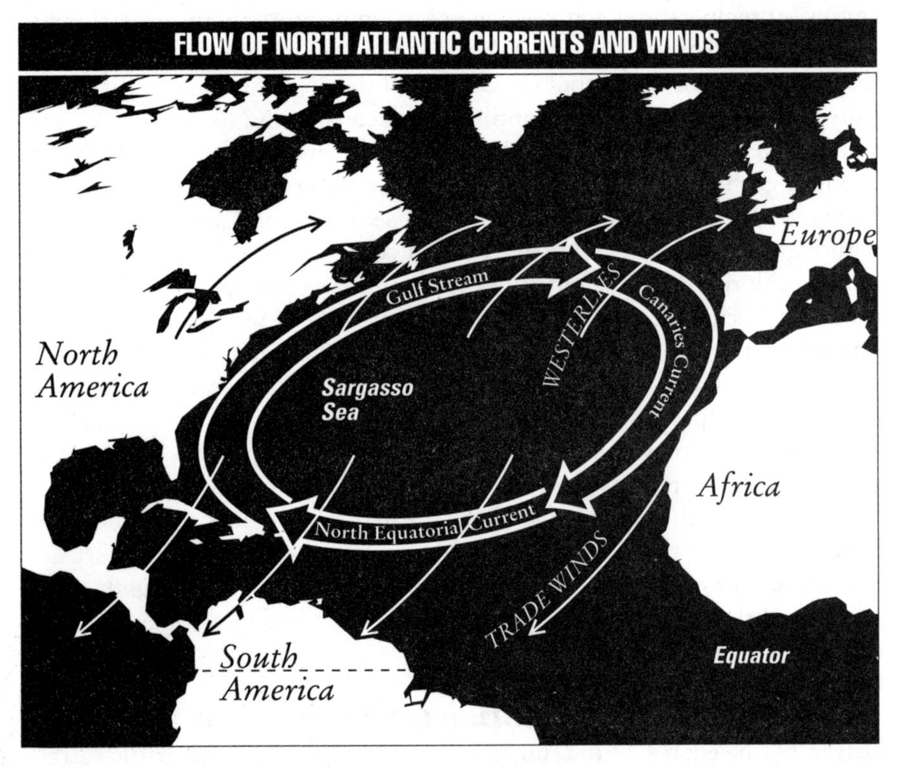

in the Northern Hemisphere tend to move toward the west, resulting in fast-moving currents along the western boundaries of oceans (Gulf Stream, Kuriosho) and producing slower currents on the eastern side (California Current). On the western side of an ocean basin (e.g., on the eastern seaboard of the continental U.S.) this causes a sharp break between coastal and open sea circulation. On the eastern side of the west coast the separation is more diffuse. Wind-driven currents move the ocean horizontally, as do certain currents (known as **geostrophic**) resulting from the distribution of water density, which is regulated by the salinity and temperature of the water. Geostrophic currents occur mainly below the level reached by the wind. Other density-controlled currents cause vertical water movements that supply deep and bottom water masses to the ocean basins. It is vertical circulation (called **thermohaline**) that moves deep waters from the North Atlantic and the Antarctic through the ocean basins and returns them to the surface. Thermohaline circulation is driven by differences in density that result from the warming of water at the equator and the cooling of it at the poles and from the evaporation and freezing of seawater.

The Gulf Stream carries the warm waters from the equator northward through the North Atlantic. Near Greenland they are cooled to below 32 degrees Fahrenheit. As the surface waters reach critical density, they sink and begin to flow as a mass along the bottom of the ocean, particularly along the western side of the basin. Rather like the cold-water return of a household heating system, cold ocean water is returned by surface and subsurface currents back toward the equator. Across the equator the water is moved almost entirely by deep-ocean currents, which generally flow more slowly than those on the surface. The warm and cold currents influence the climates of coastal regions near their flow. For example, without the Gulf Stream system, which transports fifty times more water than do all the world's rivers, the climate of the eastern seaboard of the U.S. and the Canadian Maritime Provinces would be similar to that of the sub-arctic.

Why Tides Move West

Tides are caused in large part by the powerful gravitational attraction of the moon and sun on the oceans. Although the mass of the sun is 27 million times that of the moon, the moon's power over the tides is more than twice that of the sun, because the moon is much closer to the earth.[5] The pull is stronger on the side of the earth facing the moon, causing the water on this side to be pulled toward the moon and creating "bulges" in the water that stay aligned with the moon as the earth turns. As the earth rotates eastward on its axis, the tides move in the opposite direction — westward.

Relative to the moon, it takes the earth twenty-four hours and fifty minutes to rotate. Within this time span the ocean will experience two tidal highs and lows. The sun also generates tides on opposite sides of the earth, but their effect is masked by the lunar tides except at new or full moons. At these times, when the sun and moon are nearly lined up with the earth, the solar and lunar gravitational pulls are cumulative, producing exceptionally high tides with maximum rise and fall occurring roughly every fourteen days. These are known as **spring tides**. When sun and moon are at right angles to each other relative to the earth, the oceans have relatively weak (**neap**) tides.

In the open sea, tidal currents are relatively weak, but in shallow waters and estuaries they can create large waves that travel upstream or move large amounts of sediment that might block harbors. In addition to responding to lunar and solar attraction, tidal movements are influenced by barometric pressure, offshore and onshore winds, depth of water, contours of the shore, and other variables. Because of its currents and tides, the open sea is often considered to be one vast, interconnected ecosystem. You can get an idea of the different force of tidal currents by comparing two locations on the east coast of North America. Into Passamaquoddy, a small bay on the shore of northern Maine, tidal currents carry 2 billion tons of water twice each day; into the Bay of Fundy, a mere 50 miles away, the volume of water is fifty times greater.[6]

Rock and Roll

Many of us know from personal observation that tides vary greatly within relatively short distances along a shore. To some extent, this is a consequence of local topography — the width of a bay's entrance, the depth of the channel through which the water flows, or the slope of an ocean's bottom. **Tidal oscillation** — the rocking up and down of water — takes place around a central (tideless) node, which varies from place to place. Nantucket Island, for example, is located near the node of its basin, where there is little motion, and thus has a tidal range of about 12 inches. The Bay of Fundy, some 350 miles to the north, lies at the end of an oscillating basin with a natural twelve-hour oscillation that coincides with the period of the ocean tide. This sustains the movement of water within the bay, which is enormously increased by the ocean tide and produces tides of more than 50 feet — the highest in the world.[7]

Making Waves

Although beachgoers and seafarers have observed waves from the beginnings of human existence, it was only in the last century that the first scientific descriptions of waves were recorded.* Even as recently as D-Day in World War II, there was, as one naval officer put it, "an almost desperate lack of basic information" about how waves function.[8] Waves are created by the steady push of the wind or gravitational forces associated with the moon, which give water its rising, falling, and rolling motion. As the surface of the water moves up and down, the waves move forward. They have little impact a hundred feet down, but as they move into shallow water near the shore, friction on the ocean floor makes them rise higher, until they tip forward in an arc and break. The breaker rushes up the beach until its energy is expended. Seen from a distance, waves seem to move the ocean from place to place in a recognizable shape of crest and trough. In reality the fluid particles that make up the trough and crest move elliptically. When observed up close with fluid tracers, the wave form is seen to pass through the fluid, causing the water particles to move up and down and forward and backward in the trajectory of ellipses. This motion is considered by oceanographers to be a manifestation of wave energy and explains how waves (or more accurately, swells†) can pass through each other with no apparent change of form or motion.

It is this wave energy manifest in the wave form that is transmitted from the wind to the water in the case of wind waves, or from deep-water impulses such as subsurface landslides, earthquakes, or volcanic eruptions in the case of tidal waves or tsunamis, which may travel across the ocean for thousands of miles from their point of origin, building up to great heights over shallow waters. When you understand that wave forms are the embodiment of energy, it is easy to recognize how the energy of the wind can be transferred to the water

* The first person to measure the force of a wave was Thomas Stevenson, father of Robert Louis. With a wave dynamometer, he found that in a winter gale off the coast of his native Scotland the force of a wave might be as great as 6,000 pounds to the square foot.

† Swells are ocean waves that have traveled out of their generation area.

to generate waves that correspond in energy. The height of wind waves is limited only by the velocity of the wind, the length of water over which it is blowing (the **fetch**), and the ocean bottom topography.

Life in the Ocean

The environment of life in the ocean is markedly different from that on land. On land we live, in effect, at the bottom of an "ocean" of air. Even the highest flying birds come down to earth to find food, shelter, and to mate. The tallest trees reach only about 350 feet in the air, and few land creatures penetrate more than 3 feet deep in the ground. By contrast the open ocean is virtually boundless, with life existing at all depths from sunlit surface waters to the dark ocean floor. The density of seawater is much greater than that of air, making the effective weight of marine organisms only a fraction of what it would be on land. Consequently there is no need in the ocean for the heavy internal skeletons of animals and the rigid cellulose structures of plants developed to offset the pull of gravity.

Marine organisms are generally adapted to survive in a layer of water that satisfies their particular requirements of density, light, and temperature. Because most of these organisms are denser than the seawater they inhabit, they tend to sink unless they can swim or have other means of maintaining buoyancy.* Changes of pressure in the sea are much greater than on land, restricting many organisms to surface waters, where pressures are lower, while others adapt to life at greater depths. However, some animals — sperm whales and certain seals — can function at both levels, diving to great depths and returning to the surface with ease. Ocean temperatures are generally less variable than those on land, due primarily to seawater's large heat-storage capacity. Although surface-water temperatures vary from latitude to latitude, daily and seasonal fluctuations are much smaller than in corresponding land areas. Thus marine organisms are rarely subjected to great or rapid temperature changes as are many land forms. The majority of sea creatures are cold-blooded, which means that their internal temperature is basically the same as that of the surrounding water. Generally speaking, such organisms are able to tolerate cooling (which makes them less active) more easily than heating (which can cause death). In the open sea most animals have body fluids whose salinity is close to that of the water. Thus, unlike land animals, they do not have to develop special mechanisms to prevent water and salt loss. However, in estuaries and brackish water, where the salt content is variable, animals and plants are distributed according to their tolerance for salinity.

Classifying Marine Communities

The three main types of marine organism are described functionally as **drifters, swimmers,** and **attached.** Drifters (plankton) include bacteria, phytoplankton (one-celled plants), and zooplankton (animal plankton). Swimmers

* Buoyancy is what makes it hard for you to stay at the bottom of a swimming pool.

or nekton include most adult fishes, squid, and whales. Benthos are bottom-dwelling plants and animals. Attached plants such as seaweeds and sea anemones require sunlight and usually grow on reefs or rock substrate in the shallower parts of the ocean floor. In temperate regions shallow-water ecosystems, which are subject to seasonal climate and salinity changes, tend to have comparatively few different species but many individuals of each. On the other hand, deep-sea and tropical environments, having more stable temperatures and salinity, support a greater variety of organisms. In cold-water communities of polar seas, marine metabolism is slower than that in other oceans. In these regions the surface of the water is filled with **copepods** (tiny, shrimplike organisms) and swimming snails that attract herrings and mackerel, flocks of seabirds, seals, and whales. Shrimplike **krill** are another important group of crustaceans. Dense swarms of these small creatures — they grow up to 5 centimeters (nearly 2 inches) long — feed on diatoms* and themselves constitute the main food of many fishes and filter-feeding whales.

The Copepod Coefficient

Ranging in size from 0.3 millimeters (.0012 of an inch) to 8 millimeters (0.32 of an inch), copepods are found throughout the ocean. They may be among the most numerous creatures in the world. In the surface waters of the northwestern Pacific they average 15,000 individuals per cubic yard, and in the Arctic waters there may be nearly twice that volume. Depending on temperature and availability of food — they usually eat diatoms — large copepods can double their numbers several times a year; smaller species reproduce even more frequently. Daily food consumption for older larvae ranges from 50 percent of body weight to much more when food is plentiful.

(Source: M. Grant Gross, *Oceanography*)

The Marine Food Web

The open sea covers some 90 percent of the ocean's total surface area yet produces only one-tenth of its plant and animal life. The abundance of life concentrates in the shallower coastal areas partly because energy-producing light penetrates only a few feet beneath the water's surface. Below the lighted zone of the sea most of the available nutrients come from organic material that drops to the ocean floor. Another productive marine environment is found in regions of **upwelling**, where cool, nutrient-rich deep water is drawn up to replace surface water moved away from the coast by wind-driven currents. In 1951 Rachel Carson referred to a catch of a billion pounds of sardines a year on the west coast of the United States at such a site. "The fishery could not exist except for upwelling, which sets off the old, familiar biological chain: salts, diatoms, copepods, herring."[9] However, the productivity of this particular region subsequently declined as a result of overfishing.

* Diatoms (from the Greek *diatomos* — "cut in two") are minute plants usually classified among the yellow-green algae because they contain granules of yellow pigment. They exist as single cells or in chains.

In the ocean's top layer, marine plants and minute animals capture energy from the sun through photosynthesis and convert it to a form that can be used by other organisms. Most of this energy conversion is carried out by phytoplankton, which are eaten by zooplankton (including copepods), which are then consumed by anchovies, sardines, and other small surface-feeding fish. These in turn are preyed upon by larger fish such as mackerel and tuna as well as by marine mammals and birds in the food-chain process we have described in the Introduction.

A Proliferation of Plankton

- Plankton is to the sea as grass is to the land — the basic food. Derived from a Greek word that means "wanderer," the term is applied collectively to all the minute plants and creatures that live at or near the surface of oceans or lakes. Some are passive, drifting with the currents, others swim actively in search of food. Most common are the diatoms, plants of colder waters that provide almost nine-tenths of the food in the ocean.
- During their infancy many sea animals — most fishes, starfish, jellyfish, crabs, and bottom-living clams, for example — are temporary members of the plankton family.
- Marine biologists estimate that 10,000 pounds of diatoms must be eaten to make 1 pound of copepods, 1,000 pounds of which produce 100 pounds of smelts. These, when eaten, give 10 pounds of mackerel, which as food, make 1 pound of tuna. Caught, canned, and consumed, 1 pound of tuna will increase your body weight by only 0.1 pound!

Deep Ecology

At the ocean floor is the pitch-dark and near-freezing **abyssal zone**, supporting some 98 percent of the ocean's different species (many of them decomposer bacteria).[10] Most of these survive either by eating dead plants and animals or by coming up to shallower waters to feed. In 1977 scientists discovered a new type of highly productive ecosystem deep within the abyssal zone. Vents in the ocean floor give forth very hot hydrogen-sulfide gas that is converted by specialized bacteria into chemical energy without the presence of sunlight through a process called **chemosynthesis**. In the only ecosystem on earth that does not depend on photosynthesis and light, these bacteria support a hitherto hidden world of extremely large worms, clams, blind white crabs, and other abyssal creatures.[11]

Seasons in the Sea

Although less apparent than on land, there are seasonal differences in the sea. A naturalist sampling plankton at different times of the year can find contrasts between spring, summer, fall, and winter almost as striking as those on land. In winter, when waters are well mixed from top to bottom and their temperatures are almost uniform, plankton is scarce and marine life diminishes. In spring, with longer days and more sunlight, plankton and other organisms begin to multiply. With warmer temperatures the colder, heavier water

sinks, displacing warmer layers below and bringing a rich supply of minerals to the surface.

Just as land plants depend on minerals in the soil for their growth, every marine plant is dependent upon the nutrient salts or minerals in the seawater, wrote Rachel Carson, describing how the spring sea "belongs at first to the diatoms and to all the other microscopic plant life of the plankton."[12] This is matched by a similar multiplication of the small creatures of the plankton — copepods, glassworms, pelagic shrimp, winged snail — feeding on the abundant plants and themselves devoured by fishes, crabs, mussels, and other larger animals. Then, as spring advances into summer, plant growth slackens due to the enormous amount of grazing and a reduction of minerals in the now warmer, sunlit zone of the ocean in which alone the plants can multiply. The minerals that are used up by the plants in the upper zone are now not replaced by any mixing with the lower, cooler waters, because of the difference in density between them. The plants that are not eaten die and sink, carrying with them to the bottom the nutritive salts that were once present in the upper zones. Down below, the supplies of minerals are to some extent built up by their return from dead animals broken down by bacterial action. As summer wanes and temperatures drop, the upper zone of the ocean becomes cooler and there is a new mixing of waters, which becomes richer in minerals brought up from below toward the surface. This creates once more a fertile layer where the autumnal sunlight is still strong enough to encourage photosynthesis, producing a second seasonal outburst of phytoplankton. This fall peak of bioactivity is less spectacular than that of spring and not as long-lived. As the light dwindles and winter storms stir up the water, temperatures again become uniform, and the nutrient minerals spread more or less evenly throughout the different layers of water. Reproduction of plants and animals is reduced to a minimum, completing the cycle.

Oceans at Risk

Until relatively recently it was assumed that, because of their size and the huge quantities of water they contain, oceans had an infinite capacity to absorb wastes by dilution and the natural process of regeneration. However, in the 1950s and '60s Rachel Carson's writings and Jacques Cousteau's films and books alerted the public to a growing threat to the ocean's ecosystems. The thinning of seabirds' eggshells by DDT, described by Carson, was the first of a long series of reports on degradation of the marine environment. In 1970 the newly formed U.S. Council on Environmental Quality (CEQ) issued a major warming: Pollution had already closed one-fifth of the nation's commercial shellfish beds; increasing levels of ocean dumping had created serious environmental damage; and there was a vast new threat from municipalities using the oceans "as a convenient sink for their wastes."[13]

Overfishing; dumping of garbage, human sewage, and industrial wastes; oil spills; runoff of chemicals from agriculture and transportation; heating of rivers and bays from nuclear reactors and conventional power plants; acid rain; dredging; channelization; and sandblasting of ships: These are some of the ways in which aquatic ecosystems are now subjected to damage and destruction comparable with that of Superfund toxic-waste sites and air pollution from carbon and sulfur dioxides, CFCs, ozone, and other substances.

Overfishing

Overfishing is a global problem threatening many species of marine life and the systems of which they are part. Since the 1980s the world's yearly catch of fish has exceeded what the United Nations considers to be a maximum sustainable yield.[14] This is the result of a sharply increased world demand for fish, the development of new factory-fishing technologies, and government encouragement for building ever bigger fishing boats and using nets with smaller mesh sizes that trap fish too young and small to be eaten.

Catch as Catch Can

In the late 1980s the fishing of capelin, a tiny sardine used in animal feed and commercial fats, was banned because of an interrelated chain of events. The catching of capelin off the shores of Norway deprived cod of their natural food and forced them to eat their own young. With the disappearance of capelin and cod, the seals were deprived of *their* main source of food. Then hundreds of thousands of starving seals invaded the Norwegian coast in search of food, depleting coastal fish stocks and destroying salmon hatcheries in the fjords. At the same time, sixty thousand seals were accidentally caught and drowned in Norwegian fishing nets.[15]

Natural Causes

Natural causes can also contribute to a decline in marine life. A recently identified climatic phenomenon that recurs every three to eight years, known as *El Niño*,* prevents the upwelling of cold water to the ocean's surface in the Eastern Equatorial Pacific, depriving it of nutrients, causing thousands of seabirds to starve and destroying huge numbers of fish and other marine organisms. Along with overfishing, *El Niño* is believed to have contributed to the failure of the anchovetta catch off Ecuador and the collapse of the Californian sardine fisheries in the mid-1960s and early 1970s.[16]

Ocean Dumping

Dumping of sludge, dredged and radioactive materials and burning of wastes at sea represent serious threats to marine ecosystems. **Sludge**, a mudlike, semiliquid end product of sewage treatment, often contains lead, mercury, zinc, chromium, cadmium, copper, pesticides, and disease-causing microorganisms and pathogens. These harmful substances can kill marine organisms and reduce species vitality and growth. Nutrients contained in the sludge, while useful on land, can overfertilize ocean waters, resulting in eutrophica-

* "The Christ child," because its impact is first felt in December.

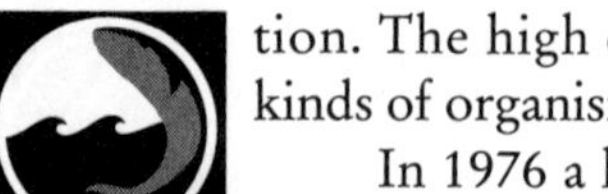

tion. The high organic content of sludge can drastically change the kinds of organisms that can live on the bottom of the ocean.

In 1976 a huge fish kill was discovered off the New Jersey shore just south of a sewage sludge dump located within a few miles of New York Harbor. "I went out to a shipwreck and it was completely dead — starfish, eels, lobsters, all sizes of crab — everything was dead," said Pat Yanaton, a microbiologist and environmental committeeman for the Eastern Diveboat Association.[17] In 1991, by order of Congress and its own state legislature, New Jersey ceased dumping treated sewage sludge into the Atlantic Ocean. New York City and its neighboring Westchester and Nassau counties were given a June 1992 deadline to stop dumping their annual 8 million tons of sewage sludge into the Atlantic 106 miles east of Cape May, New Jersey, where the marine environment and the ecological balance of the ocean might already have been impacted.[18]

Unfortunately, even when an ocean dump is phased out, it continues to endanger marine ecosystems for decades because, as in New York's case, it leaves behind millions of tons of solid wastes, industrial residue, and other contaminants accumulated over long periods of time. In addition to polluting the area where they are dumped, many of the toxic materials are carried by currents to threaten vital plant and animal habitats hundreds and sometimes thousands of miles away.

The alternatives to dumping cost money. New Jersey invested more than $200 million on new processing equipment and another $26 million a year for shipping the sludge by truck and train to a short-term landfill 1,450 miles away in Texas.[19] On a longer-term basis, different disposal methods must be found because there is growing nationwide resistance against trucking waste to out-of-state landfills. There is also increased opposition to incineration of waste because of air pollution, building costs, and disposal of the toxic ash.* At an international meeting held in January 1991 at the Woods Hole Oceanographic Institution in Woods Hole, Massachusetts, a number of proposals for delivering sewage sludge to the deep-sea bottom without contaminating the upper levels of seawater were discussed. In one of them an 18-inch hose several miles long with a 40-ton nozzle at its lower end would be lowered from a sludge ship to dump its load close to the bottom. Another idea was to link 55-gallon drums of waste into "trains" with a heavy nose cone that would sink at 50 miles per hour, penetrating deep into the bottom ooze, which in some regions is thousands of feet deep. Concerns were raised, however, that too little is known about the long-term fate of such sludge and its effects on marine life, especially deep-dwelling organisms.[20]

The most environmentally sound solution is to reclaim the sludge by removing the heavy metals and other contaminants and use it as a fertilizer rather than dumping it in the ocean or in a landfill. However, even this option can be difficult to implement on a large scale. In March 1992 New York City announced that it was proceeding with a billion-dollar plan to build eight plants to convert the 13,000 tons of sludge it produces daily into fertilizer, soil

* For more on sludge dumping, see *Design for a Livable Planet*, pages 11-17.

conditioner, or landfill cover that the city could use in parks and housing-project grounds. Residents and business people in the boroughs where the plants would be built expressed opposition to the plan because they feared noxious odors and greatly increased truck traffic. At least one of the sites was reported to have heavily contaminated soil. However, according to Albert Appleton, New York City's commissioner of environmental protection, the plants would be "completely odor-controlled." The volume of trucks would be reduced by the use of barges; as for the contaminated site, "We can clean it up and bring it back into the city's land-use inventory," he promised.[21]

In June 1992, one year after a Congressional deadline, New York City ceased dumping sewage sludge in the Atlantic, but there are more than one hundred licensed ocean dump sites in the Gulf of Mexico, the Atlantic, and the Pacific that receive materials dredged primarily from harbor and river channels. These materials are a mix of water and sediments often containing heavy metals, PCBs, oil, grease, and other pollutants. However, even "clean" dredged materials can harm marine life, because they bury marine organisms and increase the level of suspended sediments. The bulk of this dredging is carried out by the U.S. Army Corps of Engineers, whose task it is to maintain free passage for ports and harbors. If the corps' dredging program continues, there will be an even greater threat to marine ecosystems in coastal and ocean waters.

Radioactive Waste

The Ocean Dumping Act expressly bars high-level radioactive waste dumping at sea.* But between 1946 and 1970 the United States was allowed to dump more than 110,000 packages of plutonium and cesium into its waters, most of it close to major metropolitan areas — the Farallon Islands 30 miles west of San Francisco, Massachusetts Bay just outside of Boston, and two dump sites within 3 miles of Newark, New Jersey. According to a 1990 NOAA study, about a quarter of the 47,500 barrels of the atomic waste dumped in the Gulf of the Farallones National Marine Sanctuary have ruptured, threatening Pacific herring, Dover sole, rockfish, sablefish, and Dungeness crab commercially fished in the area. Plutonium, which can remain toxic for hundreds of thousands of years, and cesium have been found at levels "possibly more than 1,000 times the level expected to occur naturally."[22] The Farallon Islands have the largest population of seabirds south of Alaska and an abundance of fish, invertebrates, and marine mammals. In January 1991 NOAA began a $900,000 study to determine the extent of the damage from the radioactive material to the richest marine habitat in the West.

Ocean Incineration

Burning wastes aboard ships is also regulated by the Ocean Dumping Act, because it releases waste and toxic by-products into the marine environment.

* For further details, see chapter 9, page 276.

The law permits burning only liquid wastes of a certain composition, representing roughly 8 percent of the 250 million metric tons of hazardous wastes produced every year in the United States. Yet even this "small" percentage (some 2 million metric tons) of concentrated toxic and carcinogenic material is lethal to marine life immediately or through long-term contamination; it also can affect humans who eat the poisoned fish and seafood. Another problem is spills and leaks during the transportation of the wastes to the incinerator ship or during the actual burning at sea, which in turn causes additional problems of water pollution and ecosystem damage.

What You Can Do to Protect the Oceans

1. Find out from your congressional representatives the status of legislation affecting the oceans. The key laws include the Marine Mammal Protection Act, the Coastal Zone Management Act, the Clean Water Act, and the Endangered Species Act. (For more on these laws, see chapter 9, pages 275-277 and pages 286-291.)

2. To get information about oceans from Congress, contact:
— *U.S. House of Representatives*: Committee on Merchant Marine and Fisheries; and Committee on Public Works and Transportation.
— *U.S. Senate*: Committee on Commerce, Science and Transportation (responsible for national ocean policy study); and Committee on Environmental and Public Works (responsible for ocean dumping and solid waste disposal). For details on these committees and their subcommittees, see the Directory, pages 353-354.

3. Work with your state and local organizations concerned with ocean ecology and cleanup. You can often have more impact there than at the federal level.

4. Support the efforts of the three leading direct-action groups that are campaigning on behalf of ocean wildlife: Earth Island Institute, Greenpeace, and the Sea Shepherd Conservation Society. (For details, see the Directory.)

5. Support the programs of the Center for Marine Conservation, the American Littoral Society, the Global Coral Reef Alliance, NRDC, EDF, and other environmental groups concerned with ocean issues. (For details, see the Directory.)

6. Contact the National Audubon Society's Living Oceans Program, 306 South Bay Avenue, Islip, NY 11751; (516) 277-4289. Get a copy of their primer on *Conserving Marine Resources* and find out how you can help build a strong constituency on behalf of marine resources.

7. Write to the Izaak Walton League for a free copy of *A Citizen's Guide to Clean Water*, providing detailed information about water quality issues under the Clean Water Act.

8. As a consumer, buy only environmentally sound products made without the use of CFCs and other ecologically harmful ingredients. Take care to prevent plastic products from getting into the oceans, where they are especially harmful to marine life.

9. If you are a boat owner or mariner: Minimize the amount of nondegradable products on board; see that no trash is discarded, washed, or blow overboard; where possible, retrieve trash found in the water.

RESOURCES

1. *International Organizations*

— *International Association of Hydrological Sciences*, 2000 Florida Avenue, NW, Washington, DC 20009; (202) 462-6903.
— *International Maritime Organization*, 4 Albert Embankment, London SE1 7SR, England.
— *Law of the Sea Treaty*, Secretariat, United Nations, Room 1827A, New York, NY 10017.
— *Regional Seas Activity Center*, United Nations Environmental Program, Palais des Nations, 1121 Geneva 10, Switzerland.

2. *U.S. Government*

— *Department of Commerce:*
National Oceanic and Atmospheric Administration;
National Marine Fisheries Service;
National Ocean Service.
— *Department of Energy*, Sub Seabed Disposal Program, Geologic Repositories Office, 1000 Independence Avenue, SW, Washington, DC 20585.
— *Department of State*:
Oceans and International Environmental and Scientific Affairs Bureau;
Oceans and Fisheries Affairs;
Office of Ocean Law and Policy.
— *Department of Transportation*, U.S. Coast Guard.
— *Environmental Protection Agency*
— *National Advisory Committee on Oceans and Atmosphere*
— *National Ocean Policy Study*
(For further details including telephone numbers, see the Directory, pages 353-354.)

3. *Environmental and Conservation Organizations*

In addition to the groups mentioned in this chapter, there are many other organizations working to protect the Pacific, Atlantic, and Arctic oceans and the Gulf of Mexico. For further information, see the Directory.

4. *Research, Educational, and Trade Organizations*

— *Antarctica Project*, 1845 Calvert Street, NW, Washington, DC 20009.
— *Center for Marine Conservation*, 1725 DeSales Street, NW, Washington, DC 20036; (202) 429-5609. Dedicated to protecting marine wildlife and their habitats and to conserving ocean and coastal resources, the center conducts research, promotes public education and citizen involvement, and supports domestic and international laws and programs for marine conservation.
— *Citizens for Ocean Law*, 1601 Connecticut Avenue, NW, Washington, DC 20009.
— *Marine Environmental Research Institute*, 235 West 76th Street, New York, NY 10023; research center, P.O. Box 61, Route 176, Blue Hill, ME 04614.
— *National Ocean Industries Association*, 1050 Seventeenth Street, NW, Washington, DC 20036.
— *Society of the Plastics Industry*, 1275 K Street, NW, Washington, DC 20005.

5. *Further Reading*

— *The 1985 Citizen's Guide to the Ocean*, edited by Michael Weber et al. Washington, D.C.: Center for Environmental Education, 1985.
— *Citizens Guide to Plastics in the Ocean*. Washington, D.C.: Center for Marine Conservation, 1991.
— *The Cousteau Almanac*, Jacques-Yves Cousteau. New York: Doubleday, 1981.

— *The Ecology of the Seas*, D. H. Cushing and J. J. Walsh. Philadelphia: Saunders, 1976.
— *Federal Conservation and Management of Marine Fisheries in the United States.* Washington, D.C.: Center for Marine Conservation, 1991.
— *The Forest and the Sea*, Marston Bates. New York: Random House, 1960.
— *The Frail Ocean*, Wesley Marx. New York: Ballantine Books, 1970.
— *Life in the Oceans*, Joseph Lucas and Pamela Critch. New York: E. P. Dutton, 1974.
— *The Life of the Ocean*, N. J. Berrill. New York: McGraw-Hill, 1966.
— *The Living Ocean: Understanding and Protecting Marine Biodiversity*, Boyce Thorne-Miller and John G. Catena. Washington, D.C.: Island Press, 1991.
— *Marine Biological Diversity Strategy and Action Plan.* Washington, D.C.: Center for Marine Conservation, 1992.
— *Oceanography: A View of the Earth, M. Grant Gross.* Englewood Cliffs, N.J.: Prentice-Hall, 1982.
— *The Oceans: Our Last Resource*, Wesley Marx. San Francisco: Sierra Club Books, 1981.
— *The Open Sea*, Alister Hardy. Boston: Houghton Mifflin, 1956.
— *Plastics in the Ocean*, K. J. O'Hara, N. Atkins, and S. Iudicello. Washington, D.C.: Center for Environmental Education, 1987.
— *A Primer on Conserving Marine Resources*, Carl Safina. New York: National Audubon Society, 1992.
— *The Sea Around Us,* Rachel Carson. New York: Signet, 1961.
— *Under the Sea Wind,* Rachel Carson. New York: Oxford University Press, 1941.

NOTES

1. Foreword to Boyce Thorne-Miller and John G. Catena, *The Living Ocean* (Washington D.C.: Island Press), p. xiii.
2. Ibid., p. xiv.
3. G. Tyler Miller, *Living in the Environment* (Belmont, Calif.: Wadsworth, 1988), p. 95.
4. M. Grant Gross, *Oceanography* (Englewood Cliffs, N.J.: Prentice-Hall, 1982), p. 126.
5. See "The Moving Tides" in Rachel Carson, *The Sea Around Us* (New York: Signet, 1961), pp. 142-55.
6. Ibid., p. 142; see also Gross, *Oceanography*, p. 249.
7. Carson, *The Sea Around Us*, p. 145.
8. Ibid., p. 109.
9. Ibid., p. 138.
10. G. Tyler Miller, *Living in the Environment*, p. 108.
11. Ibid.; see also Gross, *Oceanography*, pp. 403-4.
12. Carson, *The Sea Around Us*, p. 42.
13. Council on Environmental Quality Report, cited in Beth Millemann, *And Two If by Sea: A Citizen's Guide to the Coastal Zone Management Act and Other Coastal Laws* (Washington, D.C.: Coast Alliance, 1988), p. 63.
14. For further information see "The Impact of Modern Fishing" in the excellent book by Edward Goldsmith et al., *The Imperiled Planet* (Cambridge, Mass.: MIT, 1990), pp. 179-85.
15. Ibid. pp., 180-81.
16. Edward Goldsmith and Nicholas Hildyard, *Earth Report* (London: Mitchell Beazley, 1988), p. 139.
17. Cited in Charles Kaiser, "A Huge Fish Kill Found Off Jersey," *New York Times*, July 8, 1976.
18. Nina Maria Sankovitch, coastal project attorney, Natural Resources Defense Council, letter to the *New York Times*, August 2, 1989.
19. Allan R. Gold, "New Jersey Ends Practice of Dumping Sludge in Sea," *New York Times*, March 18, 1991.
20. Walter Sullivan, "Schemes Are Debated for Dumping Sewage Deep Beneath the Sea," *New York Times*, January 15, 1991.

21. Joseph P. Fried, "Despite Foes, New York City Presses Its Plan for Sludge," *New York Times*, March 15, 1992.

22. "Atomic Waste Reported Leaking in Ocean Sanctuary Off California," *New York Times* special report, May 7, 1990; Katherine Bishop, "U.S. to Determine Danger from Barrels of Atomic Waste in Pacific," *New York Times*, January 20, 1991.

Part 2
THE LAND

GRASSLANDS, CHAPARRAL, AND TUNDRA

"The Grass was the Country as the Water is the Sea."

— Willa Cather

Plains, Prairies, and Pastures[1]

When the first Europeans explored the lands west of the Appalachians, they came across something they had never seen before, an immense (200,000 square miles), treeless, grassy region. They named it *prairie*, the Old French word for "meadow land." Stretching from mid-Canada to Texas, the prairie blended into what became known as the Great Plains, extending as far west as the Rocky Mountains. It was one of the great grassland regions of the world, comparable with the South American pampas, the South African veld, and the Asian steppes. If you count only large, unbroken grasslands, grasses still cover some 18 million square miles, or more than 30 percent of the earth's land area.[2] But little of these grasslands remains intact in North America.

In the early nineteenth century the North American grasslands formed a coherent ecosystem, supported by a diverse variety of plants and animals and able to maintain itself against erosion and encroachment by trees. In Illinois — the Prairie State — the grass grew so high that early settlers had to mark their paths with piles of stones so they could find their way back. They told of waving grass stretching to the shimmering horizon.

Most grassland regions have flat or rolling terrain, slightly alkaline soil rich in organic matter and very fertile, an abundance of grazing and burrowing animals, and a climate marked by high rates of evaporation and periodic droughts. These regions are found mainly where there's too little rain for forests and too much for deserts. From time to time grasslands need fire to maintain and renew the grasses and to inhibit invasion by trees. Grassland fires spread rapidly, due to

The Grasses Are Greener

- Known scientifically as *Gramineae* or *Poaceae*, grasses are the third largest family of flowering plants on earth.*
- Grasses are divided into approximately fifteen **tribes**, which include the wheat (wheat, barley, and rye), rice, and oat tribes. Another is the **fescue** tribe, from which come many lawn grasses.
- Grass tribes are subdivided into about 600 genera and then into species, of which about 7,500 have been identified.
- Grasses are herbaceous, not forming woody tissue nor increasing in girth as do trees.
- Grasses can be distinguished from two similar families — sedges and rushes — by their hollow, round stems and by their small flowers arranged either in vertical clusters (spikes) or in many-branched clusters (panicles).
- When fully grown, grass plants have extensive, fibrous root systems, sometimes reaching a depth twice the height of the shoots above ground.
- Most native prairie grasses are long-lived perennials, but when they are stripped from overgrazing or trampling by animals, annual species often replace them.

(Sources: *Grasslands*, Audubon Society Nature Guide; *Grasslands and Tundra*, Time-Life Books)

* Only asters and orchids have more species.

the dense cover of dry stalks, the usually high winds, and the scarcity of surface water. The use of grasslands for livestock grazing and other human purposes has led to a widespread effort to prevent and control naturally occurring fires, with ecological consequences that we will examine later.*

"Unfit for Cultivation"

In 1820 Congress commissioned Major Stephen Long to explore the lands between the Mississippi River and the Rocky Mountains. "The Great American Desert," he termed them, "wholly unfit for cultivation, and, of course, uninhabitable by a people depending on agriculture for their subsistence."[3] He wasn't quite on target,† because these grasslands now provide an annual harvest of some $200 billion and enough crops to feed 250 million Americans and 25 million Canadians, with a lot left over to export to other nations. About three quarters of the crops — notably wheat, corn, barley, oats, millet, and sorghum — are grasses. Today little remains of the original prairie ecosystem that took thousands of years to evolve. Within fifty years of the settlers' arrival, the unbroken spaces and rich soils were transformed into a "breadbasket." Existing plants and animals were replaced with wheat, cattle, and other species cultivated by humans or that could adapt to the conditions created by them. Except for a few areas, the original grasslands have vanished.

How the West Was (Really) Won[4]

Before the arrival of Euro-American settlers, an estimated sixty million American bison (commonly called buffalo) roamed the prairies. For centuries they and the Native American inhabitants had coexisted in ecological balance. The bison supplied just about everything the Plains Indians needed: food from meat and milk, cooking oil from fat, shelter and clothing from hides. Their dung and carcasses returned nutrients to the prairie earth. The buffalo depended on grass, which maintained itself with naturally occurring fires, ignited by lightning. On a smaller scale, fires were set by Native Americans to attract buffalo to the fresh growth of short, green grass.[5] The settlers quickly changed this long ecological cycle with a different agenda. They brought in their own grazing animals — cows, sheep, and goats — which cropped the grass more closely, leaving less fuel for fires. They also cleared and fenced in large areas of land for farming and to serve as firebreaks that would protect their newly-built settlements. Systematically the settlers exterminated the buffalo in order to deprive the Indians of their livelihood. Those Indians not starved to death or killed by the whites were herded into reservations. By the turn of this century the bison population had shrunk from sixty million to fewer than five hundred animals. Today, thanks to the efforts of conservationists, there are some sixty-five thousand, protected in nature preserves, wildlife refuges, and national parks.‡

* See page 179 and pages 249-251.

† In the 1860s the development of the multibladed American farm windmill made it possible to pump water from 150 feet below the ground and thus irrigate the land for crop cultivation as well as provide water for the new transcontinental railroads.

‡ Restoration of grasslands and other ecosystems is discussed in chapter 10, pages 308-313.

Types of Grassland

In North America grasslands are found where precipitation averages between 10 and 40 inches a year. At their eastern limits, they consist of **tallgrass prairie** (see map), which merges into **mixed prairie** in the Great Plains, taking in most of the Dakotas, much of Nebraska and Kansas, central Oklahoma, and parts of the Texas Panhandle. As it approaches the

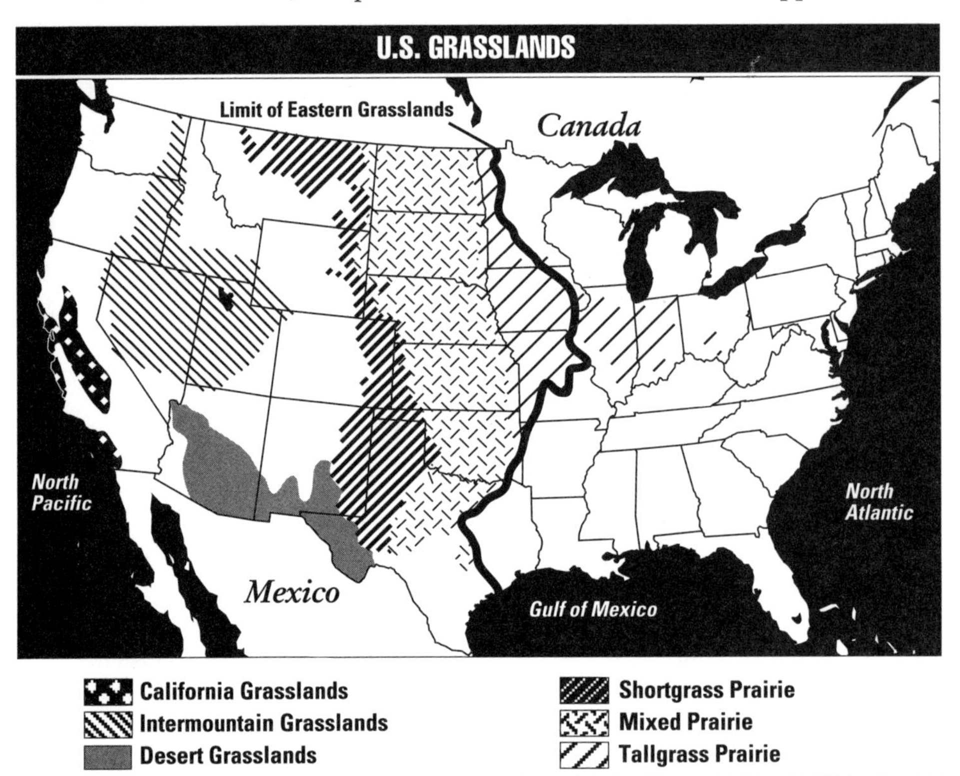

Rocky Mountains, the mixed prairie blends into the **shortgrass prairie**. Between the Rockies and the Sierra Nevada and the Cascade ranges are **intermountain grasslands** covering parts of Utah, Nevada, California, Oregon, Washington, Arizona (north of the Colorado River), Idaho, and Wyoming. Farther west are the **California grasslands**, and extending from northern Mexico into southeastern Arizona, southwestern New Mexico, and parts of Texas are the **desert grasslands** sometimes known as **desert plains**.

The *tallgrass, mixed,* and *shortgrass prairies* are not separated by clearly defined boundaries but merge into one another in zones that shift over periods of time depending on a number of factors, the most important of which is change in weather conditions. In prolonged dry spells, shortgrass species expand into the tallgrass domain. In wet times, the process is reversed. The soils of the mixed-grass prairie, receiving less precipitation than the tallgrass regions, are drier and less fertile. In the 1930s what was left of the mixed-grass prairie was severely hit by a prolonged drought, causing the topsoil, which had been stripped and laid bare by excessive crop cultivation and overgrazing, to blow eastward, creating the Dust Bowl, one of the world's worst ecological disasters.

PlainSpeaking

- The Great Plains begin where the annual rainfall drops below 20 inches, the minimum standard for dependable farming.
- The ten Great Plains states cover 20 percent of the landmass of the lower forty-eight states but have less than 3 percent of the total U.S. population.
- The region has the nation's coldest winters, hottest summers, fiercest droughts and blizzards, and worst hail, locusts, and range fires. It also has the shortest growing season. Its farm, mineral, and energy economies are weak. Its towns are emptying as farms and small businesses fail.
- The level of the Ogalla Aquifer, a nonrenewable source of groundwater that nourishes more than 11 million acres of land in the region, has, because of overdrawing by farmers, dropped from 6 feet below the surface to 60 feet in some parts of Kansas.

Desert grasslands are found usually at elevations of 4,000 to 8,000 feet on the edges of the deserts of the Southwest. Because they receive more sunshine and less precipitation than other grasslands, these are the most arid. The dry climate produces species of grasses that are low in nutritional value as well as a dominance of mesquite and cacti and other succulent plants that survive primarily on their ability to store water in their stems. The cattle boom of the 1880s and '90s radically changed the character of the desert grasslands. Widespread overgrazing by the cows stripped away the grass, exposing the soil, which became impacted by their heavy trampling. Over the course of twenty years this led to considerably increased runoff of rainwater and snowmelt and the formation of deep gullies in what previously had been slow-running streams. This phenomenon, known as **trenching**, caused the water table to drop, favoring the growth of mesquite, which has deeper roots than grass, taking up even more water, lowering the water table further, and inhibiting the regrowth of grasses. Many landowners are now trying to restore the grasslands so that they can increase the size of the rangelands for their cattle. Among the techniques they are using are rotating cattle to allow fields to rest and periodic burns to replicate the natural cleansing by lightning fires.

Intermountain grasslands are something of a misnomer, because most of the region, covering millions of acres, is dominated by sagebrush, which is a woody plant in the daisy family. The powerful turpentine odor it leaves after a rainstorm is familiar to anyone who has walked through the dry, open lands of the West. Many of these lands are owned by the federal government, which uses some of them for military purposes and others for rangelands. The climate in the intermountain west is harsh. It is generally very dry (8- to 12-inch annual average rainfall), with most of the precipitation coming in fall and winter and a short growing season from late spring to early fall. Summer temperatures often exceed 100 degrees Fahrenheit during the day.

The Plant-Animal-Soil Connection

Throughout the plains and prairies there is a very close relationship between soil, plants, and herbivores, or grazing animals. Among the larger herbivores

still remaining are mule deer, elk, and pronghorn, who eat shrubs as well as grasses. Many grassland animals, including jackrabbits, jumping mice, and grasshoppers, have strong hind legs that enable them to rise above the grass for better visibility and to outrun predators. Two of the most common creatures of the grasslands, the prairie dog and the harvester ant, are burrowers whose mounds dot the landscape, especially in those regions where the tallgrasses have been replaced by shortgrasses. In their burrowing they turn over and mix large amounts of soil, helping to create and maintain the grassland. Living in dog towns, prairie dogs provide a habitat and even housing for a host of other animals including harvester ants, crickets, burrowing owls, rattlesnakes, and coyotes.[6]

Gopher It

- A North American pocket gopher can dig up to 300 feet of tunnel in just one night.
- A jackrabbit has a top speed of 45 mph.
- Pronghorns can cruise for miles at 45 mph and reach 60 mph when severely pressed.
- In 1900 a prairie-dog town in Texas covered 25,000 square miles and housed an estimated 400 million animals.
- Kansas has 300 species of grasshoppers.

(Source: *Grasslands and Tundra*)

The Konza Prairie

> *"We get up in arms over the destruction of the [Amazonian] rain forest, when the natural habitats in North America have been all but eliminated."*
>
> — Buzz Hoagland, professor of biology, University of Kansas

The last reserve of the native tallgrass prairie that once covered 250,000 square miles of North America is in the Flint Hills region of eastern Kansas, a rough ellipse 40 miles wide and running some 200 miles from northern Oklahoma into southern Nebraska. Protected from the sodbuster's plow by its flinty subsoil, the Konza Prairie became a tallgrass pastureland as cattle replaced the bison. Much of what remains of tallgrass ecosystems is scraps, fragments, and old railway rights-of-way. Here and there along the roads are prairie islands — patches of the namesake tallgrasses: big bluestem, Indian grass, and switch grass, growing as tall as 10 feet. They are part of a complex system that supports at least 250 other plant species, including Mead's milkweed and the western prairie fringed orchid, both threatened. Among the 400 species of animals identified by biologists are Swainson's hawks, greater prairie chickens, and bobcats.

The chief threat to the tallgrass is cattle grazing, which has reduced plant and animal diversity and increased soil erosion. Other threats, coming from a million acres of surrounding farmland, are herbicides and the introduction of exotic plants that compete with native species. Increased suppres-

sion of fires by humans has accounted for the steady encroachment of scrub oak, which provides undesirable shade, reducing the diversity of the native plant life. This tallgrass region represents a classic conflict between conservationists, who want to set aside 320,000 acres in the Flint Hills as a national park, and ranchers, who vigorously oppose the idea. "We take care of the Flint Hills better than any government agency could....We follow sound soil-and-water conservation practices we've either learned ourselves in agricultural college or that researchers and extension folks tell us about," said one veteran rancher.[7]

THE KONZA PRAIRIE

NE
Lincoln
Konza Prairie
Kansas City
Lawrence
KS
FLINT HILLS
MO
Tallgrass Prairie Reserve
Tulsa
OK
AR
TX

In November 1990 the largest remaining stretch of pristine prairie — the 80-acre Elkins Prairie near Lawrence in northeast Kansas — disappeared under the plow, despite attempts by the Nature Conservancy and local environmentalists to buy it.[8] At stake in this conflict is how to balance urban growth and environmental protection. Since 1980 the town of Lawrence grew by 25 percent to 65,000 people, but development was tempered by strong community support for ecosystem protection. Conservationists prevented the building of a suburban shopping mall, closed a riverfront promenade in winter to protect bald eagles, and rerouted a highway to save the habitat of a rare frog. But the Elkins Prairie stood in the way of the next stage of development — a highway bypass that was approved by voters. In defense of the farmer who bought the land, many residents agreed with his neighbor and fellow farmer, Larry Warren: "I question the wisdom of plowing up good prairie, but I would defend his right to do it."

Taking an opposite position, Buzz Hoagland, a professor at the nearby University of Kansas, argued that people have a responsibility to preserve the environment, even at their own expense. "We get up in arms over the destruction of the [Amazonian] rain forest, when the natural habitats in North America have been all but eliminated. We don't seem to get upset about destroying the things in our own backyard." For its part, the Nature Conservancy said it could not have raised its offer to buy the Elkins Prairie above $3,500 an acre because "We can't unjustifiably enrich someone using the funds of a nonprofit organization."[9]

Tallgrass Prairie Preserve

Since 1990 the Nature Conservancy has run the Tallgrass Prairie Preserve 55 miles northwest of Tulsa. The aim of the preserve is to restore a functioning tallgrass prairie ecosystem. The cornerstone of this effort is recreating the his-

toric interaction of fire and bison. This involves a prescribed burning program to remove accumulated vegetation and regenerate native grasses. The schedule allows the prairie to rest from livestock grazing for up to three years before reintroducing 1,800 bison, whose natural herding and grazing behavior benefits the grasslands. Other plans under way include the rebuilding of a scenic gravel drive running for 50 miles through the preserve, which the Conservancy hopes will attract tourists. Mineral rights in Osage County, where the preserve is located, are owned by the Osage Indians, who will continue to drill for oil and gas. The Osage operate fifteen thousand oil wells in the county, generating nearly $18 million in revenue annually.

Note: Additional reports on prairie restoration will be found in chapter 10, Restoring the Earth, *pages 308-313.*

Don't Fence Me In

> *"The rangers working now were trained to produce cows, not healthy ecosystems."*
> — Tom France, director, Northern Rockies Wildlife Federation[10]

Today some 25,000 private ranchers have federal permission to graze their livestock on 270 *million* acres of public land in the sixteen western states,* an area greater in size than the entire eastern seaboard from Maine to Florida. It is an arrangement that has existed since 1934, when Congress ended unregulated grazing, but which is coming under increasing scrutiny by environmentalists, the media, and even the government itself. The degradation of these lands and the escalating cost to taxpayers of administering the program are uniting a coalition of conservationists, hunters, and sports enthusiasts, who share a common interest. They want to cut back or even do away with the publicly-supported grazing system by which the U.S. Forest Service, subsidizing ranchers who produce only 2 percent of the nation's beef, took a loss of $25 million in 1989.[11]

According to critics of the system, cattle are destroying the rangeland environment by overgrazing the land and eroding the soil, by polluting the water with their wastes, and by trampling down stream banks and often causing streams to disappear, which lowers the water table. **Riparian** areas — the narrow corridors of green vegetation alongside rivers and streams — are a particularly important source of survival for 75 to 85 percent of the wildlife. Using government reports to document their case, the National Wildlife Federation and the Natural Resources Defense Council filed an administrative appeal against the Bureau of Land Management (BLM) and the Forest Service for not conforming to their own standards, using questionable means of measuring consumption of grass by cattle and having inadequate plans to address shortcomings. On the other side of the (barbed-wire) fence, ranchers claim that their use of the public lands is important to the region's and the nation's economy and that claims of ecological destruction are grossly exaggerated. According to one rancher, "The real people in this country know

* Arizona, California, Colorado, Idaho, Kansas, Montana, Nebraska, Nevada, New Mexico, North and South Dakota, Oklahoma, Oregon, Utah, Washington, and Wyoming.

that with logging, or mining, or grazing livestock... you're going to have some things that don't look exactly the way you want them. It doesn't mean the land is destroyed; it's the way things have to be."[12]

Somewhere in the middle of the dispute, the Forest Service, formerly a staunch ally of the ranchers, is slowly shifting gears in the face of mounting public pressure. To stop overgrazing and save riparian habitats, the agency is considering cutting back the numbers of cattle allowed on the rangeland. One way to do this would be to raise the grazing fees to market levels, forcing ranchers to run less cattle and providing the Service with more funds to enforce the law. However, repeated attempts in Congress to raise the fees have been rebuffed by legislators from the western states. There is a moral question in the case: Who really pays for the grazing? As George Wuerthner, a former BLM botanist, puts it, "We pay a false price for beef as it is. Every filet mignon is a spring dried up, every steak is a riparian zone, every hamburger is another dead wolf."* [13]

The government's delay in taking action is masking the long-term price that will be paid as range conditions get worse. "We can't reduce vast areas of range to monoculture and have it continue to function," says Stephanie Woods, a Montana ranger. "You keep pulling out plugs and eventually you pull out the link that holds the whole system together." Removing plants in Upper Ruby, Montana, or eliminating species of fish, birds, or soil microorganisms will ultimately lead to its collapse — "just like the Amazon rain forest."[14]

Rivaling and perhaps even surpassing ranching as a problem on public lands is trespass grazing. For example, roughly half of all national parks in the West are trespassed "more or less regularly by livestock from adjacent public and private lands, or from National Park Service allotments themselves."[15] In Big Bend National Park, a designated World Biosphere Reserve in southwest Texas, trespassing livestock, mainly from Mexico, so heavily degrade the Rio Grande Canyon that in many riparian areas cottonwood regeneration is virtually nonexistent. The solution is clearly stated in one of NPS's own guidelines: "Where grazing is permitted or its continuation is not in the best interest of public use or maintenance of the park ecosystem, it will be eliminated."[16] It still remains to be implemented by the Park Service and other federal agencies.†

California Grasslands

Just over a hundred years ago grasslands covered a quarter of the state of California. But like other North American grasslands, they have been radically changed by human activities. You can see what remains of them on hillsides along the coast from San Francisco to southern Oregon and, to a lesser extent, in the flatlands of the Central Valley, the world's richest agricultural area. Because coastal grasslands are often located on steep slopes, they are less accessible for intensive cultivation. However, as suburban housing creeps inexorably over the hills, these ecosystems too are facing the threat of obliteration.‡

* The government allows federal hunters to kill wolves and other predators that harass livestock.

† For more on the effects of cattle grazing, see chapter 7, pages 228-229.

‡ For a further discussion of this problem, see the report on chaparral, page 156.

Gray Ranch/San Pedro

Living in the center of New York City,* I (JN) find it hard but fascinating to imagine a field of blond grama grass *three* times the size of Manhattan.* The "field" in question is 40,000 acres of grasslands, itself a relatively small portion of the 321,000-acre Gray Ranch in the southwestern "boot heel" of New Mexico purchased in 1990 by the Nature Conservancy. Just north of the Mexican border, surrounded by desert and the rugged Animas Mountains, this represents the most extensive example of intact grasslands in North America. Perhaps the largest private conservation project in the world, Gray Ranch is located 75 miles to the east of the San Pedro River Bioreserve across the border in Arizona also run by the Conservancy.

> *"A unique place on the face of the earth."*
> — Charles J. Ault, biologist, U.S. Fish and Wildlife Service[17]

Unparalleled Biodiversity

Gray Ranch has an unequaled array of plant and animal life, harboring 360 bird species, representing nearly half of North America's bird fauna. Its river system supports a rich aquatic and streamside habitat that includes the most extensive cottonwood-willow and mesquite riparian forests remaining in the Southwest. Gray Ranch has at least 718 species of plants, of which 71 are rare or endangered, including the white butterfly cactus and the night-blooming cereus. It boasts more species and subspecies of mammals than you will find at any national park or wildlife refuge. Within its 321,000 acres there are sixty five natural communities, twelve of which are globally rare. Among its animals are the bald eagle, the Sanborn's long-nosed bat, and the ridge-nosed rattlesnake, all endangered species. Virtually all of the species at Gray Ranch are native to the region. This, Ault suggests, is because the soft contours of the land, the lack of deep erosion, and the unusually high amount of plant cover have prevented cattle from doing serious damage, even though all of the land around the ranch is dedicated to ranching.

Before the Nature Conservancy's purchase of Gray Ranch, the greatest potential danger was its being broken into smaller parcels and not being managed as a single ecosystem. Although this land is now secure, streamside woodlands alongside the San Pedro River face the threat of becoming depleted, fragmented, and destroyed due to the residential, agricultural, and industrial demands being placed on the basin's life-sustaining waters. Since the 1920s, agriculture within the river's riparian corridor has increased by more than 800 percent and copper mining continues to boom in the San Pedro River Basin.

Back at the Ranch

In acquiring Gray Ranch, the Conservancy was well aware that managing it would be a complex task, not only because of its great size and ecological diversi-

* Sometimes I would like to see the city covered with grasslands or at least with woodlands and salt marshes as it was three hundred years ago!

ty but because the acquisition touched on a whole of array of political and cultural sensitivities at local, state, and federal levels. Besides containing a wealth of natural diversity, the property includes thirteen archaeological sites nominated for the National Register of Historic Places. Although this was by far the largest purchase the Conservancy had made, they were not without experience at managing big conservation projects, having protected more than 4 million acres of terrain in the U.S. since 1951. "What is particularly interesting," said Jeff Babb, Gray Ranch's manager, "is that the property encompasses seven major ecosystems: desert scrub, semidesert grassland, evergreen/oak woodlands, Montane chaparral, chaparral, mixed broadleaf and coniferous forest, and Montane forest."[18]

"To truly insure our natural world for future generations, we are developing creative ways to safeguard much larger landscapes."
— John C. Sawhill, president, The Nature Conservancy

Land-Use Priorities

The Conservancy has designated three prime land-use priorities for the property. Conceptually (but not geographically) they can be seen as a kind of three-dimensional dart board — a core preserve within an inner ring surrounded by an outer ring. At the center of the bullseye are the highest peaks of the Animas Mountain range, which straddles the Continental Divide. Kept completely off-limits to cattle grazing or any other commercial activity, this area, which accounts for about 15 percent of the ranch, will be managed exclusively as a nature reserve. The inner ring, comprising 20 percent of the total, is designated as a conservation management area within which "best-use" practices will predominate. These will include restoration of certain areas, possible use of prescribed fire for ecosystem management, and some cattle grazing, which will be controlled by fencing. The remaining, "outer" part of the property, which is mainly rangeland, woodland, and desert, will be carefully surveyed before longer-term uses are decided. Conservancy scientists and managers are working to assemble the biological information needed to ensure that the site's natural features will not be degraded, depleted, or fragmented.

Sensitive to local interests and concerns, the new owners of Gray Ranch are trying to develop good working relations with neighboring ranchers. Cattle grazing, Babb explains, has gone on for 150 years and will be permitted in the outer ring, which comprises more than half of the property, and under certain conditions in the inner ring. Virtually all of the lands surrounding Gray Ranch and the neighboring San Pedro River watershed are devoted to ranching, agriculture, and mining. In the public sector, the Nature Conservancy works with the Arizona Game and Fish Department, the Arizona Natural Heritage program, the Cochise, Graham, and Pima county governments, the U.S. Fish and Wildlife Service, and the Forest Service. Private partners include the Centro Ecologico de Sonora, the Rocky Mountain Heritage Task Force, and the San Pedro Water Management Council. The Nature Conservancy is also seeking new public and private partners for its work in this region, Babb said. The Conservancy's success at Gray Ranch will depend to a large extent on what

they find out in their species inventories and other scientific studies, how soon they develop the right partnerships (including perhaps a greater commitment than at present from the federal agencies), and how effective they are at raising the money to offset the $18 million purchase price for an important and innovative ecological project.

Chaparral

In the semiarid regions of the western United States you come across hilly or mountainous lands covered with densely growing evergreen shrubs and dwarf trees, with occasional open, grassy meadows or patches of oak trees in the canyons. Known as **chaparral**, this type of ecosystem occurs where cool, moist winters are combined with warm, dry summers. It is found on the southern coast of California, where it is dominated by shrub oaks and chamise, in Arizona and New Mexico, dominated by Gambel oak, and in the sagebrush of the Great Basin. Although chaparral is not a major North American biome in terms of size, its ecology is particularly meaningful to the large concentrations of people who live within its boundaries. Its trees and plants are heavy seeders and highly flammable, requiring heat to germinate and the action of fire to clear away the old growth and make way for the new, as it recycles nutrients through the ecosystem. When large numbers of people moved into chaparral regions, which include greater Los Angeles, they tried to prevent the naturally occurring fires from destroying their newly built homes. Little did they realize that they were actually compounding the problem. Deprived of fire, chaparral grows taller and denser, presenting an even greater fire hazard in the dry season, when lightning or human carelessness gives rise to veritable infernos. Once chaparral is swept by fire, it returns either to lush green sprouts that come up from buried root crowns or, if there is a seed source available, to grass, and then the characteristic shrubs. These provide excellent food for deer, sheep, and cattle. As the sprouts grow, the chaparral becomes denser, the canopy closes overhead, and the litter accumulates on the ground, setting the stage for another fire. At low levels near the grasslands, burned-over chaparral, if maintained by periodic controlled burning, reverts to grass. However, in higher country, particularly at the edge of the pine forest, fires can destroy the trees and open the way for chaparral to spread.

Tundra — Arctic and Alpine

Our idea of "north" depends on where we are. To Central or South Americans, everyone who lives in the U.S. or Canada is North American, regardless of whether they come from Tallahassee, Florida, or Tuktoyaktuk in Canada's Northwest Territories. For those of us who live in the Northern Hemisphere, north can be said to begin, ecologically at least, with the coniferous (cone-bearing) forest, which is known also as

taiga (a Russian word) or *boreal* (northern) forest. Beginning where the coniferous forests end* and reaching to the polar seas at the top of the world is the tundra belt, stretching across one-tenth of the earth's surface. For much of the year it is a harsh, dark realm of bitter cold, high winds, and limited food resources.

> *"Tundra: A habitat characterized by short annual growing seasons, severe winters, low annual precipitation, and an absence of trees. Tundra is found primarily beyond the tree line in northern regions, where it is called* arctic *tundra; it also occurs above tree line on high mountains, where it is called* alpine *tundra."*
>
> — The Life of the Far North[19]

Beyond the Tree Line[20]

Depending on which dictionary you consult, the word *tundra* comes from Finnish, Lapp, or Russian, but all agree that it means "treeless plain." It may surprise you to learn that, for a region so close to the North Pole, the tundra snow is not particularly deep (annual precipitation is only 12 to 20 inches). The arctic tundra landscape is molded by frost. Its unique conditions are created by alternate freezing and thawing of the soil's sublayer (which varies from a few inches in some places to a foot or two in others) and the presence beneath that of *permafrost*, a permanently frozen layer, impenetrable to water and roots. More than 80 percent of Alaska and 50 percent of Canada and Russia have this frozen underpinning, which in some places is almost a mile deep. Since the permafrost is frozen solid, water cannot penetrate it. All the rain and meltwater from snow are held in the shallow sublayer of the arctic flatlands, which are dotted with bogs, marshes, and lakes and crisscrossed with streams and rivers. This reservoir of water on top of the permafrost is vital to the existence of vegetation in the arctic. At higher locations and areas exposed to the wind, the ground is bare or rock-covered, with sparsely scattered plants.

Tundra Vegetation

In the tundra regions plants must adapt not only to extreme cold but to the short duration of the growing season — between sixty and one hundred days — and to a lack of available water, which is a severe limitation to growth. Nutrients, particularly nitrates and phosphates, are also in short supply. Droppings from birds and animals are one source of enrichment for the soil. To survive, tundra plants must be able to carry on photosynthesis at near-freezing temperatures. Although their roots grow close to permafrost, they must withstand freezing at any stage in their growth. Flowers may be buried under snow or encased in ice for several days and, when they thaw out, produce healthy seeds. Vegetation must also function at low light intensity.

Although sometimes called the "barren grounds," the tundra has evolved plant species that survive harsh conditions. Birch shrubs may spread their branches 10 to 15 feet across the soil but raise them only a few inches

* At approximately latitude 60°.

above the ground. Despite its distinctive climate and the existence of indigenous species, the tundra does not have a vegetation type unique to itself as does a prairie or a rain forest. In terms of its ecological structure, the tundra can be described as a grassland.[21]

Compared with other biomes, the vegetation of the tundra consists of the relatively few species that are able to cope with the rigorous environment. Most of the plants are perennial, reproducing vegetatively,* not by seed. They compensate for the brief summer by keeping most of their living tissue below ground, storing large reserves of fats and carbohydrates in their roots before becoming dormant in the fall, and drawing on these reserves when they begin to grow in the spring. Because the sun in the arctic is low-angled, some plants have steeply inclined leaves to catch the twenty-four-hour daily summer sunlight. Others have especially large foliage to maximize their absorption of light. The predominant tundra vegetation — grasses, sedges, shrubs, and lichens — is simple in structure and grows close to the ground.

The Alaskan Tundra — Aside from its offshore islands and coastal regions in the southeast, where there are rain forests of giant Sitka spruce and hemlock, a large part of the Alaskan landmass, a vast terrain twice the size of Texas, comes within the tundra biome. At the first signs of spring, purple mountain saxifrage, arctic poppies, northern shooting stars, and myriad other wildflowers burst into life, attracting an incredible profusion of insects, rodents, birds, and other creatures, some of whom have hibernated throughout the nine-month winter.

Lichen Lore

- Lichens, which grow on rocks, wood, or on the ground, are an important source of food for many animals, including caribou and reindeer. In fact, the so-called reindeer moss is not a moss but a lichen.
- Lichens are primitive plants with no stems, leaves, or roots. They are actually two plants — an alga and a fungus living together in a complex relationship. The fungus forms the body in which the algal cells are embedded. Its threadlike filaments (*hyphae*) derive their nourishment from the cells while protecting them from possible destruction by bright light or extreme loss of water. Interestingly, when either component of the lichen is grown in isolation, it cannot stand the extreme conditions that are tolerated by the combined species.
- Lichens are often the first living things to grow on bare rock. They begin by growing in cracks or depressions in the rock or by secreting acids that dissolve the rock's surface. Having gained a foothold, they consolidate their position by expanding and contracting, which further disrupts the rock. Lichens can do this because they remain alive even when deprived of water. When water becomes available, they swell, increasing their weight and size ten to thirty times.

(Source: *Botany: A Human Concern*[22])

* Vegetative reproduction is the formation of a new generation of plants or animals by nonsexual means, such as the production of runners on a strawberry plant.

Among the three hundred or so bird species that come to Alaska from all over the world to nest or feed on the tundra, a few hardy denizens live there year round — snowy owls swooping on lemmings and other prey, ravens scrounging the remains of dead animals and human garbage, ptarmigans picking berries and dormant buds of dwarf willows and birches, and gyrfalcons snagging arctic hare and ptarmigan.

With its broad expanse of lakes and marshes, the arctic tundra attracts millions of waterfowl, plovers, and sandpipers, who arrive when the ice is melted and return south when winter sets in. You can see geese, ducks, gulls, songbirds, waders, and other birds in such profusion as to suggest the bird populations that must have existed on the prairies and plains of the lower continent before settlers changed the face of the landscape so radically. Many birds feed on the abundant insects, others prefer the tundra plants, aquatic vegetation, or fish. First to arrive in the spring are waterfowl, then shorebirds, and finally land birds. Some travel long distances. The arctic tern makes an annual round trip of 21,000 miles to and from Antarctica. In April caribou leave their winter habitats in the boreal forests to the south, making their way to their calving grounds in the tundra. Traveling in herds that number in the thousands, these animals follow trails so deeply worn in the tundra that you can see them clearly from the air. In the summer of 1984 ten thousand members of one herd, following their usual migration route, tried to cross a swollen river and were drowned in the raging current. Caribou eat a variety of plants and, unlike reindeer, move from area to area to find their sustenance. By reducing the risk of overgrazing, they ensure supplies of food for the next time they pass through. Aside from humans and insects, the caribou's main predators are golden eagles and bears, who prey on the calves, and wolves, who hunt the caribou in packs. Another large mammal of the tundra is the musk-ox, a year-round herbivore, living off lichens, grasses, willows, and other plants. They too are vulnerable to attack from wolves and humans.

Threats to the Tundra

For all its vast ruggedness, the tundra environment is subject to an increasing barrage of natural and human insults that threaten its ecological balance. Overgrazing by reindeer and uncontrolled hunting of musk-ox and other animals by humans are conspicuous threats. A more insidious danger comes from atmospheric pollution.

Arctic Haze

"People think of the Arctic as being clean and unspoiled. It is not really pristine anymore. It has become highly contaminated." — Dr. Jerome O. Nriagu, Canadian National Water Research Institute and Atmospheric Research Service.[23]

In the 1950s a blue-gray haze was detected in the arctic and has steadily worsened since then. At times during winter and spring, smog levels equal that found in large American cities. Arctic haze contains sulfur particles, a byproduct of smelting and fossil fuel combustion. Some of the pollutants have recently been

traced to industrial sources in Western Europe. A more localized source of the smog is the flotilla of cruise ships that pump dense smoke along the coastal areas of Alaska, especially in the area of Juneau, Sitka, Skagway, and Glacier Bay.[24] Particularly sensitive to pollutants carried by atmospheric dust and moisture, slow-growing lichens are among the first to die, harbingers of worse things to come. Lichens have also accumulated radioactivity from nuclear tests over the polar regions and from the 1986 nuclear-reactor disaster at Chernobyl. Caribou and other animals feeding on lichen pick up the radiation and pass it on to those who feed on them — eagles, wolves, and native people.

> *"We're standing in the world's greatest Arctic ecosystem...if they can drill here, where can't they drill? If we're going to develop this, we might as well go ahead and dam the Grand Canyon. You can make the same arguments for national energy needs."*
> — Tom Mahoney, the Sierra Club[25]

The first commercial exploiters of the tundra regions came in search of bears, seals, whales, and other animals that could supply them with furs and oils. But their impact pales in significance compared with the effects of the multinational corporations now exploring and developing the mineral resources locked up under the ice and permafrost. The gold rush of 1896 was the first extensive mineral exploitation in the arctic, but the mule and gold-pan technology of the time did little harm to the environment. Today it is quite another story, with bulldozers and other giant earth-moving machines. In 1987 $80 million was spent on gold prospecting in Canada's Northwest Territories alone.* Oil prospecting and drilling, as we know from the Exxon *Valdez* disaster, bring even greater ecological harm. In May 1988, a report by the Fish and Wildlife Service found that environmental damage from the Prudhoe Bay oil field and the Trans-Alaska Pipeline (which runs 800 miles from the Arctic coast to Valdez on Prince William Sound in the south) was "on a substantially greater scale than was envisaged in the Government's environmental impact statements, prepared before the projects began 16 years ago."[26] Among the findings were that 11,000 acres of vegetation used by wildlife at Prudhoe Bay close to the Arctic National Wildlife Refuge (ANWR)† were lost, almost double what was predicted, and that most bird, bear, wolf, and other animal populations in the area had declined due to oil industry operations. The oil-related projects had also created "substantially more air and water pollution" than originally predicted.[27]

The Arctic National Wildlife Refuge (ANWR)

ANWR (pronounced "an-whar") is the northernmost of all the U.S. national wildlife refuges, encompassing one of the world's most remarkable assemblages of landscapes, arctic plants, and wildlife. Sometimes called "North America's Serengeti," it is the habitat of caribou, musk-oxen, polar and grizzly

* For more on the ecological impact of mining, see pages 229-230 and pages 277-279.
† For more on the threats to the Arctic National Wildlife Refuge, see chapter 8, pages 284-285.

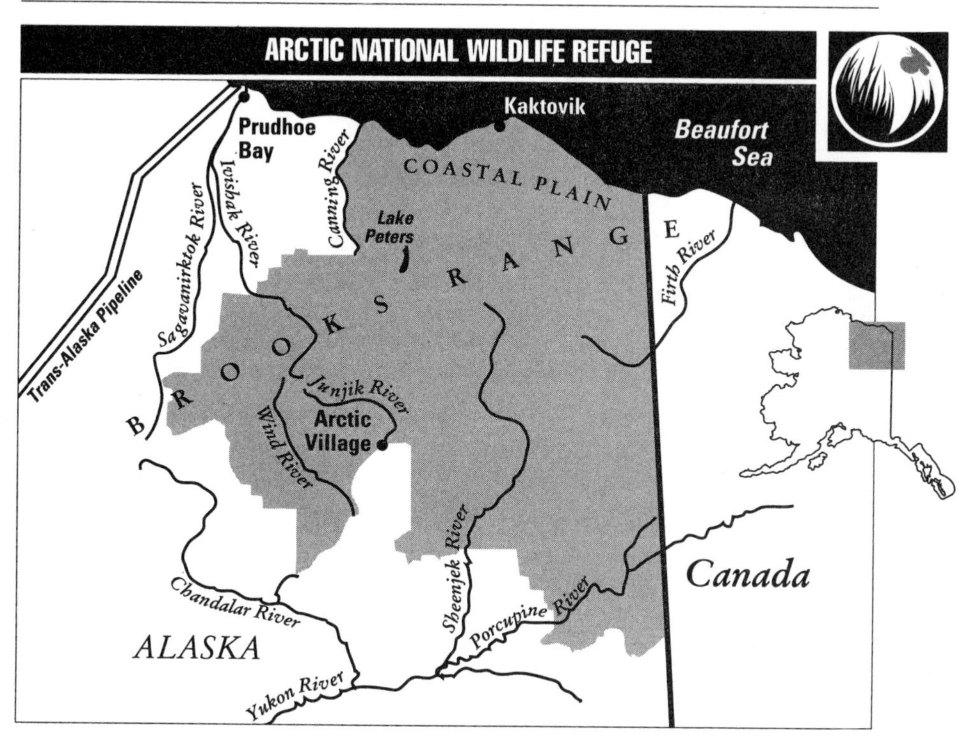

bears, wolves, and 135 species of birds, including arctic tern, long-tailed jaegers, Asian songbirds, and other transcontinental migrants.

> *"The Arctic has strange stillness that no other wilderness knows. It has loneliness too — a feeling of isolation and remoteness born of vast spaces, the rolling tundra, and the barren domes of limestone mountains.... All the noises of civilization have been left behind; now the music of the wilderness can be heard."*
>
> — Supreme Court Justice William O. Douglas[28]

The Caribou Connection

Just as great herds of bison once roamed the Great Plains, two major herds of caribou are found in the ANWR region. The central arctic herd, which now numbers about 20,000 animals, is a common sight at the Prudhoe Bay end of the Trans-Alaska oil pipeline, leading presidential candidate George Bush to proclaim in 1988 that caribou "like to snuggle up next to the pipeline."[29] Not surprisingly, oil industry scientists tend to agree: "I'm fairly confident that oil development can continue on the North Slope [of the Brooks Range] with no harm to the caribou," stated G. Scott Ronzio, manager of environmental sciences for ARCO Alaska.[30] The second major herd, the 180,000-head Porcupine* caribou herd, presents a different problem. It does its calving on the arctic coastal plain in an area where the proposed new oil development would take place. According to Dr. David R. Klein, a biologist considered to be one of North America's foremost experts

* Named after the Porcupine River, which flows through the southern edge of the ANWR.

on caribou, the development would deprive the Porcupine herd females and their young of their best feeding grounds.[31] During the brief arctic spring and summer enormous swarms of mosquitoes rise from the spongelike tundra, driving the caribou to the cooler coastal area for relief from ocean breezes. There they voraciously devour grasses and flower buds. If the animals were forced by oil and gas operations to scatter inland, they would again be harassed by mosquitoes to the point where they would be unable to eat enough food for their 1,000-mile return migration north, says Klein.[32] Depending on the caribou as their primary food source is another vital component of the coastal ecosystem — the Gwich'in people, who live in fifteen small communities just south of the Brooks Range and Canada's Yukon by hunting and fishing in the plains, forests, and tundra. Numbering about seven thousand, this is a population that will probably be destroyed and certainly will be dramatically changed if the Porcupine caribou herd diminishes.[33]

The northernmost native tribe in North America, the Gwich'in are one of the last aboriginal groups to resist assimilation. "For too long this has been an issue about environment versus energy, and no one wanted to hear about the Gwich'in," said Sarah James, chairperson of a political committee of Gwich'in villages, who lives in Arctic Village, just south of the ANWR. "We are not fighting because that place looks beautiful or because it has abundant wildlife...[but] because our way of life depends on the land. In order for it to take care of us, we have to take care of the land in return."[34] Other native people near the refuge, Inupiat who live in eight villages inside the Arctic Circle near the town of Barrow, for example, take a different but contradictory position. They are mostly whale hunters who get millions of dollars from oil drilling on the North Slope (and who would get much more if the refuge is opened up). Although they support the drilling in the refuge, they vehemently oppose an Interior Department plan to allow drilling offshore in the Arctic Ocean, because it poses a potential threat to their whale hunting.[35]

Peaks and Valleys

Nowhere else in North America is the transition from the coastal plain to the adjacent mountains as abrupt as it is in the Arctic National Wildlife Refuge. The rugged chain of spectacular peaks comprising the Brooks Range curves through the refuge, forming a continental divide through northern Alaska (see map). Deep valleys reach far within the mountains to the base of glaciated peaks. Uplifted east to west, the northern slope descends in a tundra-covered plain to the Arctic Ocean. This treeless expanse, covered with grasses, sedges, lichens, and other low plants, is cut by braided streams flowing into the sea. A narrow reef of gravel protects much of the shoreline from the arctic ice pack. In the south-sloping valleys of the Brooks Range, rivers wind serpentine courses through valley floors dotted with lakes, sloughs, and wetlands. Groves of stunted white spruce and balsam poplars stand among meadows and muskeg (bogs) and, further south, grade into dense, taller spruce forests. Winter in the ANWR is long and severe. Summer is short and intense. Plants adapted to the arctic survive even though permafrost lies 18 inches or more below the ground, which is covered by snow for nine

months of the year. In June and July the twenty four-hour sunlight produces extremely rapid leaf growth, although the yearly growth of trees and shrubs is slight. It may take three hundred years for a typical white spruce to reach a diameter of 4 inches and seventy-five years for a willow shrub to grow 6 feet tall. The result is an extremely fragile ecosystem easily disrupted by human activities.

Arctic Wildlife

Wildlife in the arctic need vast areas to ensure their survival. Following natural migration paths, generally along river valleys, the Porcupine caribou herd covers more than a thousand miles in its annual journey from wintering grounds in the Brooks Range and the Yukon to the coastal plains. From a distance you can spot Dall sheep as white specks resting or moving along the mountainsides. Moose, most common in the southern portions of the ANWR, are found primarily among the willow thickets in river bottoms and near small lakes. Musk-oxen roam the coastal plain. Having disappeared from the North Slope in the mid-1800s, largely because of overhunting, seventy of them were reintroduced to the mainland in 1969 from a herd on Alaska's Nunivak Island. The transplant was successful and today they number more than five hundred. Although not migratory, they move in response to seasonal changes, using streams and rivers as travel corridors. Some musk-oxen have wandered south of the Brooks Range, east into Canada; others have gone as far west as the oil pipeline corridor.[36] Biologists and conservationists worry that the oil companies' gouging of streams in the refuge to build roads would displace the shaggy, lumbering animals from their prime habitat, a concern the petroleum industry says is unfounded. Grizzly bears are widely scattered through the ANWR but are most commonly encountered in mountain valleys. An important link in the arctic ecosystem, grizzlies wander the tundra meadows and river bottoms seeking out ground squirrels, carrion, roots, and berries.*

Destroying the Last Great Wilderness

In 1987 the U.S. Department of the Interior estimated that there was about a 20 percent chance of recovering a large amount of oil from the ANWR. Since then the DOI and the petroleum industry (which were in virtual agreement on the need to drill in the ANWR) have contended that the refuge contains 3.2 billion barrels of economically recoverable oil, enough to supply U.S. needs for six months.[37] In response to the DOI claims, Bob Adler, senior attorney for the Natural Resources Defense Council, said that the department had "overpriced oil and underpriced the cost of development so that they could maximize estimates of how much oil there was. They want to make Congress salivate."[38]

Prodrilling partisans argue that oil production of the coastal plain would directly affect only a "footprint" of about 12,000 acres of the refuge, with little impact on the wildlife. Environmentalists such as National Audubon policy analyst Dorene Bolze refute this claim, pointing out that drilling waste and toxins at Prudhoe Bay have already contaminated tundra for miles around —

* For details on threats to the grizzly bear, see pages 248-250.

not to mention the damage from drilling pads, roads, air pollution, and noise.[39] Even if the prodrilling claims were accurate, the National Audubon Society and other conservation groups point out that the need to extract oil from the Arctic National Wildlife Refuge could be spared by the simple expedient of improving the efficiency of our automobiles by 2 miles per gallon, which would save more than 3 billion barrels of oil by 2020.[40]

The Legendary Lemming

- Biologists generally do not support the myth that lemmings are suicidal but like all animals, they are part of an ecosystem and susceptible to stresses that affect it.
- The most numerous of the tundra mammals, lemmings do not hibernate. They mate throughout most of the year, spending the winter under the snow, feeding on the green shoots of grasses and sedges.
- To survive they need to consume as much as twice their weight in food every day. Very fast eaters, they excrete most of what they ingest. When the snow melts, these wastes soak into the ground, fertilizing soil and plants.
- In winter lemmings enrich the environment by stripping the plants to get to the edible shoots, speeding the process by which dead leaves and stems decompose and return nutrients to the earth.
- Every three to six years the lemming population crashes. The wolves, weasels, and other predators dependent on them for food migrate elsewhere or die off.

In November 1991 the U.S. Senate defeated an administration-supported energy bill introduced by Senator J. Bennett Johnston (D-Louisiana) that would have opened up the ANWR to oil drilling. While environmentalists hailed the decision as an important victory, the energy secretary, James D. Watkins, warned that the vote "may end up costing us 30,000 jobs and add billions to our trade deficit."[41] In February 1992 the Senate overwhelmingly passed a different version of the bill, which excluded drilling for oil in the ANWR but permitted new drilling for oil and gas in the waters off the Florida panhandle.[42] Meanwhile powerful pro-oil lobbyists in Washington and throughout the U.S. have vowed to continue their fight for oil drilling in the ANWR and other wilderness locations.

What You Can Do to Preserve the ANWR

1. The Arctic Refuge will be at risk until it wins wilderness protection. It is important, therefore, to urge your senators and representatives to support legislation that would grant *permanent* wilderness status to the ANWR. Remind them that the refuge is an irreplaceable part of U.S. heritage that must be protected for future generations.

2. Get your legislators to work for a sound energy policy based on increased conservation, energy efficiency, and the use of environmentally friendly renewable energy such as solar and wind power.

3. Work with National Audubon, the Sierra Club, Greenpeace, the Wilderness Society, and other groups spearheading the fight to save the ANWR.

4. Get involved, for example, in Audubon's high-priority Arctic Refuge Campaign. For further details, read *Audubon Activist*'s special "Arctic Refuge" issue, April 1991.

Arctic National Wildlife Refuge — Myths and Facts*

Myth: "The oil industry has an excellent environmental record at Prudhoe Bay, the Endicott Field, and other Alaskan oil fields."

— **Fact:** In 1986 alone, 64 million gallons of toxic drilling waste were discharged directly onto the tundra by North Slope operations. The Exxon *Valdez* spilled 11 million gallons of Alaskan oil.

Myth: "The impacts of development on coastal plain wildlife will be minimal."

— **Fact:** Oil production and support activities are likely to result in a population decrease or change in distribution of 20 to 40 percent in the region's caribou population; 50 percent in the numbers of snow geese; and 25 to 50 percent in musk-ox populations. (Source: U.S. Department of Interior.)

Myth: "There's plenty of wilderness for everybody. We don't need this small piece of it."

— **Fact:** Less than 4 percent of the original U.S. wilderness remains. The Arctic refuge's coastal plain is virtually the last stretch of arctic coastline of Alaska not open for development.

Myth: "The country needs the oil in the ANWR to ensure a secure energy future."

— **Fact:** There is at best a one-in-five chance of finding an economically recoverable quantity of oil under the refuge's coastal plain. Even that amount would take at least ten years to reach production.

Polar Bears at Risk

Animals at the top of the food chain are more at risk than others from the buildup of toxins. Between 1969 and 1984 the polar bears' diet of seal meat led to a fourfold increase in the levels of polychlorinated biphenyls (PCBs) in their fatty tissue. If, as seems likely, current PCB inputs continue in the arctic, the bears' fatty tissue will by the year 2005 exceed the limit of fifty parts per million, which designates their bodies as "toxic waste." The PCB intake will probably make them infertile long before before then.[43]

The polar bear–seal connection could be seriously endangered by the setting off of explosive blasts by gas and oil prospectors. If, as expected, the reverberating sounds drive away seals, polar bear survival could be jeopardized by loss of food. Polar bears are inquisitive animals, attracted by the debris of human life that steadily encroaches on the arctic. As they sort through garbage carelessly strewn in wilderness areas, they risk injury to themselves and present a hazard to careless people.

* Reprinted with permission from *Audubon Activist*, April 1991.

James Bay—Still at Risk

> *"A hydroelectric project the size of France threatens a vast subarctic wilderness and the Indians who have lived there for millennia."*
> — Jim Gordon, *In These Times**

Located 700 miles north of Detroit, James Bay is a large, shallow inlet of Hudson Bay, bounded on the west by the province of Ontario and on the east by northern Quebec. Together the bays form the largest inland seas on the North American continent and serve as a critical link between the arctic and subarctic ecosystems. James Bay is a vast subarctic wilderness, the habitat for seals, whales, caribou, and polar bears. Its mud flats and coastal marshes are vital staging grounds for millions of migrating birds that feed on the region's abundant algae, marsh grass, and clams. James Bay is also the focus of a classic struggle between environmental values and economic growth. On its Quebec side the bay has a steeply elevated gradient that is highly conducive to the generation of hydroelectric power.

In 1985 the first phase of a colossal hydroelectric project was completed by Hydro-Quebec, a utility owned by the province of Quebec. This involved building three powerhouses with a capacity of more than 10,000 megawatts — the equivalent of six or seven large nuclear plants — on La Grande River along with the diversion into this river of water from the upper basins of the Eastmain and Kanaaupscow rivers, which nearly double the flow of La Grande.[44] In Phase 2, Hydro-Quebec plans to push north, flooding the Great Whale River and building at least three more powerhouses with a capacity of 3,000 megawatts at a cost of perhaps $10 billion. In Phase 3 the utility envisions diverting the Rupert and Nottaway rivers south of La Grande into the Broadback River, storing the water in seven new reservoirs, sending it through the turbines of eleven dams, and transforming the lower parts of the diverted rivers into dry bedrock. Ranking as one of the largest construction projects ever undertaken in human history, the James Bay project would generate the force of twelve Niagara Falls and reshape the terrain in a region the size of France.

In terms of its ecological impact, the James Bay project is seen by environmentalists as the northern equivalent of the destruction of the tropical rainforests, but to the government of Quebec it represents a golden opportunity to reduce unemployment and generate income by exporting electricity and water to the United States. Currently at issue is Phase 2, which if carried out would drastically alter eight river systems, flood thousands of lakes and 5,000 square miles of forests, and destroy shoreline ecosystems, waterfowl nesting sites, snow-goose staging areas, and caribou calving areas as well as displace numerous native Cree and Inuit communities. Among the most threatening environmental concerns is the project's plan to alter the seasonal flow of water in northern Quebec. Under natural conditions the region's rivers run lowest in winter and highest during the spring melt. But because Quebec's electricity demand is highest in winter, the volume of water released from the reservoirs

* October 31 – November 6, 1990.

to spin the generating turbines would be ten times the natural flow, while the normal spring runoff would diminish.

As the *Audubon Activist* emphasizes, the potential ecological damage from Phase 2 is enormous: "The resulting changes in the ecosystem, including the timing of water flows, the salinity of the water, and the patterns of ice formation, could have a disastrous effect on the availability

JAMES BAY

of nutrients that support the bird and animal life on the bay. The impacts from the first phase of dam-building are only beginning to be understood. One known result is that the fish from the reservoirs, a staple of Cree diet, contain 9 times the acceptable level of mercury."[45] A 1984 study of the Cree village of Chisabi, downstream from the La Grande complex, reported that two-thirds of the residents had dangerously unsafe blood levels of methyl mercury, which is believed to cause birth defects, including palsy, retarda-

tion, stillbirths, and convulsions.[46] Hydro-Quebec concedes that methyl mercury will continue to be a problem for twenty-five to thirty years at La Grande and, if it is flooded, Great Whale, but suggests that unsafe levels can be reduced by strict limits on eating trout, pike, and other fish. "Telling us that we will be O.K. if we don't eat fish is like telling us we will be O.K. if we cut our own legs off" is the response of Andrew Natachequan, a Cree elder in the village of Whapmagootstui.[47]

In 1985 the sudden release of water from one of the dams drowned ten thousand caribou. Another impact has been the disturbance of seasonal water flows into James Bay, changing salinity and destroying fish spawning grounds. The changed salinity is already harming algae, marsh grasses, and other aquatic organisms on which migratory birds feed. It is also upsetting the habitats of the ringed seals, beluga whales, and polar bears that form part of the bay's ecosystem. A further consequence of the huge project is the environmental cost of road building, which not only destroys acres of wilderness but also opens up huge areas to human intrusion.

Considerable pressure to complete the second phase of the James Bay project has been applied by Hydro-Quebec and by commercial interests in Canada and the U.S. that stand to profit from the arrangement. The main recipient of the James Bay power was to have been New York State, which was considering a twenty-year, $20 billion contract to buy 1,000 megawatts of power — about as much as a medium-size nuclear plant. Opponents of the project raised a number of disturbing questions that forced New York to reconsider its contract decision. In addition to the dire ecological consequences outlined above, there are serious legal, economic, and cultural arguments against the plan. As National Audubon and other environmental groups in the U.S. and Canada point out, the legal requirements of an environmental impact statement for James Bay were not fulfilled, nor was any legal limit placed on the number of rivers that would have to be reshaped by the project.

To the surprise of many people, New York governor Mario M. Cuomo in March 1992 abruptly cut short negotiations with the Quebec government and canceled the contract to buy electricity from the controversial James Bay project. Hailed as a victory by environmentalists and native leaders in Canada, the move was made on the grounds not of wilderness conservation but of economics. In what many observers believe to be a landmark case the decision represents one of the first times that government officials had compared the value of conserving energy against purchasing it from such a mega-scale source. New York state energy officials said that because of the move consumers could actually end up paying less for their electricity.[48]

Although the New York decision has dealt a blow to the prospects of completing James Bay Phase 2, Hydro-Quebec says that it is determined to proceed with the project. Conservationists and other members of the James Bay Coalition express concern that the winning of this particular battle may lead to a lessening of the opposition to what still could turn into one of the greatest ecological tragedies of the twenty-first century.

What You Can Do about the Future of James Bay

1. Support the efforts of the James Bay Coalition, which includes: Earth Island Institute; Environmental Planning Lobby, 33 Central Avenue, Albany, NY 12210; (518) 462-5526; Friends of the Earth; Grand Council of the Cree, 2 Lakeshore Road, Nemaska, James Bay, Quebec, Canada J0Y 380; Greenpeace; the Humane Society of the United States; National Audubon Society; the Native American Council of New York City; Natural Resources Defense Council; Rainforest Action Network; the Sierra Club; and the Student Environmental Action Coalition of New York (SEAC), 300 Mercer Street, #17B, New York, NY 10003; (212) 627-1440.
2. Continue to express your opposition to the James Bay project to the following key politicians:
The Premier of Quebec, 885 Grande-Allee East, Building J, Third Floor, Quebec, Quebec, Canada G1A 1A2; the Minister of Energy, Government of Quebec, 200 B Chemin St. Foy, Quebec, Quebec G1R 4X7; the Minister of the Environment, Government of Quebec, 3900 Rue Marly, Sixth Floor, St. Foy, Quebec, Canada 61X 4E4; the Consul General, Canadian Consulate, 1251 Avenue of the Americas, New York, NY 10020; the Governor of New York, State Capitol, Albany, NY 12224; the Mayor of New York City, City Hall, New York, NY 10007; the Chairperson of the Board, New York Power Authority, 1633 Broadway, New York, NY 10019.
3. Support: the Quebec Coalition, c/o Les Ami(e)s de la Terre, P.O. Box 804, Succ. Place d'Armes, Montreal, Quebec, Canada H2Y 3J2; and the Canadian Wildlife Federation, 1673 Carling Avenue, Ottawa, Ontario, Canada K2A 3Z1.
4. Get in touch with the James Bay and Northern Wilderness Task Force, c/o Solidarity Foundation, 310 West 52nd Street, New York, NY 10019, and the Arctic to Amazonia Alliance, Box 73, Strafford, VT 05072.
5. Work through your legislators to support programs favoring energy conservation and efficiency and environmentally sound renewable energy.
6. Save energy wherever you can — at home, at work, and in the type of transportation you use.*

Alpine Tundra

Unlike arctic tundra, which is defined by latitude, alpine tundra is created by altitudinal conditions; it is found in mountainous areas located in temperate zones — the Swiss Alps, the Andes, the North American Rockies, and the New Hampshire Presidential Range.

Alpine tundra regions have a harsh climate with strong, frigid winds, snow, and widely fluctuating temperatures. Sometimes known as islands in the sky, because they are surrounded by other kinds of ecosystems, they are

* For more on how you can save energy, see "75 Ways You Can Save Energy," in *Design for a Livable Planet*, pages 216-222.

characterized by rocky slopes, alpine meadows, and shrubs that hug the ground to withstand the buffeting of the wind and provide a warm, insulating cushion for plants and insects. There is also some alpine tundra in the high Appalachians which is less cold and windswept than that of the northern regions. Except at very high altitudes, as in western mountains and a few eastern areas such as New Hampshire's White Mountains, alpine tundra is distinguished from arctic tundra by a lack of permafrost. This

Alpine Garden, New Hampshire*

Highest of the eleven peaks that make up the Presidential Range, Mount Washington lies in the heart of the White Mountain National Forest. Here, one thousand feet down and east of the mountaintop, is a natural area known as the Alpine Garden. Two-thirds of the garden's 110 kinds of flowering plants are alpine species whose major range is far to the north. Many plant species that live here are more common in Labrador, Newfoundland, and areas in the Arctic. Alpine azalea, moss cassiope, mountain heath, alpine bearberry, and two eyebrights range from Newfoundland and the mountains of Quebec south only to the Presidential Range. Alpine bluet grows only on the islands of Saint Pierre and Miquelon and on Mount Washington....

The garden surface has been shaped by frost through the centuries. Permanent moisture and fluctuating low temperatures have created the subsurface soil known as permafrost, which has remained frozen for thousands of years. Above the permafrost is a layer of water-saturated soil that repeatedly freezes and thaws, heaving rocks and pebbles out of the ground.

Plants in the Alpine Garden contend with bitter conditions.... [Those] that have adapted to life in the tundra make the most of the natural features. Something as seemingly unimportant as the lee side of a rock may give just enough protection for a seedling to grow into a mature plant.... Most plants on Mount Washington are perennials, with taproots that penetrate more than a foot into the soil. Because they have a stable root system, they needn't perpetuate themselves every year. If conditions are especially harsh, a plant need only form a few new leaves to replenish its supply of stored food. To avoid the intensity of the wind, most alpine plants are very short, and often they are plastered against the ground....

The streams that radiate across the Alpine Garden are lined with dark green bands of vegetation. The streamsides are the most interesting of the plant communities because they contain a diversity of species. One is the tea-leaved willow, which looks like a normal willow in miniature.... Another curious plant, the alpine bistort, forms several pale pink flowers, but none of the flowers set seed. Instead, tiny balls of plant tissue, called bulblets, develop below the base of the flowers. After dropping to the ground, the bulblets can grow into new plants without benefit of pollination and fertilization.

* Extracted with permission from "Alpine Garden, New Hampshire" by Robert H. Mohlenbrock, *Natural History*, July 1985.

permits generally better soil drainage and a longer growing season, with flowering plants like sedges, grasses, and dwarf willows more common than lichens and mosses. One flowering plant, the San Francisco Peaks groundsel, found only in the Coconino National Forest ten miles north of Flagstaff, Arizona, is listed by the U.S. Fish and Wildlife Service as a threatened species.

In the West, alpine tundra is inhabited by mountain goats (actually not goats but a species of chamois), mountain sheep, elk, marmots (a winter-hibernating mountain woodchuck), and other smaller rodents including pocket gophers, whose tunneling and root-eating habits influence the pattern of alpine vegetation. Because of ever present winds, short-winged insects are more prevalent or, in the case of butterflies, fly close to the ground. Compared with what you find in warmer climates, insect development generally is slow, with butterflies taking two years to mature and grasshoppers three. Although low in nutrients because of the short growing season, cold temperatures, and slow decomposition of organic matter, alpine tundra provides a livable if rigorous environment for plants and animals specially adapted to thrive there.

RESOURCES

1. U.S. Government

The following are the main federal agencies concerned with grasslands, chaparral, and tundra. Where not given here, details of addresses and phone numbers are listed in the Directory, pages 352-353.

— *Arctic National Wildlife Refuge* (DOI), Box 20, 101 Twelfth Avenue, Fairbanks, AK 99701; (907) 456-0250;
— *Bureau of Indian Affairs* (DOI);
— *Bureau of Land Management* (DOI);
— *Fish and Wildlife Service* (DOI);
— *National Park Service* (DOI);
— *National Forest Service* (Department of Agriculture);
— *Environmental Protection Agency*, Region 10 (Alaska, Washington, Oregon, Idaho) 1200 Sixth Avenue, Seattle, WA 98101; (206) 442-5810.

2. Environmental and Conservation Organizations

See the Directory.

3. Further Reading

— *Alaska: The Fight to Save a Great Land.* Series of articles in *Sierra*, January/February 1991.
— *Alaska in the 21st Century.* Report by 17 cosponsors of the "Celebrate Wild Alaska!" conference. Washington, D.C.: National Audubon Society, 1991.
— *Altars of Unhewn Stone: Science and the Earth*, Wes Jackson. San Francisco: North Point Press, 1987.
— "The Arctic." Report in *Imperiled Planet*, Edward Goldsmith et al., editors. Cambridge, Mass.: MIT Press, 1990.
— *Arctic Dreams*, Barry Lopez. New York: Charles Scribner's, 1986.
— *Audubon Perspectives: Fight for Survival*, Roger L. DiSilvestro. New York: John Wiley, 1990. Companion book to Audubon-TBS television series. Includes in-depth coverage of the ANWR.
— *Fish and Wildlife Resources of the Arctic Coastal Plain*, Arctic National Wildlife Refuge,

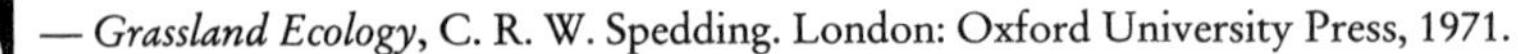

Fairbanks, AK 99701.
— *Grassland Ecology*, C. R. W. Spedding. London: Oxford University Press, 1971.
— *Grasslands and Tundra*, editors of Time-Life Books. Alexandria, Va: Time-Life Books, 1985.
— *Grasslands of the United States*, Howard B. Sprague, editor. Ames: Iowa State University Press, 1974.
— "The Great Alaska Debate: Can Oil and Water Mix?" Timothy Egan. *New York Times Magazine*, August 4, 1991.
— *Grizzly Years*, Doug Peacock. New York: Henry Holt, 1990.
— *How Grasses Grow*, R. H. M. Langer. Baltimore: University Park Press, 1979.
— *Journal of a Prairie Year*, Paul Gruchow. Minneapolis: University of Minnesota Press, 1987.
— *Konza Prairie: A Tallgrass Natural History*, O. J. Reichman. Lawrence: University Press of Kansas, 1987.
— *Lands Beyond the Forest*, Paul B. Sears. Englewood Cliffs, N.J.: Prentice-Hall, 1969.
— "Last Refuge," Paul Rauber. *Sierra*, January/February 1992.
— *The Life of Prairies and Plains*, Durward L. Allen. New York: McGraw-Hill, 1969.
— *The Life of the Far North*, William A. Fuller and John C. Holmes. New York: McGraw-Hill, 1972.
— *The Time of the Buffalo*, Tom McHugh. New York: Knopf, 1972.
— *The Wake of the Unseen Object: Among the Native Cultures of Bush Alaska*, Tom Kizzia. New York: Henry Holt, 1991.

NOTES

1. Much of the background information in this section is based on Lauren Brown's *Grasslands* in the Audubon Society Nature Guide series (New York: Knopf, 1989).
2. *Grasslands and Tundra* (Alexandria, Va.: Time-Life Books, 1985), p. 20.
3. Ibid., p. 45.
4. An excellent account of the ecological impact of replacement of bison with cows is given in Janine M. Benyus, *The Field Guide to Wildlife Habitats of the Western United States* (New York: Simon & Schuster, 1989), pp. 136-38.
5. Edward Goldsmith et al., *Imperiled Planet* (Cambridge, Mass.: MIT Press, 1990), pp. 117-27; see also Brown, *Grasslands*, pp. 49-51.
6. Brown, *Grasslands*, p. 78.
7. *Grasslands and Tundra*, p. 151.
8. "Ignoring Pleas of Environmentalists, A Kansan Plows His Virgin Prairie," *New York Times*, November 23, 1990.
9. Ibid.
10. Quoted in Elizabeth Royte, "Showdown in Cattle Country," *New York Times Magazine*, December 16, 1990, p. 66.
11. Documented in *Rangeland Management*, a report to Congress by the U.S. General Accounting Office, GAO/RCED-88-80, June 1988.
12. Royte, "Showdown," p. 66.
13. Ibid.; see also Wuerthner's "Ranchers and Refuges: 3 Case Studies," a well-documented report in *The Earth First! Reader* (Salt Lake City: Peregrine Smith, 1991), pp. 64-78.
14. Ibid.
15. Dale Turner and Lynn Jacobs, "Livestock Grazing on the National Parks: A National Disgrace," in *The Earth First! Reader*, p. 121.
16. Cited in ibid., p. 126.
17. Quoted in Bruce Selcraig, "The Secrets of Gray Ranch," *New York Times Magazine*, June 3, 1990, p. 52.
18. Personal communication with the author.
19. William A. Fuller and John C. Holmes, *The Life of the Far North* (New York: McGraw-Hill, 1972), p. 225.

20. This section is based on material from a variety of sources including: Fuller and Holmes, *The Life of the Far North; Grasslands and Tundra* (Alexandria, Va.: Time-Life Books, 1985); Barry Lopez, *Arctic Dreams* (New York: Charles Scribner's, 1986); Dyan Zaslowsky, *These American Lands,* especially chapter 6 (New York: Henry Holt, 1986); and "The Fight to Save Alaska," a special issue of *Sierra*, January/February 1991.

21. Robert Leo Smith, *Elements of Ecology and Field Biology* (New York: Harper & Row, 1977), p. 426.

22. David L. Rayle and Hale L. Wedberg, *Botany: A Human Concern* (Philadelphia: Saunders College, 1980), pp. 293-94.

23. "Tracing Arctic Haze," Science Watch, *New York Times*, January 15, 1991.

24. "Alaska Liners Creating Smog, Block Views," *New York Times*, December 22, 1991.

25. Cited in Goldsmith et al., *Imperiled Planet*, p. 229.

26. Philip Shabecoff, "Alaska Oilfield Report Cites Unexpected Harm to Wildlife," *New York Times*, May 11, 1988.

27. Ibid.

28. William O. Douglas, *My Wilderness* (Sausalito, Calif.: Comstock Editions, 1960), p. 11.

29. Cited in Timothy Egan, "The Great Alaska Debate: Can Oil and Wilderness Mix?" *New York Times Magazine*, August 4, 1991.

30. Ibid.

31. Ibid.

32. Ibid.

33. Roger L. DiSilvestro, *Fight for Survival: A Companion to the Audubon Television Specials* (New York: John Wiley, 1990), p. 191.

34. Tom Kizzia, "Tradition Ties Caribou People to Wild Alaska," *Audubon Activist*, April 1991, pp. 1 and 6.

35. See also DiSilvestro, *Fight for Survival*, pp. 174-77.

36. U.S. Fish and Wildlife Service.

37. Egan, "The Great Alaska Debate."

38. Paul Rauber, "Last Refuge," *Sierra*, January/February 1992, p. 43.

39. Fred Baumgarten, "America's Arctic Refuge — At the Crossroads," *Audubon Activist*, April 1991, p. 4.

40. *Audubon Activist*, March 1991, p. 4, and April 1991, p. 4; see also Robert Watson, "Looking for Oil in All the Wrong Places," Natural Resources Defense Council, 1991.

41. *Ecology USA*, vol. 9, no. 22 (November 4, 1991), p. 213.

42. *Ecology USA*, vol. 21, no. 4 (February 24, 1992), p. 34.

43. Edward Goldsmith et al., *Imperiled Planet*, p. 232.

44. Details of the James Bay project can be found in several position papers and reports published by the National Audubon Society in 1990 and 1991; in Steve Turner and Todd Nachowitz, "The Damming of Native Lands," *The Nation*, October 21, 1991, and Sam Howe Verhovek, "Power Struggle," *New York Times Magazine*, January 12, 1992.

45. Jennifer Hansell, "Quebec Eyes Bay for Power," *Audubon Activist*, September/October 1989.

46. Verhovek, "Power Struggle," p. 20.

47. Ibid.

48. Sam Howe Verhovek, "Cuomo, Citing Economic Issues, Cancels Quebec Power Contract," *New York Times*, March 27, 1992.

6
FORESTS

"I want to tell what the forests were like. I will have to speak in a forgotten language."

— W. S. Merwin

The Nation States of Trees[1]

Endless forms most beautiful and wonderful [that] have been and are being evolved," is how Charles Darwin described the forests that even today cover a third of the earth's land surface.[2] Traveling slowly around the world in sailing ships, the naturalists of the eighteenth and nineteenth centuries brought back to their European colleagues the then-astonishing news that the plants of the world were organized by continents and geography into something like nation states. "Over enormous blocks of territory there spread formations of plants whose members were all of the same shape."[3] Voyaging northward to Scandinavia, Russia, or Canada, the naturalists found "whole nation states of forests given over to the cone, to the dark green brooding ranks of needle-leaved conifers."[4] In other parts of the globe they came across deciduous and rain forests, grasslands, and other plant formations. Each one, they observed, occupied a distinct region of the earth similar in topography and climate to another region with similar kinds of plants. It became clear to the naturalists that the formations follow regular patterns according to latitude and altitude. Below the arctic tundra line, extending from the Pacific to the Atlantic are broad bands of conifers. Some 400 to 600 miles south they give way to deciduous trees in the temperate regions that make up most of the United States. Still further south, in Central America, are the tropical rain forests that reach deep into Amazonia.

Forest Formations

- A forest is a large tract of land covered with trees and underbrush. It usually requires an annual average precipitation of 30 inches or more to support various tree species and smaller forms of vegetation.
- A **closed forest** is a land area with an almost complete cover of trees. An **open**

forest is a land area only partly covered by trees. **High forest** is a mature woodland, usually composed of tall trees whose tops form a closed canopy.

• The three main kinds of forest cover are **coniferous, deciduous,** and **tropical.** The coniferous forest, also known as **boreal** (northern) **forest,** is confined to the Northern Hemisphere; it is typical of regions with long, cold winters and short summers.

• **Evergreen coniferous forests** are characterized by pines, firs, redwoods, hemlock, spruce, and other *gymnosperms* (plants whose seeds are borne in cones). They usually occur as dense stands with one or a few species dominating. Abundant in northern latitudes, coniferous forests occur in large areas of Canada and in the higher mountain regions of the western U.S., through the Sierra Nevada and Rocky Mountains and even into northern Mexico. They are found in areas with cold winters and growing seasons of five months or less. In the southeastern U.S., where winters are milder than those in the western mountains and Canada, large areas are covered by evergreen pine forests. If undisturbed, these forests are gradually replaced by oak, hickory, and magnolia. Pine trees persist because they regenerate quickly after fires, whereas oak, hickory, and magnolia do not.

• **Deciduous forests**. Unlike evergreens, which drop their leaves as new ones are produced, deciduous trees — oak, maple, and beech, for example — sprout a new set of leaves each growing season and lose them all at once with the onset of winter. Over much of the eastern half of the U.S., where winters are less severe than in the north, different species of deciduous trees intermix and form the dominant vegetation. Many temperate forests are made up mainly of one or two kinds of trees that predominate. You find other trees there but less frequently.

• **Mixed hemlock-deciduous forests.** North of the deciduous forest, in the region around the Great Lakes and eastward through New England to the Atlantic coast, only a few dominant deciduous species, notably beech and maple, remain and mix with the evergreen hemlock tree.

• **Tropical rain forests** require warm, frost-free temperatures all year and abundant rainfall. Seventy-five million years ago tropical rain forests covered most of what is now the United States. Today in the U.S. they are found only in Hawaii, Puerto Rico, and Texas, as well as in Central and South America and other parts of the world. Because of the warm, moist conditions in rain forests, dead organic materials decay rapidly and minerals are rapidly washed out of the ground. This is why tropical regions generally have poor soil and cannot support the kind of agriculture practiced in temperate climates. (Unlike temperate forests, where nutrients are stored in the soil, tropical forests store nutrients in the trees themselves.) It is also why tropical soils are red. Along with the plant nutrients, the rains have taken out even the silica, the flinty rock that gives the gray tints to northern terrains.

• **Temperate rain forests**. South of mainland Alaska the coniferous forest differs from that found in the north due to a combination of climate and topography. Moisture-laden winds blowing inland from the Pacific meet the barrier of the Coast Ranges and Cascade Mountains of the Pacific Northwest and rise abruptly. Cooled by the upward thrust, the moisture in the air is released as rain or snow in amounts of up to 250 inches a year. Even in sum-

mer when the rainfall is low, the cool air brings in heavy fog, which collects on the forest foliage and drips to the ground, adding as much as 50 inches more moisture. The mix of abundant moisture, high humidity, and warm temperatures supports a community of lavish vegetation dominated by Douglas fir, Sitka spruce, western hemlock, and western red cedar. On the California coast to the south, because of similar conditions, you find the **coast redwood forest**.

● If forest formations are geographically patterned by weather and topography, they are also structurally stratified by light, temperature, and humidity. As we noted earlier, the topmost layer of a forest is the canopy, which receives the fullest impact of sun, wind, and rain. The forest floor is shaded throughout the year in most coniferous and tropical rain forests, and in late spring, summer, and early autumn in deciduous forests.

Although different types of forest support different species of animal life, you find the greatest concentration and diversity of life on and just below the ground layer. Other creatures live in various layers from low shrubs to the canopy. The flora and fauna of tropical rain forest are found from above the canopy down to the forest floor, with many animals and plants existing strictly in the trees. Whatever the forest, its living environment of plants and animals is greatly determined by the trees of which it is composed.

The Succession Process

Nothing in nature stands still. Every biological vacuum is eventually filled with living organisms. Left to natural processes, a community of plants is gradually replaced by others as the vegetation progresses toward a **climax**. This last stage of succession takes place when a community becomes mature, self-sustaining, and self-reproducing. It sustains itself when there is a balance between energy captured from sunlight and energy released by decomposition of organic material, and between the uptake of nutrients by plants and the return of nutrients to the soil by the fall of litter. The climax community is a well-developed ecosystem with a complex food web and a wide diversity of plant and animal species.[5] In general, ecologists recognize two main types of succession: primary and secondary. Which occurs in a given area depends on the conditions there at the beginning of the process.

> *"Any good local botanist can tell you the date a farmer quit farming merely by looking at the plants that are now growing on the land."*
> — Paul Colinvaux, ecologist[6]

Primary succession occurs in soilless areas — rocks exposed by retreating glaciers, sandbars deposited by a shift in ocean currents, or strip-mined terrain from which all topsoil has been removed. On these barren surfaces, succession from bare rock to a mature forest can take hundreds to thousands of years.

Secondary succession, more common than primary, takes place in terrain where the natural vegetation has been removed or destroyed but where the soil is left more or less intact. For example, if a Douglas fir forest in Idaho burns down, aspens and other **pioneer** trees will move in for twenty to thirty years, but eventually the Douglas firs will come back. When farmland that was

once a maple and beech forest is abandoned, it will revert eventually to the same type of forest. If you live in an area dominated by trees, you can see secondary succession taking place nearby, in an abandoned field, city lot, or any patch of earth where weeds can grow. First to arrive are **annual plants**, scattering their tiny seeds far and wide. **Perennials**, plants whose root systems take hold of the field year after year, come next, followed by shrubs, then the first trees of the returning woodlands.

We can't tell for certain that succession will end with the primeval forest, because "scientific" humans have not been around long enough to record the entire process. But, as Colinvaux says, it seems like a rather safe conjecture. Because the dynamics of forest succession are almost universal, you will generally find the same progression of forest patterns throughout the same latitudes. While such universal patterns do exist, making it easy to predict the past history of land use in many cases, scientists are coming to appreciate the importance of natural disturbances — wind, fire, flood, et cetera — in setting succession back to earlier stages of development. Before a forest reaches climax a wildfire might burn it down, and the process must begin anew. Annual flooding may keep a streamside forest "held up" at any early stage indefinitely. Disturbances can range in magnitude from the toppling down of a single old-growth tree to catastrophic fires such as those that burned millions of acres in and around Yellowstone National Park in 1988.* Because forests are subject to so many disturbances both large and small over long periods of time, ecologists recognize that "unpredictable" events are at least as important as any "predictable" successional process. In some ecosystems the climax stage may never be reached simply because disturbances are constantly resetting the successional clock. Nature is not as orderly as many scientists first thought!

The Coniferous Forest

Across the North American continent south of the treeless arctic tundra is a 400- to 800-mile-wide zone of coniferous forest. It stretches in an almost unbroken belt from New England across northern New York and southern Canada westward through the Rocky Mountains to the Pacific Ocean. The forest reaches north into Alaska and the arctic and south through the Sierra Nevada into northern Mexico. On the East Coast from Quebec southward, the coniferous forest extends south along the crests of the Appalachian Mountains through West Virginia, Virginia, North Carolina, Kentucky, Tennessee, Georgia, and Alabama. This biome is dominated by a relatively few species of evergreen, cone-bearing trees such as fir, spruce, pine, and larch. Their needle-shaped, wax-coated leaves conserve heat and water during the long, cold winters when air temperatures are low and the soil is frozen. Beneath dense stands of these trees, a carpet of fallen needles and litter covers the soil, which is poor in nutrients because of the high amounts of tannic acid in the needles slowing down their decay during the brief summer growing season.

* For more on the Yellowstone fire, see pages 249-251.

Life in the Coniferous Zone

The predominant vegetation consists of needle-leaf evergreens — balsam fir and black and white spruce (across the northern stretches to Alaska, in the Adirondacks and the White Mountains), red and white pine, hemlock, and white cedar (in the Great Lakes region), Fraser fir and red spruce (in the southern Appalachians), Douglas fir, Sitka spruce, western hemlock, and western red cedar (in the temperate rain forest of the northern Pacific coast), redwoods (in northern California), and jack pine (in dry or fire-scorched areas). Although low seasonal temperatures slow the growth rate throughout most of this biome, these coniferous forests rank among the world's most productive lumber-producing regions. The animals in this biome are well adapted by their fur and shaggy coats, their capacity to store fat, and their ability to hibernate to survive the long, cold winters. They include caribou, moose, wolves, and grizzly bears, as well as weasels, minks, and martens, which hunt smaller mammals, birds, and insects. The predominant birds are hawks, owls, chickadees, and woodpeckers. Among the many insect species that thrive during the warm summer months are pine beetles, spruce budworms, and other tree pests, as well as mosquitoes and flies fed upon by birds that fly in from the south.

The Deciduous Forest

Deciduous trees invade the southern fringe of the coniferous forest biome in ever-growing numbers. Together they form a coniferous-deciduous ecotone, which covers most of the eastern half of the United States excluding northern New England and New York, Appalachia, and the subtropical forests of southern Florida. Characterized by a moderate climate and deciduous trees, this biome is also the most extensively affected by human activity. Receiving between 30 and 50 inches of precipitation a year, it has high humidity during the growing season, which can last for six months or more. Because deciduous trees shed their leaves and lose their ability to photosynthesize, they do not grow during the winter months. Each fall a single acre of forest may be covered with more than ten million leaves, which together with other decaying material decompose rapidly on the damp ground to produce a rich layer of humus. The soil found in a typical deciduous woodland is known as brown-earth or brown-forest soil; it is formed when the downward drainage of rain water or melting snow is counterbalanced by the upward flow of moisture in the ground. This process helps to circulate nutrients and keep them within the soil rather than wash them out, as happens in tropical rain forests, for instance. Rich soil, ample moisture, and the long, warm growing season support a variety of plants that grow to different levels in the forest. An upper canopy is formed by taller deciduous trees (e.g., ash and maple) whose broad leaves receive the sun's full strength. Beneath these are smaller trees, often of the same species, whose leaves still allow some sunlight to filter down to smaller understory trees (e.g., dogwood and red bud), under which grows a layer of shrubs. On the forest floor, receiving only about 5 percent of noonday sunlight, are ferns, mosses, and other small plants that flower in late winter or early spring before becoming shaded by new leaf growth overhead. Deciduous trees provide food for a very large number of consumers, including elk and white-tailed deer, the most abundant species of deer in North America. Tree bark, buds, and seeds supply food year-round for many birds and animals including finches, chickadees, rabbits, deer mice, and beavers, all of whom are important components of the deciduous forest ecosystem.

Feeding the Forest

> *"Viewed from the air the forest is a formless mosaic, a quilt of ghostly white clearcuts."*
> — Lee Green, *Audubon.*[7]

Like all ecosystems, forests require nutrients, especially oxygen, carbon, hydrogen, and nitrogen, which are captured, retained, and recycled by trees (and other plants) in different ways. Nitrogen from the atmosphere is converted by bacteria in the soil into compounds that are taken up by the plants' root systems. Rainfall is a major supplier of nutrients, which a tree either intercepts and stores as the rain filters down or absorbs from moisture on the ground through the tree's millions of tiny rootlets. Then, as water evaporates from the leaves, more moisture and nutrients are drawn upward from the soil through the tree's trunk into the leaves, where they combine with carbon dioxide from the air to produce carbohydrates by the process of photosynthesis. In all forests, fungi, earthworms, termites, and other decomposers break down fallen leaves and other debris, recycling nutrients back into the soil. Certain types of fungus living among a tree's root hairs extract minerals from the decomposing litter and transfer the nutrients to the roots, which in turn supply the fungus with carbohydrates. This symbiotic relationship has been shown to increase the growth rate of a young pine by 20 percent.[8] Clear-cutting often deprives the soil of these specialized fungi, preventing regeneration of trees and slowing down the forest's regrowth.

In forests the accumulation of nutrients is considerable, both in trees and in forest soils. Some trees such as Douglas firs are known as **accumulators**, because they remove elements in the soil in amounts large enough to upset the short-term nutritional budget of the ecosystem by slowing down the biological cycling of minerals. When this happens, the rate of cycling is maintained only by the pumping of nutrients from deep reserves in the soil, the weathering of rock, and the eventual decay of the Douglas fir itself.[9] The nutritional gains to a forest ecosystem from precipitation and other sources are offset by losses from water often draining more mineral matter away from forests than the rainfall supplies. There are additional losses from the removal of timber. The absence of trees to take up the water results in a greater amount passing through the ecosystem, rapidly dissolving the nutrients and carrying them in solution. When trees are removed or die, the amount of root surface is reduced. This means that smaller portions of the

nutrients in the soil will be removed by the roots and also that there will be relatively little vegetation left to take up the nutrients. Loss of vegetation leads to increased surface temperatures, which may speed up the conversion of organic matter and its being dissolved and taken away by the runoff of soil and water.[10] Regardless of how rich or sterile the soil may be, the supply of nutrients to trees, and indeed to all plants, must be continuously accessible. Prolonged stoppage at any point can cause the ecosystem's function to cease. To keep it healthy, the nutrient sources — rain, runoff, decomposers, and recycling organisms — have to be maintained at a relatively stable equilibrium.

The Effects of Clear-cutting

Clear-cutting — a logging technique that removes virtually all of the forest canopy trees — is practiced extensively in western forests because it makes life easy for timber operators and, in the short term, the Forest Service. It saves the cost of selecting trees to be cut and eliminates the need to spare trees that will not be cut. And it allows forest managers to do less maintenance in that it usually requires only one entry into the area to be cut. The environmental impact of clear-cutting is devastating:

- It destroys the trees that forests depend on.
- It allows wind to penetrate the remaining forest, drying it and damaging the root systems of trees formed in less wind-stressed environments. This leads to soil erosion.
- The loss of canopy caused by clear-cutting also creates soil erosion by exposing naked soils to flooding and runoff from rain.
- Clear-cutting decreases the trees' interception of snow, thus covering the food sources of elk, deer, and other animals.
- A clearcut acts as a barrier to species that can't disperse across open areas.
- Species that live in trees and those that need shelter and moderate climate can't survive the harsher climate created by clearcuts.
- Deprived of shade normally provided by trees, streams become too warm in summertime, often exceeding the temperature limits of cool-water species of insects and fish.

(Source: Elliot A. Norse, *Ancient Forests of the Pacific Northwest*)

Adapting to Stress

In their defense against strong winds, floods, temperature extremes, drought, fire, low soil fertility, and other natural or human ravages, different types of vegetation respond in different ways. For example, when the temperature rises above 90 degrees Fahrenheit, some plants consume stored nutrients as a defense against excessive heat. This results in less productivity and less growth. Pine trees withstand fire by means of their insulating bark and their ability to regrow in open, disturbed areas. Spruce, fir, and hemlock have developed needles with less surface area than leaves of other species, so that they are less exposed to cold and moisture loss through transpiration.

A single old-growth tree may have as many as 70 million needles and a total of 50,000 square feet of leaf surface. These needles collect astonishing amounts of moisture and chemical nutrients from the atmosphere. Catherine Caulfield reports that when forests were cut around the watershed from which Portland, Oregon, gets some of its water, U.S. Forest Service scientists expected more water to enter the reservoir. Instead, water levels dropped. Researchers then discovered that almost a third of the water had never come from rain but from the no-longer-standing old-growth trees, which had collected it from passing clouds and fog banks.[11]

Too much stress may result in reduced numbers of species. As we noted previously, the cold winter and short growing season of coniferous forests result in fewer plant and animal species than are found in more southerly forests. Adaptation to stress also applies to entire ecosystems, whose components are interconnected. Not all components are affected in the same way. In a prolonged drought, for example, certain trees may die but not others; and even if trees die, the understory herbs and shrubs generally survive and help the forest community grow back. A forest, like all ecological communities, is a combination of myriad plant and animal species, each with its particular requirements and functions responding to a shared environment.

Old Growth: Going, Going...

"A land full of the largest forest... with as much beauty and delectable appearance as it would be possible to express," wrote the explorer Verrazano describing the East Coast of North America in 1524, and that was very much how the continent remained until the early 1800s when commercial logging got under way.[12] By 1850 production of lumber had become the leading manufacturing industry of the United States. Twenty years later the great hardwood and white pine forests of the Northeast were exhausted. Next to succumb to axes and saws were the pine forests, the cypress swamps, and the live-oak stands of the South, followed by the forest ecosystems of the Great Lakes. A similar process took place in Canada from New Brunswick to Quebec and inexorably on to the West Coast. The Pacific conifer forest, stretching 2,000 miles from San Francisco Bay north to the Alaska Panhandle and inland to the Cascade and Coast ranges, was the last frontier.

Forests that have been allowed to develop over long periods of time without catastrophic disturbance are known as **old-growth** or **ancient forests,** because of their untouched, primeval quality. They differ from younger forests in species composition and structure and in their cycling of energy, nutrients, and water. The trees in the old-growth forests of the Pacific Northwest — Douglas fir, spruce, cedar, hemlock, and redwood, many of them 300 feet or taller and 50 feet around — are much larger than their counterparts in younger forests. They are among the largest and oldest trees on earth, some of them more than a thousand years old; many were there before Verrazano.

According to the U.S. Forest Service's Old Growth Definition Task Group, an old-growth forest is defined by the following characteristics:

- Two or more species of trees with a wide range of tree sizes and ages;
- Six to eight Douglas fir or other coniferous trees per acre, at least 20 inches

in diameter or at least two hundred years old;

- At least 10 tons per acre of fallen logs, including at least two log sections per acre that are at least 24 inches in diameter and 50 feet long;
- A multilayered canopy;
- Two to four snags (standing dead trees) per acre at least 20 inches in diameter and at least 15 feet tall.

(Source: *The Oregonian*)

"The Last Great Buffalo Hunt"

— Lou Gold, environmental activist

U.S. FORESTS: AN HISTORIC LANDSCAPE

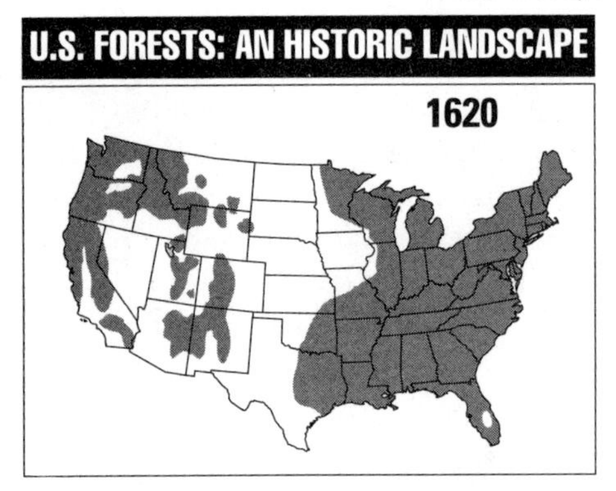

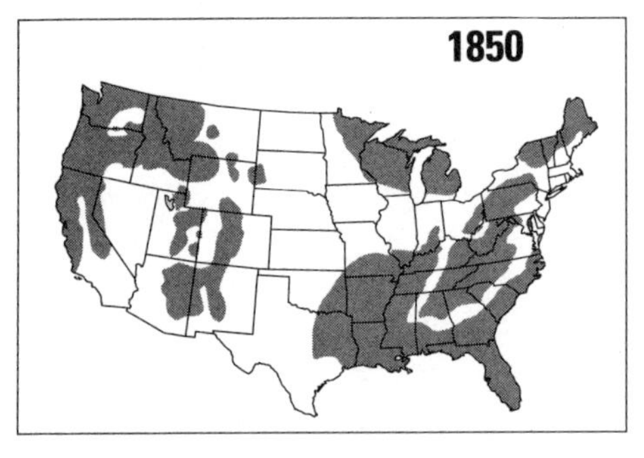

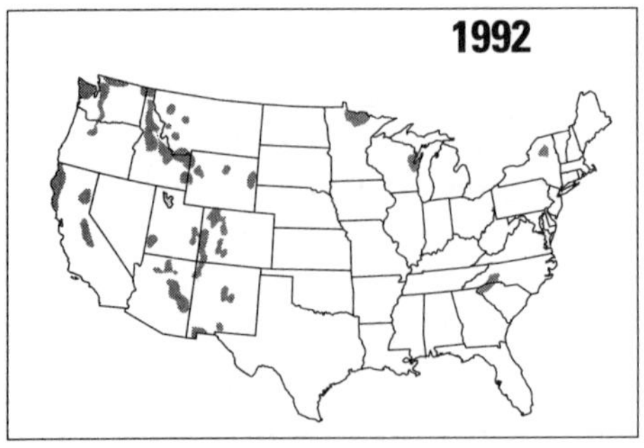

As you can see from the accompanying maps, the ancient forests are dwindling fast. Today less than 5 percent of them remain. In the Pacific Northwest more than 2 square miles of the oldest and largest trees are clear-cut each week. Even more alarming is the fact that less than 1 percent of ancient forests in the U.S. are protected from logging. Almost all the remaining unprotected groves are scheduled for elimination under current U.S. Forest Service and Bureau of Land Management plans, which subsidize the timber industry to log on public lands.*

The federal "management" of national forests includes slash-burning† and spraying chemical herbicides to kill the weeds that grow naturally in a clearcut. These weeds are needed to hold and replenish the soil during its periods of convalescence and, as dead matter, provide necessary nutrients and fertilizer for subsequent groves of trees. As the authors saw on a hike through Willamette National Forest in Oregon in early 1992, the Forest Service and the timber industry do not replant forests. They plant tree farms, which have only one or two species and bear as much relationship to forests as do cornfields to natural prairies. These plantations are logged

* For more on the Forest Service's subsidizing of clear-cutting, see "Voodoo Economics," page 195.

† Less than half of all the trees cut down are actually used. Those that are left are burned, producing 40 percent of all carbon dioxide (CO_2) emissions in the Northwest. CO_2 is the major "greenhouse" gas causing global warming. Newly planted trees will take two hundred years to reclaim from the atmosphere the amount of carbon dioxide released when an ancient forest is cut.

again in forty years, preventing the regrowth of a natural forest ecosystem, which would require two to three hundred years to develop. Such a long-term perspective runs counter to the policy of short-term economic exploitation currently favored by the federal government and a timber industry intent on maximizing quick profits.

Saving Siskiyou[13]

To increase the amount of timber cut on its land the Forest Service has built 360,000 miles of logging roads in the national forests, compared with a total of 44,000 miles of roads in the entire U.S. highway system. Nowhere is this program more threatening than in the Siskiyou National Forest, a landscape of exceptional beauty and diversity in southwest Oregon. Extending west to east from a few miles from the Pacific Ocean inland to the 7,000-foot glaciated summits of the High Siskiyous, the region adapts the richest forest gene pool of the Northern Hemisphere to abrupt changes in microclimate and soils. More than 1,400 plant species have been identified, many of them rare and endemic. Animals such as fisher and wolverine, extinct in most pacific northwest forests, are found here, and there are more salmon and steelhead in Siskiyou's rivers than in any other national forest outside of Alaska.

> *"The forest is much more than trees. It's many, many different organisms. Disrupt the web anywhere and you'll touch off disturbances throughout the system."*
> — Lou Gold[14]

Like other national forests, the Siskiyou has been logged and roaded over most of its land base. However, due to its remoteness, adverse terrain, legal obstacles, and other factors, logging on large areas of the forest have been deferred. Now the Forest Service plans to enter and log all of its remaining roadless areas within the next five years. At the forefront of a growing grass-roots opposition to the opening up of roadless forest in the Siskiyou and other national forests is the Siskiyou Regional Education Project (SREP) founded by Lou Gold in 1983. A former professor of political science, Gold is now a full-time environmental activist who works with Earth First!, the Alliance for the Wild Rockies, Save America's Forests, and other action-minded groups from his camp on a 3,800-foot Siskiyou peak called Bald Mountain.

Part of the answer, SREP advocates, is to create more national parks and to expand old ones to follow ecological rather than political boundaries. One such initiative would be the preservation of at least 300,000 acres of the forest as Siskiyou National Park. However, the Forest Service gave the idea short shrift, and when it issued its final plan in 1989, it not only excluded plans for a national park but called for large-scale logging — as much as 160 million board feet* a year, much of it in remnant old-growth forest.[15] In addition to halting logging on public land, SREP calls for reduced timber exports, especially of unmilled logs, and more efficient use and recycling of wood products.

* One board foot is 1 foot square and 1 inch thick. See also "Board Facts," page 192.

Some loggers blame environmentalists like Gold for trying to destroy the economy of their timber towns. "But we don't want to destroy the economy, we want to change it," says Gold. He calls sawmills "economic dinosaurs," because they were built to process old-growth trees. When those trees are gone, the associated jobs will be lost. "What it looks like to me is that they're trying to organize the last great buffalo hunt."[16] The best strategy, he says, is to work for legislation that preserves and restores forests as whole ecosystems, and makes way for a life-sustaining economy. "We need to make way for a diversified and earth-constructive vision in which economy and ecology work together — for the road builder to work on rehabilitation, for a revitalized salmon fishing industry, for the multimillion-dollar-a-year wild mushroom crop that comes out of our Pacific Northwest forests."[17]

What You Can Do to Save Siskiyou and Other Forests

1. Support the work of the Siskiyou Regional Education Project, P.O. Box 220, Cave Junction, OR 97523; (503) 592-4459. For $10 you can order its Ancient Forest Action Pack, which includes articles, graphics, and many organizing ideas. For a $35 contribution you can get "Lessons from the Ancient Forest: Earth Wisdom and Political Activism," a 53-minute video that conveys a powerful message with well-documented facts, beautiful images, and wry humor. This is one of the most effective and remarkable presentations we have seen in a long time. Also don't miss Lou Gold live if he's speaking anywhere near where you live.

2. Call the Western Ancient Forest Coalition, a Washington, D.C.– based grass-roots lobby, for the latest update on legislation affecting western and other forests.

3. Press your congressional representatives to stop the Forest Service and the BLM from logging the last ancient forests and to support such legislation as the Forest Biodiversity and Clearcutting Prohibition Act (H.R. 1969), sponsored by Representative John Bryant (D-Texas), and the Pacific Northwest Community Recovery and Ecosystem Conservation Act (S 1536), introduced in the Senate by Senator Brock Adams (D-Washington).

Unkindest Cut

In a June 18, 1990, front-page story datelined Pend Oreille, Washington, the *Wall Street Journal* reported that the Plum Creek Timber Company was "chainsawing a trail of profits and cut-over forests" over a 1.4 million-acre empire that had been a gift of President Lincoln to the railroads as an incentive to build lines to the Pacific. It had cut to the very banks of pristine trout streams, bulldozing a road into one wilderness after another. By the company's own figures, it is logging the forest at perhaps twice the rate that it can grow back. Ironically, reported the *Journal*, Plum Creek's advertising was based on the theme "For Us, Every Day Is Earth Day," showing a forester tenderly planting a seedling with the caption "Our Roots Are Here."[18] Plum Creek is one of many similar operations taking place throughout North America.[19] In

Verrazano's time old-growth forests covered 70,000 square miles (450 million acres) of the United States and Canada. Today less than 40 percent of the Canadian and less than 10 percent of the U.S. forests have survived, almost all on public lands, and most of these forests are scheduled to be cut for lumber, plywood, and pulp, primarily for export to Asia. What remains is going fast. In the U.S. national forests alone, some 50,000 acres of ancient forest each year are being cleared to feed the saw and pulp mills of Oregon and Washington. The U.S. Bureau of Land Management sells off another 20,000 acres annually. Conservationists estimate that the last of the old growth on state and private lands will be gone within fifteen years.[20]

Thriving on thin, nutrient-poor soils and often on steep hillsides, the Pacific forest ecosystems have been described by the forest ecologist Jerry Franklin as a triumph of life over adversity.[21] As he explains, forests have a natural life span, reaching maturity in about two hundred years, after which most of their energy goes into sustaining themselves. As decay slowly sets in, it can take many more centuries before the trees actually die and fall over. However, to loggers and foresters, it makes little sense to leave a tree in the ground after it has reached its peak of wood production. Some professional foresters claim that old-growth forests should be cleared and replanted with healthy young trees as soon as possible. However, this argument is refuted by Franklin and his colleagues, who demonstrate that the postmature, old-growth phase is in fact the richest, most complex stage of the forest's life.

The Life and Times of an Old-Growth Forest

Old-growth forests require at least 175 years to develop. Their multilayered canopies produce heavily filtered light, resulting in moderately dense, patchy layers of shrubs, herbs, and seedlings. Many snags, rotted stumps, and large logs in various stages of decay are strewn over the forest floor and lying across streams. Although smaller streams may be choked with organic debris, larger ones usually have clear water flowing over gravel beds and plunging over log dams into catch pools. Even after their death, trees continue to influence the forest system, first as standing snags and then as downed logs. Snags act as wildlife habitats, providing food sources, nesting places, and stages for courtship rituals.

The Redwood Heritage

Hundreds of millions of years ago a moister, warmer Northern Hemisphere was circled with redwoods.* Gradually the Cascade and Sierra Nevada mountains were pushed up, creating a drier, cooler climate that was less friendly to these mighty trees. Their feathery leaves do not have a waxy coating, and their

* Botanists recognize three separate genera of redwood, each of which consists of a single species: *Sequoia sempervirens* (commonly called the Coast Redwood) and *Sequoiadendron giganteum* (Giant Sequoia), both evergreens and native to California; and *Metasequoia glyptostroboides* (Dawn Redwood), brought from China to the U.S. In contrast to its California cousins, the dawn redwood is deciduous.

pores too easily release water vapor to the air. Nor did the trees develop root hairs, which help other species take up water from the soil. Consequently they were confined to a 450-mile-long coastal strip from southern Oregon to the Santa Lucia Mountains at the southern tip of Monterey County, California, which provides the high humidity, moist soil, and moderate temperatures they require. Redwoods are limited to within 50 miles of the Pacific Ocean, where onshore breezes bring saturated clouds of coastal fog to reach the trees, wrapping their branches in a blanket of moisture. Droplets of water condense on their needles and drop to the ground, adding the equivalent of 7 to 12 inches of rainfall a year and bringing minerals that seep into the trees' roots. This has been particularly important to the redwoods' survival during the recent prolonged drought in California.

The need for water also explains why many of the most magnificent redwood groves develop in stream canyons or along riverside flats that are subject to periodic flooding and deposition of mud and silt. Few other trees are able to withstand this buildup of silt, which helps maintain the purity of the redwood stands. The moist wood and thick, nonresinous bark of older redwoods make them resistant to fire and attack from insects or fungal diseases. Even when bark beetles attack fire-scarred trees, they do little damage. Injured redwoods display an amazing capacity to recover. If their tops are blown away by a storm, the trees quickly grow new ones. When branches are seared off by fire, a feathery column of new growth sprouts from the trunk. As we discovered from a recent visit to Muir Woods National Monument some 20 miles northeast of San Francisco, sprouts also appear from stumps of fallen-down redwoods, putting down their own roots and eventually forming a circle of younger trees around the parent. For the redwood, as for many other long-lived conifers, the chief agents of destruction are humans, who prize their lumber for building houses and other uses.

When the first settlers arrived, the redwood forests covered much of what is now San Mateo, Marin, and Santa Cruz counties as well as east San Francisco Bay and the hills behind Oakland. Today more than 95 percent of the original redwood forest has been cut at least once and less than 5 percent remains as it was. Of this, half is preserved in federal and state parks (notably Redwood National Park, Muir Woods National Monument, and Humboldt Redwoods State Park); the other half is privately owned and steadily disappearing as the trees are hauled out on logging skidders. As trees are removed from private forests adjacent to parks, the runoff from rains can no longer be controlled by redwood roots. This accelerates flooding to a level that even the alluvial-grown redwoods cannot survive.

With the loss of the trees comes an inevitable loss of wildlife. According to the *Field Guide to Wildlife Habitats*, 193 species of wildlife use redwoods for food, cover, or to fulfill their special habitat requirements during their lives.[22] In the forest canopy you can spot such birds as western flycatchers, Vaux's swift, and the varied thrush. On the tree trunks you find brown creepers and pygmy nuthatches in search of insects, hairy and pileated woodpeckers attacking the bark, as well as northern spotted owls and northern flying squirrels that nest in the holes left by the woodpeckers. In the branches close to the trunk live red

tree voles. Seeking nuts and berries fallen on the ground are blue grouse and chipmunks, shrews, and other rodents from underground burrows. Residing inside logs are moisture-loving salamanders and hosts of bacteria, fungi, and insects.* Among the larger creatures residing in the redwoods are mountain lions, marten, elk, deer, mountain beaver, and black bear.

Saving the Redwoods

The first state redwood park was created in 1902 at Big Basin in California's Santa Cruz County. Five years later Muir Woods in Marin County was dedicated as a national monument, and in 1917 the Save-the-Redwoods League was organized. One of its first actions was to recommend establishing a national park of some 64,000 acres in the lower Klamath River area. Congress turned down this proposal and several later ones, recommending a park of more than 2 million acres that would include most of the then-existing redwood forests. By 1964, a National Park Service survey revealed, only 300,000 acres of old-growth timber remained, with 50,000 of those protected in state parks. In 1968 Congress finally established Redwood National Park, albeit on only 30,000 acres, which were expanded by another 48,000 acres ten years later. The Save-the-Redwoods League, which played an important role in getting the legislation passed, works with the California Department of Parks and Recreation and the National Park Service for the long-term protection of this scenically and ecologically important forest ecosystem.

Taking Out the Giant Sequoias

Giant sequoias once dominated the forests of North America, but today only seventy-five groves remain, all in California and half of them in Sequoia National Forest. But even this doesn't guarantee their safety. After enduring hundreds of thousands of years of fire, storms, drought, and flood, the giant sequoias are being taken out by the agency designated to protect them. In 1982 the United States Forest Service, under pressure from the timber industry, quietly abandoned its long-standing policy of keeping out of the sequoia groves, reclassifying them as eligible for logging. Without public notification or debate, thirty-six groves of giant sequoias came under the ax.

> *"Sequoias are not just an attractive part of nature, they are the biggest trees on Earth, the planet's largest living things."*
> — Lee Green[23]

Sequoia and Redwoods

- The giant sequoia (*Sequoiadendron giganteum*) is often confused with its close relative, the taller but slenderer coast redwood (*Sequoia sempervirens*), which is the world's tallest living thing.
- The tallest redwood, in California's Redwood National Park, measures 368 feet high.

* For a more detailed description of the life within a log, see page 204.

- The giant sequoia General Sherman Tree in California's Giant Forest is thought to weigh 1,250 tons. It measures 105 feet around at its base and is still growing.
- Redwoods flourish only in a fog belt, growing in a narrow, 500-mile-long, discontinuous strip of Pacific coast from southern Oregon down to San Luis Obispo County, California. The thickest, tallest forests are in California's Del Norte and Humboldt counties, where redwood groves average 300 feet tall.
- Coast redwoods are among the plant world's most efficient photosynthesizers. They can grow in a location so shaded that only 1 percent of the incoming sunlight reaches their leaves.
- Despite their great height, redwoods have roots that are shallow, penetrating only 10 to 15 feet deep but spreading out matlike for 100 feet or more.
- On mature redwoods the bark runs from 6 to 12 inches thick. This thickness and its tannin content make the trees resistant to fire and attack by insects and fungi.
- Fire benefits the redwoods' ecological health by clearing the forest floor so that redwood seeds can reach the mineral soil. Fire suppression by humans has had a negative effect on the redwood ecosystem.
- Coast redwoods reproduce by sprouting from burls* as well as by seeds. This gives these trees a competitive edge over other conifers and other trees that reproduce only by seeds.

(Sources: U.S. Forest Service; National Park Service)

Why did the stewards of a 1.1-million-acre public forest violate their own trust and mandate? We did it for the good of the groves, answered the Forest Service. The young giant sequoias, they said, were being crowded out by pines, white firs, and cedars, which posed a wildfire threat to the big trees. What they didn't tell us was that sequoias generally grow in spots surrounded by the majestic specimens of old-growth ponderosa and sugar pines, which are commodities of high value to the timber industry. The Forest Service action does not make sense to botanists in particular and conservationists in general. Isolating the sequoias, they warn, will have dangerous long-term consequences, making them vulnerable to windthrow, soil erosion, and disease, as well as disrupting the forest's nutrient cycles. Representing a broad coalition of opposition, the Sierra Club filed suit in federal court, charging the Service with violating the National Environmental Policy Act (NEPA) by failing to prepare environmental-impact statements† on nine timber sales, four of which involved giant sequoias. In late 1989 the Service agreed to an out-of-court settlement enjoining them from further cutting in the contested areas until the environmental effects were properly assessed. But with more than 1,000 acres of giant sequoia habitat already partially logged, irretrievable damage has been done.

* A burl is a dome-shaped growth on the trunk of a tree.

† According to NEPA, whenever a government agency plans to initiate, finance, or permit a "major" action, it must prepare an "environmental-impact statement" (EIS) assessing the project's environmental effects. As part of this process, the appropriate agency must hold hearings to take public comment and provide a public review period of a draft EIS. Until a final EIS is released, the project cannot proceed. For more on this, see chapter 9, page 294.

Who Runs the Forests?

When the Forest Service was founded in 1905 by the conservationist Gifford Pinchot,* it was entrusted with 86 million acres of forestland to administer. The forests were established with two major goals in mind: protecting watersheds and managing public timber in a more responsible way than the cut-and-run practices of the nineteenth century. Public forestland, said Pinchot, "is to be devoted to its most productive use for the permanent good of the whole people, not for the temporary benefit of individuals or companies."[24] Under his administration, conservation of species and *sustainable production*† of renewable resources were accorded highest priority.

> *"We are doing the same thing to our rain forests the Third World countries are doing to theirs."*
> — Jeff DeBonis, former U.S. Forest Service timber sales planner

To control overgrazing, which threatened watersheds in the West, he instituted a system of permits and fees; he also restricted logging to meeting local timber needs. But Pinchot gave the Forest Service an orientation toward **multiple use**, which in principle means reconciling the interests of all forest users — timber growers and cutters, hikers, campers, hunters, skiers, and others — "from the standpoint of the greatest good of the greatest number in the long run."[25] In practice, things worked out differently. For the first three decades of its existence the Forest Service developed a "comfortable, almost symbiotic relationship"[26] with the timber industry. After World War II the Service embarked on a program of large-scale logging operations that brought in clear-cutting, widespread road building, and the inevitable destruction of habitat. The pressure to produce timber from the national forests intensified. By 1970 the annual harvest had quadrupled. Most heavily cut were the Pacific forests, highly prized for the quality and volume of their trees.

Forest Service mismanagement and overcutting of the national forests are widely documented.[27] In 1969 one of its own studies predicted that, due to overcutting, its Douglas-fir harvests would drop by almost a half once the old-growth trees were gone. At the same time the agency was criticized for ignoring its legal obligation to protect all the resources of the forest, and in 1975 the Fourth Circuit Court of Appeals banned clear-cutting in the Monongahela National Forest in West Virginia.[28] In response to this emergency, Congress in 1976 adopted the National Forest Management Act (NFMA) ordering the Service to limit the timber cut to an amount that each forest could sustain in perpetuity. It also directed the agency to "provide for diversity of plant and animal communities."[29] Yet twelve years later the Wilderness Society described the recently issued Forest Service's long-term

* For more on Pinchot and the issue of conservation vs. preservation, see page 258.

† Sustainable production is economic development that can continue indefinitely because it is based on the use of renewable resources, causing insufficient environmental damage to pose an eventual limit.

Board Facts

- Board feet measure the amount of wood in a tree that can be used by humans. One board foot is 1 foot square and 1 inch thick.
- It takes about 10,000 board feet of timber to build an average single-family house.
- One large, old Douglas fir may contain 10,000 board feet; a giant fir might contain 30,000.
- Since 1979 an average of 12 billion board feet of timber a year have been cut in national forests, about a quarter of which comes from the Pacific Northwest.
- In 1989 Oregon and Washington cut 14.8 billion board feet of raw logs, 3.7 billion of which were exported, mainly to Japan, Korea, and Taiwan.
- Because at least 60 percent of all Pacific Northwest lumber is exported *unfinished*, the 10 percent increase in wood taken from national forests since 1977 has been accompanied by a 15 percent drop in jobs in logging and wood processing.

(Sources: U.S. Forest Service; Native Forest Council)

management plans for twelve national forests in the West* as "nothing less than a prescription for disaster." The plans would increase logging by 15 percent, destroy critical wildlife habitat, greatly exaggerate the amount of old growth that actually exists, eliminate nearly half of the unprotected roadless areas within the next fifteen years, drastically reduce recreational opportunities and scenic beauty, and degrade water quality and fisheries.[30] In 1989 the Society revealed that the Forest Service was selling timber at prices well below its costs and subsidizing the lumber industry and timber-dependent communities to the tune of more than $400 million a year.† "The costs of this policy are staggering. In 1985 and 1986, the Forest Service recovered less than one cent in timber payments for every dollar it spent on timber sales and road construction. Between 1977 and 1986, total losses in the Tongass were more than one third of a billion dollars."[31]

The Fragmented Forest

Fifty years of clear-cutting, road building, and Forest Service mismanagement have left North America from the southern Appalachians to the Pacific Northwest with severely fragmented forests. A poignant example, given by Catherine Caufield in her *New Yorker* report "The Ancient Forest," is southeastern Alaska's Chichagof Island: "Almost every watershed of this once-perfect landscape of forests, rivers, white sand beaches, and snow-capped mountain ridges has been violated, largely as a result of the Forest

* These are: (Washington) Olympic, Mt. Baker-Snoqualmie, and Gifford Pinchot; (Oregon) Mt. Hood, Willamette, Umpqua, Rogue River, Siuslaw, and Siskiyou; (California) Six Rivers, Shasta-Trinity, and Klamath.

† See also *Design for a Livable Planet*, pages 115-133.

Service's allowing loggers to search out and take the biggest trees, which are scattered in small stands — a process known as high-grading."[32] These old-growth stands, Caufield explains, are a small part of the forest but the most profitable to cut. They are also the heart of the forest ecosystem, supporting "both a mass of timber and a mass of life." Many of the remaining clearcuts are separated from each other only by narrow strips of standing trees that are supposed to serve as wildlife refuges. But they are inadequate to support the animals displaced by the logging.

As Caufield reports, the Forest Service claims that only 12 percent of the forest in question has been logged. While this may be literally factual, the logging has concentrated on the old-growth forest and the clearcuts are spread over almost every one of the region's watersheds. "Gone are dense, low-elevation forests where deer sheltered and fed in winter, and riparian forests where brown bears lived and in whose waters millions of salmon spawned each spring."[33] There are already 200 miles of logging roads on the northeastern 400 square miles of Chichagof Island, with another 200 miles planned by the Forest Service. Almost certainly the roads will bring an influx of hunters and tourists, destroying the isolated coastal community of Tenakee Springs, its neighboring Tlingit Indians, and the environment of which they form part.

Forest Facts

- A mature Douglas fir can store more than a thousand gallons of water in its sapwood.
- An Alaska yellow cedar can live more than 3,000 years.
- Every day the timber industry logs 170 acres of old-growth forest — the equivalent of 129 football fields.
- Less than 5 percent of America's old-growth forests remains today.
- Air pollution is the suspected cause of forest decline in at least fifteen states.

(Sources: the Sierra Club, the Wilderness Society, Native Forest Council, *Technology Review*)

Saving America's Rain Forests

The images of burning rain forests in the Amazon are so vivid for many people that they often overlook the rainforest massacres being perpetrated in our own backyards — in Alaska, British Columbia, the Pacific Northwest, Hawaii, and Puerto Rico. Yet, as the following case histories show, these are regions where we can have a direct influence through the legislative process, by supporting the efforts of conservation and environmental groups, and, when necessary, with demonstrations, boycotts, and other forms of citizen action.

Trashing the Tongass

The nation's largest national forest, the Tongass covers 25,000 square miles (16 million acres) along the rugged coastline of southeastern Alaska from north of Juneau to south of Ketchikan. This is a scenic region of islands, fjords, coves, and snow-capped mountains, warmed by an offshoot of the

Japanese Current, which produces mild winters and considerable rainfall. The trees are often festooned with hanging mosses, reminiscent more of Florida than Alaska. The forest is dotted with **muskeg** areas (low marshy bogs) that produce a wide variety of flowering plants, including the Alaska water lily and the marsh marigold.

TONGASS NATIONAL FOREST

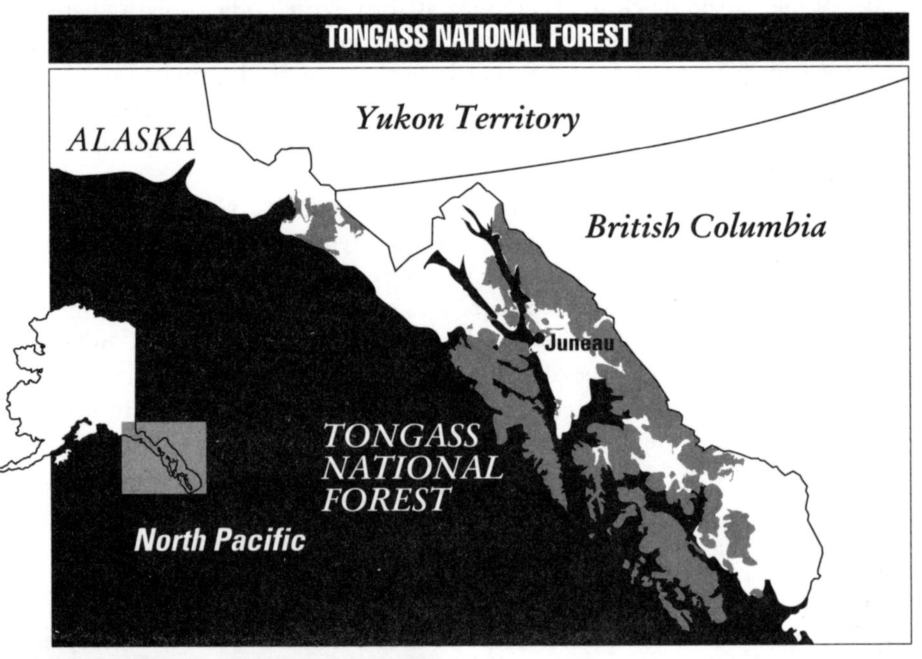

> *"A place of endless rhythm and beauty."*
> — John Muir

A thousand square miles larger than the state of West Virginia, the Tongass National Forest is home to the greatest concentrations of grizzly bears* and bald eagles left in North America. Its streams provide the spawning grounds for a salmon fishery that is vital to the economy of Alaska. Its spectacular wildness is a magnet for a rapidly growing number of tourists. It is a priceless, irreplaceable, and critically endangered resource. The Tongass, which is the northern extension of the coastal rain forests of the Pacific Northwest and British Columbia, shelters great stands of Sitka spruce and Western hemlock in one of the world's last temperate-zone rain forests. Much more than a collection of eight-hundred-year-old trees, these stands are an integral part of a complex ecosystem. Their upper branches are covered with several centimeters of organic "soil," which supports entire communities of plants, birds, and other tree-dwelling creatures. Like grizzlies and eagles, Sitka black-tailed deer depend on the old-growth stands, especially in winter. In times of heavy snowfall, large branches in the forest canopy intercept enough snow to maintain snow-free patches on the forest floor, allowing deer to feed on small plants.

* One island in this region, Admiralty Island, contains nearly twice as many grizzlies as in all the lower forty-eight states, where we are spending millions of dollars trying to save the last of the great bears.

Because the Tongass's remote location makes logging uneconomical and because 95 percent of southeast Alaska is federally owned, its destruction could be brought about only by federal action. In 1947 Congress authorized the Forest Service to sign fifty-year contracts with private timber companies who promised to build pulp mills and create jobs. These mills, one Japanese-owned, have thus been able to buy Tongass timber at prices averaging about $2 per 1,000 board feet, compared with two to six *hundred* times that amount on the open market. In addition to selling the timber at cut-rate prices, the Service facilitates the logging by building roads into the backcountry, causing serious soil erosion and damaging salmon fisheries. At around $150,000 a mile, the cost of these roads often exceeds the value of the timber made accessible by them. As the *New York Times* commented in a 1989 editorial, "The Forest Service has been selling 500-year-old trees for about the price of a cheeseburger."[34]

Voodoo Economics

Since 1982 the Forest Service has spent an average of $55 million a year to subsidize clear-cutting in the Tongass, while only $550,000 annually has been returned to the U.S. Treasury in timber receipts, a return of less than 2¢ on the dollar, or put another way, a loss of over 98¢ for every dollar invested. A common justification for this program is that it provides a stable employment base for residents of southeast Alaska. Yet during 1980-1988 the number of directly timber-dependent jobs dropped from 2,700 to 1,781, and furthermore, approximately 40 percent of the loggers are not true residents of Alaska but maintain homes and families in Washington and Idaho.

Almost all wood products taken from the Tongass are destined for the Far East. Southeast Alaska residents who need building materials must buy expensive "imported" lumber from Washington State. The prime beneficiaries of the huge Tongass logging subsidy are two multinational logging companies — Alaska Pulp Co., a Japanese consortium, and Louisiana-Pacific Ketchikan, the largest purchaser of federal timber in the U.S. As holders of unprecedented fifty-year contracts, these companies have a virtual monopoly of logging in the Tongass, excluding smaller, independent timber operators.

(Source: *Inner Voice*)

This heavily subsidized, environmentally disastrous logging is not winning friends in Alaska's fast-growing tourist industry, which feeds well over a billion dollars a year into the state's economy, employing more people than either logging or fishing. In the southeast corner of the state the number-one tourist attraction is the Inside Passage, the marine highway that leads cruise ships along a series of straits, sounds, corridors, and canals all running through the Tongass forest. "People don't come here to see clearcuts," warned David Cline, National Audubon Society's vice president for Alaska.[35] Serious as the loss of taxpayer and tourist dollars may be, the longer-range tragedy of the Tongass is the lasting destruction of Alaska's unique rain forest.

British Columbia at Loggerheads

"British Columbia's lumber lords suffer few checks on their enterprise, and scoff at local protest. Without international support, the people of the province can only watch as their forests — some of the world's most magnificent — are decimated." So begins a cover story in *Sierra* magazine.[36] The forests are being cut at an annual rate of over 600,000 acres, more than is logged in a year from all the national forests in the United States. At the current rate the world will lose its largest temperate rain forest within fifteen years. At stake are stands of hemlock, cedar, and Sitka spruce that support one of the richest ecosystems on earth. One of the most depleted areas is Vancouver Island, where mile after mile of forest has been turned into a barren stumpland where 250-foot-high trees used to stand.[38] In the battle to save what's left, forestry experts and conservationists emphasize the environmental value of the ancient forests: The trees wring moisture from the clouds and help keep the coastal climate cool; clear-cutting and slash-burning cause flooding of the land in winter, pollution of the air in summer, and may lead to global warming; and it is hypocritical for British Columbia to promote the tourist attraction of its natural wonders while allowing the forests to be cut down.

> *"I tell my guys if they see a spotted owl to shoot it."*
> — Jack Munro, Canadian president, the International Woodworkers of America[37]

As *Sierra* reports, Vancouver Island is being logged faster than any other part of the province. Out of eighty-nine of its largest watersheds only six remain uncut. Around the village of Kyuquot in the north of the island the environment that provided a plentiful livelihood for fishermen and Native inhabitants is rapidly giving way to massive clearcuts, erosion scars, clogged spawning streams, and muddy estuaries. One creek just south of the village is slated for logging despite fears that its steep slopes will give way, covering salmon-spawning grounds with mud and debris.[39] The impact of the wholesale tree cutting spreads far beyond individual villages like Kyuquot. It can shake the farthest reaches of the forest food web. Logging along the British Columbia coast is wreaking havoc on streams, causing salmon populations to drop dramatically. The salmon's decline leaves local grizzly bears and bald eagles at risk and may even wipe out orcas ("killer" whales) accustomed to feasting on salmon at the river's mouth.[40]

The Olympic Crown

The Olympic National Park and surrounding National Forest and state lands on the Olympic Peninsula in western Washington are filled with what *The Big Outside* describes as "the crowning temperate rain forest on Earth." Douglas fir, western red cedar, Sitka spruce, and western hemlock, reaching almost 300 feet high and over 8 feet in diameter, abound. The 140 inches of precipitation a year allow many **epiphytic** plants to flourish. Found primarily in old-growth forests, epiphytes (from the Greek *epi* ["upon"] and *phyte* ["plant"] grow nonparasitically upon other plants, drawing their nutrients directly from raindrops and particles in the air. Not only do they cause no harm to their "host" trees, they in fact contribute significantly to the forest's

fertility in different ways. When pieces of one species, the lungwort (so called because it looks like the inside of a human lung), break off and fall to the ground, they are eaten by browsing animals or decompose to add nitrogen to the soil. This is another example of **symbiosis** — a close and mutually beneficial association of organisms of different species that often cannot live independently.*

A recent study by a University of Washington ecologist found that mats of epiphytic mosses, lichens, and liverworts along the branches and trunks of bigleaf maple and other deciduous trees covered a network of their roots, enabling the trees to tap into water and nutrients from "arboreal cupboards in their own crowns."[41] As we have seen, old-growth forests are complete ecosystems composed of organisms that have evolved together over centuries into complex natural communities. Even if we allowed them to grow back, the growth in the clearcuts would not even begin to approach the *beginning* stages of old growth for 250 to 300 years.

Trouble in Hawaii

> *"Why should America's last big tropical rain forest be sacrificed so all the new hotels can run their air conditioners?"*
> — Russell Ruderman, biologist[42]

By Brazilian standards Wao Kele O Puna (green forest of Puna) is not big. But its 27,000 acres, located on the eastern end of the island of Hawaii, constitute the United States' largest remaining tropical rain forest. Having survived sugar plantations, logging, and urban development that leveled virtually all of Hawaii's other rain forests, Puna is now imperiled by a plan to turn the eastern rift of the Kilauea Volcano into one of the world's largest geothermal power plants. At this point only a few roads have been cut into the Puna forest, a rich preserve of giant ferns, full-canopied Ohia trees, and plants most Americans see only in botanical garden greenhouses. However, the state has announced plans to develop 500 megawatts of electricity at Puna — far more than is needed by the island of Hawaii — most of which would be transmitted by undersea cable to the other, more populous islands.

Many residents in the Puna area, where it rains more than 100 inches a year, collect water from rainfall and generate electricity from solar energy cells on their rooftops. The billion dollars that might have to be spent on the cable would be enough to put photovoltaic panels on every home in Hawaii, said Bill Reich, a Democratic Party official in Puna. Opposition comes from those who fear that building roads, power plants, and transmission lines will destroy the ecological web that keeps the forest alive. "We've got to be cautious with what is left here," said U.S. Fish and Wildlife Service botanist James Jacobi, who warns that many original species would be wiped out by the project.[43] Native Hawaiians formed the Pele Defense Fund, challenging the project as an attack on their freedom to practice their religion and claiming that drilling

* See also "The Mushroom Connection," page 200.

would violate their goddess Pele, who is said to reside in the volcano's steaming caldera. In January 1991 a lawsuit brought by the Sierra Club Legal Defense Fund in support of the Wao Kele O Puna rain forest forced the U.S. Department of Energy to agree to prepare an environmental impact statement before the geothermal project could proceed.

World's Wettest

With an average rainfall of 480 inches a year, Alakai on the Hawaiian island of Kauai is the world's wettest rain forest. Sometimes called a swamp, sometimes a wilderness, Alakai actually consists of four interconnected ecosystems. "At the highest altitudes, the Alakai is a barren place marked by 8-inch Ohia trees, some of which might be hundreds of years old. In the center of its impermeable plateau, a bog drains out towards the cliffs, like a fairway through a golf course. Surrounding this bog and spilling down the cliffs is a dense shroud of vegetation."[44] There are no native mammals in Alakai, no amphibians or reptiles, but an enormous diversity of plant and insect life, most of which is found nowhere else in the world. The preceding description comes from a report by John Nielsen in *National Wildlife* on a maverick botanist, Michael Doyle, who risks his life to study a unique family of tropical plants known as *Gunnera* or, to Hawaiians, *'apee*. Brilliant green, with umbrellalike leaves that are often 8 feet wide, these plants have grown on the edges of forests such as Alakai for millions of years, evolving only slightly. They are the largest herbs in the world with no known relatives, extinct or living. Blue-green algae living in their stems convert nitrogen from the air into usable form, replacing nitrogen leached from the soil by the incessant rain. Fungi in their roots draw nutrients from the soil. According to Doyle, such an arrangement is unknown elsewhere in the world of flowering plants. For that reason, he believes, *Gunnera* might prove of great value in the search for self-fertilizing crops or plants able to survive in rough terrain or poor soil.

Species Loss

Media interest in pandas, whales, and other large animals often overshadows the fact that the health of an ecosystem often depends on smaller and less charismatic species. When a forest is fragmented by logging, for instance, some changes become more immediately obvious. The number of larger animal species may decline within a couple of years or sooner. Aside from the loss of trees, the harm to vegetation may take longer to show. The forest suffers wind damage at the edges. Trees get blown down, leaf litter dries out and, taking longer to decay, piles up under the trees that are left. This increases the risk of fire and can lead to nutrients becoming locked up in the litter rather than being recycled into the ecosystem. Even less evident than the changes in vegetation are the losses of insect species, but they may be more damaging in the long run. Certain bees and other insects that play a key role in pollinating trees have been shown to be unable or unwilling to cross a forest area that has been cleared. Their loss may have unforeseen consequences because of the

intricate relationship between species within an ecosystem. For example, botanists have recently observed that certain kinds of tree may survive as rare isolated individuals only by providing birds and insects with the means of pollinating other (similar) trees in the forest. If these rare specimens are destroyed, it is possible that the entire forest will suffer. And even when a species manages to survive the fragmentation of its habitat, its numbers may be so reduced that it becomes unable to maintain a breeding population.[45] This was what happened to the California condor and what might well become the fate of the northern spotted owl.

*The Owl and the Forest**

> *"If the spotted owl goes, it is hard to imagine hunters responding to an open season on mice."*
> — David Kelly, *Audubon*[46]

Although a reservoir of old-growth habitat is believed to be essential for the survival of the pileated woodpecker, the marten, and at least a dozen other identified species, the northern spotted owl (*Strix occidentalis caurina*) is the only species whose continued existence has been *proven* to depend on old growth. Why, may we ask, is saving this bird (one pair of which requires at least 2,000 acres to survive and of which there are perhaps 2,000 to 3,000 pairs left in the western United States) so crucial?[47] The dodo, the great auk, and within living memory, the passenger pigeon have become extinct without any noticeable effect on the world we inhabit today. Is the brouhaha over spotted owls, snail darters, and other small creatures just a reaction of bleeding-heart environmentalists? Or are these threatened creatures canaries in the coal mine, sending out warning signals we should take seriously? To answer these questions, we need to understand the association between old-growth forests and a 16-inch-tall, gentle brown bird rarely seen by the environmentalists who want to protect it or by the loggers who feel threatened by its existence.

Why Northern Spotted Owls Need Old-Growth Forests

- Because they do not build free-standing nests, they need the broken treetops and cavities found in large, old trees for their nesting sites.
- The dense, multistoried canopy of old growth provides thermal cover that protects spotted owls from extreme summer heat and winter cold.
- The dense cover also protects the spotted owl from its predators, the great horned owl and the northern goshawk.
- Snags and other decaying matter on the forest floor are ideal homes for small rodents and other spotted owl prey.

Northern spotted owls require substantial areas of old-growth forest to survive. Based on data from radio collars fitted on to individual birds, biologists estimate that each owl pair uses between 2,300 and 4,200 acres.[48]

* In this chapter we discuss the northern spotted owl's ecological role. The legal and economic aspects of this endangered species are examined in chapter 9, pages 289-291.

Because of this dependency, the amount of remaining old-growth acreage is critical. According to the Wilderness Society, less than 3 million acres of suitable spotted owl habitat are left in the Pacific Northwest. Ninety percent of this acreage is controlled by the USFS and the BLM, who consider well over half of it suitable for timber production.[49] As we have noted, the Forest Service policy of excessive logging is severely reducing old-growth and mature forests. It is now generally agreed that unless the Service dramatically reduces logging rates, most northern spotted owl habitat on public land will vanish within twenty years, leaving only isolated patches of forest fragments and roadside or riparian buffer strips.[50] This habitat loss contributes to inbreeding as well as less nesting and foraging ground for the owls. It also makes them more vulnerable to predators, forest fires, and other external threats. The Endangered Species Act (ESA) defines an endangered species as "any species which is in danger throughout all or a significant portion of its range."[51] Within this definition and given the current rate of logging throughout its habitat and the low success rate of juvenile owl dispersal in a fragmented landscape, the northern spotted owl may well become extinct within the next decade.[52]

The Mushroom Connection

Recent studies by Chris Maser, a wildlife biologist formerly with the BLM in Oregon, Jim Trappe, a mushroom expert, and other researchers explain the importance of the northern spotted owl to old-growth forests: The bird's basic role is to cull and keep healthy the forest's population of small mammals, principally the northern flying squirrel, which is its favorite prey.[53] In preserving this ecological balance, the owl helps maintain a vital connection between the roots of the conifer trees and certain soil fungi (mushrooms). As they grow, the fungi wrap themselves tightly around the trees' rootlets, penetrating the outer cellular layers. Acting as sponges, the fungi absorb minerals, nitrogen, and water from the soil and make them available them to their hosts. They also produce growth-regulating chemicals that induce the trees to grow new root tips and make them live longer. Squirrels, voles, and mice eat the fruit of the fungi — truffles — then excrete the spores at new sites in the forest, helping produce a new generation of fungi, which can then work in partnership with more trees. Edge-dwelling chipmunks, deer mice, and heather voles move the spores farther away into clearcuts, burns, and meadows. The rodents perform another function: spreading nitrogen-fixing bacteria throughout the forest.[54]

These functions have evolved by natural selection over millions of years in what Maser and Trappe describe as a "massive symbiosis, a ballet of interaction between hundreds of plant and animal species in the "rotten" wood of the forest floor." It all seems to have one end: providing food for the forest and streams. Commenting on the connections between old-growth forests, spotted owls, and other predators, David Kelly wryly observes, "It is true that the forest survives without the wolf, which has been replaced in the system by the guys with the blaze-orange vests and the 30.06 rifles. Possibly it could survive without the secretive mountain lion, which is currently hanging on in a range already technically too small for its requirements. But if the spotted owl goes, it is hard to imagine hunters responding to an open season on mice."[55]

Can Forests Be Saved Piecemeal?

In trying to save old growth, conservationists emphasize its importance in sustaining *long-term* forest production and providing high-quality water and wildlife habitat. On the other hand, the Forest Service proposes preserving only *pockets* of trees for spotted owls and other endangered species. But trying to save a species in this limited way may not keep old growth intact over the long run. "We just can't save islands of it," says Rick Brown, a National Wildlife Federation specialist in Oregon.[56] He advocates a reserve system that spans age groups of forests, including young, unmanaged stands that will become future old growth. A similar position is taken by Maser, who warns that shortsightedness could be disastrous. "By converting old growth to young stands, we're redesigning the forests," he says. "We can't duplicate what nature has been doing for centuries. And we cannot have a sustainable timber industry without a sustainable forest."[57]

Jobs Versus Owls

> *"We cut down trees, they grow back. What's the problem?"*
> — Matt Anderson, ex-logger[58]

Forks, Washington, in the foothills of the Olympic Mountains 100 miles due east of Seattle, is the self-proclaimed "Logging Capital of the World." It is where commercial logging began in the 1860s. But in 1991 its loggers cut a third less timber than in 1990 — the lowest level in a decade. Forks has lost so many timber jobs that its unemployment rate stands at 20 percent — triple the rate of 1989. According to a front-page report in the *Wall Street Journal*, there is a hard realization that the lost timber jobs are gone forever and Forks is one of many such logging towns headed the way of textile, mining, and steel towns.

Defining the causes of this decline is a matter of bitter contention. Many people agree that there is a complexity of factors at work, including economic change, automation, the recession, and the dwindling number of trees. But in the minds of most of the 2,500 inhabitants of Forks, the underlying reason is owls. In the window of a now defunct clothing store there is a drawing by the seven-year-old son of a logger with the caption, "An owl needs 2,000 acres to live. Why can't I have a room to live."[59]

Environmentalists view the problem differently: "Charred clearcuts the size of a dozen football fields stretch across the flanks of mountains... down the side of one mountain runs the scar of a landslide hundreds of yards long, brought on by rains after the trees were stripped. The slide has clogged Pistol Creek, once a fine little trout stream, with mud and stumps. Trout don't live there anymore."[60] The aim of the environmentalists is not simply to save an endangered species but to preserve the habitat, ancient trees, and vital watercourses that compose the ecosystems of which the owl is a part.

How many jobs are at stake? The actual numbers are hard to calculate. The Northwest Forest Resource Council, a timber industry lobby group, claims that the listing of the spotted owl as an endangered species will eliminate

130,000 wood products industry and service jobs.[61] On the other hand, conservationists scoff at these estimates, emphasizing that today's highly automated lumber mills employ 16,800 fewer workers than in 1978 and that productivity improvements will eliminate nearly 25,000 jobs in Oregon and Washington by the year 2030.[62] According to the National Wildlife Federation and the Wilderness Society, declining timber harvests will cause only 8,200 lost jobs by the year 2030, and old-growth protection will cause only 2,300 jobs to be lost by that time.[63] In January 1992 the Fish and Wildlife Service put the number of jobs that would be lost in the effort to save the spotted owl at 33,000, although it said many of those jobs would have been lost anyway due to market forces.[64] In *its* plan to save the spotted owl, the Forest Service put the job loss at 20,700. Regardless of which plan is accepted by the secretary of agriculture, neither the forestry products industry nor the environmental movement is likely to be appeased. "We do not feel the science is there to warrant taking such drastic action," said Chris West, vice president of the Northwest Forestry Association, an industry group based in Portland, Oregon. "We don't think the owl is threatened, we think environmentalists are putting us out of business."[65] At least one government agency seems to support the industry position. If logging cannot proceed in western Oregon, stated Robert W. Nesbit, a lawyer with the Bureau of Land Management, "It is far more likely to result in homeless people than homeless owls."[66]

The Fish and Wildlife Service, which has acted as protector of the owl, disagrees with the BLM, maintaining that because the demand for timber is down, no jobs would be lost by not cutting the old-growth trees. Environmentalists argue that logging accounts for only 2 percent of employment in Washington and 6 percent in Oregon and that the proposed reductions would have little impact. On the other hand, some state officials in Washington and Oregon say that the 50 percent cut advocated by a panel of federal biologists could lead to record unemployment rates in small towns like Forks that are dependent on national-forest logging for their economic well-being. Conservationists and many others believe that these towns need to move from dependence on logging to other sources of income such as increased tourism and the development of new industries. The present crisis, they say, is the result of years of overcutting by excessively greedy forest products interests. "We're talking about an industry with record harvests, record profits, record exports, at a time when workers' wages have been rolled back and jobs lost to automation," said Larry Tuttle, Oregon representative of the Wilderness Society.[67]

The northern spotted owl is only one of several indicator species whose decline warns of a greater threat — that North America's few remaining stands of old-growth forests are in danger of disappearing as distinct and valuable ecosystems. In addition to campaigning on behalf of the owl, conservationists are demanding protection for the red-cockaded woodpecker, a striking zebra-backed bird that lives in loblolly and longleaf pines from Texas to Virginia; the northern goshawk in Arizona and New Mexico; the grizzly bear in the northern Rockies; and other species threatened by logging across a wide swath of the national forests. As the battle to save the ancient forests is joined in the courtrooms and corridors of government, victories are won and lost on both sides.

In 1990, after years of campaigning by conservationists, Congress passed the Tongass Timber Reform Act, which closed off almost a quarter of that forest's old growth. But according to critics, the Forest Service is ignoring the law and preparing to sell Tongass timber even faster than before. And in August 1991 a federal judge refused to issue a restraining order to halt the cutting of old-growth ponderosa pines in six national forests of the Southwest, which environmentalists claimed was endangering the northern goshawk. What is needed, they claim, is for the government to identify ecologically significant old-growth forests on a nationwide basis and to protect these areas from logging and other kinds of environmentally disruptive exploitation. This calls for legislation such as the Ancient Forest Protection Act that would ban money-losing federal timber sales and force the Forest Service to act in a more accountable way environmentally.*

In April 1990 a federal task force determined that the spotted owl was imperiled and that 3.8 million acres of its habitat on federal land should be protected. Shortly after that, the Fish and Wildlife Service proposed logging restrictions on 11.6 million acres, but the Bush administration delayed acting on the plan, convening a special panel known as the God Squad and empowered to overrule the Endangered Species Act.† The panel, chaired by the secretary of the interior, has the authority to allow a species to become extinct if saving it creates significant regional or national hardship.

In February 1992 in Portland, Oregon, U.S. District judge Helen Frye issued a preliminary injunction blocking all logging in old-growth forests on BLM-managed land, including forty-four timber sales being considered for exemption by the God Squad. At the same time the secretary of the interior, in an action denounced by conservation groups as undercutting the federal government's commitment to save the spotted owl, announced the creation of yet another task force to develop ways to save the owl while limiting job losses. "Every time they've come up with a situation where they have had to follow the law, they've refused to and either violated it or tried to circumvent it," commented Larry Tuttle, director of the Wilderness Society's Portland office.[68]

New Forestry

In the light of mounting evidence that excessive logging is causing irreparable damage to forest ecosystems, new management approaches are being considered. These include a technique being tested by a team of botanists at the University of Wisconsin based on preserving large and continuous parcels of land to protect fast-disappearing ecological communities.‡ New forestry is based on the growing knowledge of how forest ecosystems evolve over millions of years and how they recover from disturbances such as hurricanes or

* Such legislation has been proposed by Representative Jim Jontz (D-Indiana) and other legislators. For further information, see "Pending Legislation," pages 357-363.
† For more on ESA, see chapter 9, pages 286-296.
‡ For further information on this proposal, see *Design for a Livable Planet*, page 118.

fires. "Most forests are driven by disturbance," said David Perry, professor of ecology and forestry at Oregon State University. "So the idea of disturbing a forest to take out commodities is not necessarily a contradiction in terms of maintaining a natural forest."[69] In line with this thinking, loggers should mimic natural disturbances as they cut, leaving behind 20 to 70 percent of the living trees of various species (including some of the largest specimens that could otherwise yield enough lumber to build an average house) along with standing dead trees and downed logs on the forest floor. This contradicts the current logging practice of clear-cutting large forest tracts, which is invariably followed by burning and large-scale application of herbicides to prepare the site for the monocultivation of tree species selected only for their economic value and planted in straight rows for ease in future cutting.

> *"A little chaos can be a wonderful thing."*
> — Jerry Franklin, professor of ecosystems studies, University of Washington

Eco-Logs

Our dictionary describes a log as an unhewn length of the trunk or large limb of a felled tree, but it gives no hint of the role logs play in the forest ecology. Recent research, however, shows that logs are not isolated elements in a woodland but *active* participants in its many functions. In the humid Northwest forests, large logs, taking as much as five hundred years to decay, serve as germination sites for conifers, shrubs, ferns, mosses, lichens, fungi, and other spore-bearing plants. When a tree falls down, its status as a habitat changes dramatically. It becomes host for an army of bacteria, yeasts, and fungi accompanied by carpenter ants, mites, and other insects. Spiders, centipedes, and salamanders prey on the insects. Worms, millipedes, and beetles devour the remains. If the log lies on a slope, it traps plant matter and soil, both ideal for burrowing creatures whose digging and defecation help plants become established. During this process some plant seedlings work their way through the insect tunnels into the log itself.

Certain logs, called "nurse logs," develop lush garden communities of their own. Absorbing large amounts of water, logs cycle energy and nutrients that otherwise might be washed away by rain or streams. These resources, which include significant amounts of nitrogen fixed by bacteria, are stored and released slowly into the forest system or saved for times of major disturbance. When the nitrogen bound up in leaves and branches is vaporized by fire, it is replenished from a supply held by the logs. Logs also influence what the *Audubon Nature Guide* calls the "sociology" of the forest by channeling wildlife traffic — offering pathways for smaller creatures and blocking them for larger ones. When they fall across streams, trees create pools where organic debris is deposited, forming the basis of the food web as they support the microorganisms that in turn support fish and other vertebrates. Log dams create habitat diversity and thus increase the numbers of water and land creatures along stream corridors.

A key to new forestry is that after fires and other natural disturbances certain **remnant plants** survive as threads of continuity for forest recovery. The remnants include a few live trees, other trees that are dead but still standing, and a wealth of logs and other debris on the forest floor. These biological legacies, as Franklin and his colleagues call them, provide the shade and nutrition required by seedlings, help control erosion, and provide habitat for creatures that are vital links in forest regeneration. Experiments along these lines have begun in the Siskiyou National Forest in southern Oregon and the Willamette National Forest just to the north. Dr. Franklin acknowledges that initially a new-forestry site might look even worse than a conventional clearcut, adding, however, that biologically "a little chaos can be a wonderful thing."[70] A cautionary view is expressed by James Montieth, former director of the Oregon Natural Resources Council, a Portland-based environmental group: "New Forestry would be great if it were coupled with a reduction of the cut so we reach a place that is really sustainable.... We're concerned that it not become a placebo for the real problem, which is, we're just cutting too fast in the national forests."[71]

The Eastern Forests

Ranging 2,000 miles from north to south, the forests of eastern North America extend from the southern shores of Hudson Bay to Florida. From subarctic to subtropics, they form part of many forested ecosystems from mountaintop to seashore. As we have noted, the most northerly of the eastern forests is the coniferous or boreal forest biome, stretching in an unbroken belt from the northern Atlantic coast to Alaska. In southern Canada and the northern U.S. the boreal forest mixes with the deciduous to form a **transition** forest — a 150-mile-wide belt reaching from the Maritime Provinces to Minnesota. Much of this region is covered with a system of large lakes, but in Quebec, New England, and New York, it reaches high elevations of mountain landscapes. The dominant tree in the transition zone is the sugar maple, crowding out fir, spruce, pine, and other conifers but codominant with such deciduous species as beech, birch, poplar, and basswood. Spreading from the Great Lakes region to the edge of the Deep South, a distance of some 500 miles, is the **mixed deciduous** forest biome, in which the sugar maple is joined by dozens of other species, including oak, hickory, sycamore, yellow birch, and, farther south, magnolia.

Along the southern and eastern borders of the mixed deciduous forests on the coastal plains of the Atlantic and the Gulf stretch the southern pinelands, large expanses of longleaf, loblolly, and slash pines whose crowns are singed by the periodic fires that hold back competing oak, hickory, and other deciduous trees. Pines survive all but the hottest fires, because their bark is well insulated from the heat. Adapted to grow best in open sunlight, they (as well as aspen in the north) spring up before other tree species do after fire. On the Florida coast and in parts of southern Texas, Mississippi, and Louisiana, temperate and subtropical forests grow together. From Lake Okeechobee south to Key West, the vegetation

becomes almost exclusively subtropical, with "islands" and hummocks of tropical hardwood species.*

Species Richness

The Eastern deciduous forest is made up of many microhabitats — slopes facing south for species that need the sun; ravines; and, facing north for organisms requiring lower temperatures, open areas, ponds, streams, ridges, and mountaintops. In this complex environment, trees compete for space to push their broad leaves toward the sun's light. This vigorous growth results in a cornucopia for numerous organisms coexisting throughout the forest. All of this action, including both production and consumption of food, requires huge amounts of energy. One scientific estimate is that the total energy used by plants and animals on an acre of beech-maple forest every year is equivalent to that needed to supply an average home with electricity for nearly fifty years.[72] Because the climate in this region is varied from winter cold to summer heat and because the deciduous forest is well provided with food and shelter, it represents a biological crossroads, with many species of insects, birds, and animals passing through the forest from one habitat to another. Among the long-distance travelers are migratory birds including the spectacular broad-winged hawk, which winters mainly in South and Central America and summers in the eastern U.S. and southern Canada. Some creatures — gray squirrels, for example — have shorter ranges. Others adapt to temperature extremes by burrowing underground. These include raccoons, red foxes, Eastern cottontail rabbits, and the ubiquitous woodchucks. According to one estimate, woodchucks move more than 1.5 million tons of New York State soil to the surface every year.[73] (Quite a bit of this was from our own garden!) In California my coauthor and other gardeners wage a neverending war against pocket gophers.

As a result of logging and other human activities, the once abundant eastern elk and bison have disappeared, but in areas where these activities are restricted, as in Virginia's Shenandoah National Park, black bear are making a comeback, finding larger trees to den in than when the mountains were newly logged. With more than 100 million people living within a day's drive of the eastern forests of Canada and the U.S., the human impact on wildlife is enormous. This is due primarily to the steady encroachment of agriculture and urban development, removing much of the natural habitat, the slaughter of untold numbers of animals, inadvertently by automobiles and deliberately by hunting, and the indirect effects of air and water pollution, acid rain, and global warming.

The Northern Forest

Stretching from the Great Lakes to the Atlantic Ocean, the 26-million-acre (60,425-square-mile) Northern Forest is one of the largest tracts of continuously forested land in North America. Although substantially changed since the advent of the first Euro-Americans, the forest retains many features that

* See the sections on the Everglades and Okefenokee Swamp in chapter 2, page 64 and pages 71-72.

were there when the newcomers arrived. Much of its immense area is still undeveloped, although it is bounded both north and south by dense human populations who depend on it for open space and threaten to annihilate the breathing room it provides them. The Northern Forest is first and foremost a biological resource. The transitional spruce-fir and hardwood forests are unique ecosystems. It is a land covered by spruce-fir thickets and hardwood stands, interspersed with wetlands, bogs, ridges, and windblown mountain peaks. Carved up by numerous lakes and rivers — the Hudson, Mohawk, Connecticut, Kennebec, Penobscot, and Saint John have their headwaters there — it contains a remarkable complement of wildlife species that have adapted to these conditions over millennia.

The soil and climate of the region make it especially suitable for growing trees. Already supplying a significant share of the nation's timber and firewood, the importance of this resource will grow as the era of oil and nuclear power diminishes. Geographically, the North Forest system extends into the Canadian provinces of Ontario, Quebec, and New Brunswick. Eleven million acres of it, including Lake Champlain and much of the Adirondacks, are designated as a United Nations' International Biosphere Reserve. But what distinguishes the North Forest above all is that it is 84 percent privately owned, a far greater percentage than in other forested areas of the nation.

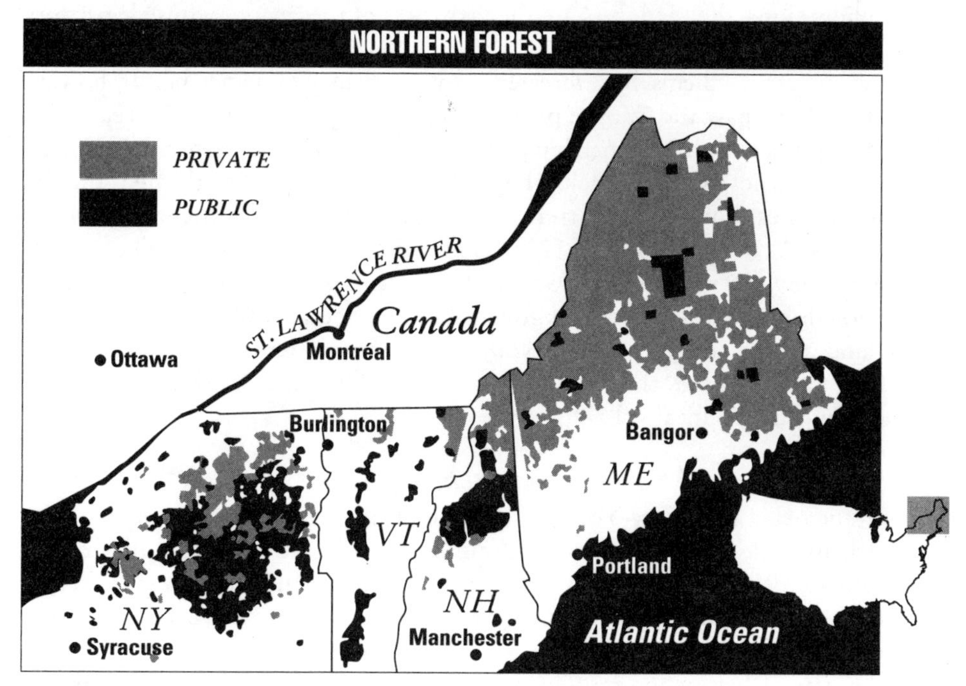

Whose Land Is It, Anyway?

Uses of privately owned forestland depend on who the owner is. On 60 percent of the private land — more than 13 million acres in some tracts larger than 5,000 acres — commercial forestry is the dominant use. Most of the commercial forest is owned by fewer than fifty companies and their families. Industrial landowners own wood-processing facilities and forestland totaling

about 10 million acres. Ninety percent of these owners are paper manufacturers; the others are companies with sawmills or specialty-product plants. Several of the paper companies also own sawmills.

The environmental record of the timber and paper companies is, to put it bluntly, appalling. Many if not most of the lakes and rivers of the region are or have been heavily polluted for over a century by wastes from these industries. In 1992 Maine's Penobscot River was listed by American Rivers as one of the United States' ten most endangered rivers. Bad as the record is, even greater ecological damage is likely to result from changes in land ownership. Depressed economic conditions have made it increasingly difficult for landowners to hold on to lands that had been generally open to hunting, fishing, and other public uses. The boom of the 1980s and the extension of interstate highways into the northern forests exerted pressure for the development of recreational housing and facilities, especially on land near water or with scenic views. This pushed up the price of forestland to the point where holding parcels for traditional uses put owners at a financial disadvantage. The process of change was intensified in 1988 by the divestiture of 1.5 million acres of forestland owned by the Diamond International Company — the folks who brought us, among other benefits, the wooden safety match.

As speculators make huge profits on the sale and resale of the land for subdivisions, the big losers are the taxpayers of the region, whose legislators belatedly bought a meager 60,000 acres of the land for public use, and the forest ecosystems themselves, threatened by the change in ownership. If, however, as Foreman and Wolke propose in *The Big Outside*, such areas were transferred to public ownership and closed to logging and motorized vehicles, the wilderness would quickly begin to reestablish itself with the return of wolverines, lynx, pine marten, caribou, and other wilderness-dependent species. At the 1989 price of $200/acre for timber company land, all of northern Maine's 10 million acres could be bought for $2 billion. There is more than enough money in the federal government's Land and Water Conservation Fund to buy land in northern New England for wilderness restoration and a new national park, they maintain, adding, "All that is lacking are popular demand and political will."[74]

Adirondacks at Risk[75]

In the belief that New York's economic future depended on preserving the Adirondack watershed and maintaining the navigational reliability of the Hudson River, the state in 1885 created the Adirondack Forest Preserve, declaring it would remain as "forests forever wild." Seven years later it was reconstituted into Adirondack Park, which now occupies a region larger than the state of Massachusetts. The park is a complex mixture of publicly owned wilderness consisting of alpine mountains, forests, lakes, river valleys, wetlands, and privately owned forests and farmlands. Nowhere else in the eastern United States does such a large area exist in a natural state, nor is there a region so rich in diversity of native species, including 90 percent of all the plant and animal species of the Northeast. Of the park's 6 million acres, 3.7 million are owned privately. Providing a year-round home for 150,000 residents, the region is visited annually by nine million hunters, campers, hikers, anglers, boaters, skiers, and tourists, all of whom have interests that, combined with those of developers and conservationists, make the park an arena of widely conflicting concerns and objectives.

"Last Clear Chance"

In May 1990 a commission appointed by New York governor Mario Cuomo recognized that the park was the victim of "an era of unbridled land speculation and unwarranted development that may threaten the unique open space and wilderness character of the region."[77] Asserting that the state faced a "last clear chance" to preserve this unique natural resource, the commission urged the state to buy more than 650,000 acres — more than 1,000 square miles — as wilderness and to impose strong rules to limit development elsewhere in the park. While environmentalists generally embraced its recommendations, many residents of the 104 communities within the park criticized the report for ignoring their economic concerns. Protection of Adirondack land was set back in November 1990 when a New York State bond proposal that would have raised nearly $1 billion for such purposes was narrowly defeated at the polls. Within three months two large parcels (amounting to more than 100,000 acres) of the land the state wants to preserve as wilderness were put up for sale by private developers. The potential sale for "acreage homesites" put further pressure on the state to include funds for the land in its budget, but the economy-minded governor and legislature deferred taking action on this critical issue.[78] While the political wrangling continues, the lake, stream, and forest ecosystems of the Adirondacks and other Eastern forests are imperiled by a threat less visible than development but potentially as harmful.

> *"My acreage contains a minimum of 156 miles of public road frontage, 40 miles of river frontage, 225 acres of ponds and several majestic waterfalls."*
> — Land speculator Henry Lassiter describing his Adirondack properties to potential buyers[76]

Ill Winds

Because so many factors affect forest ecosystems, it is not always simple to determine which ones are responsible for killing large numbers of trees. However, we know from a decade of research in Germany that air pollution in one form or another is a leading cause of *waldsterben* — forest death. Polluted air can injure vegetation either acutely, as it does in areas around fossil fuel–burning power stations and smelters, or gradually, at longer distance and in diluted concentrations. In recent years very tall smokestacks have been built to reduce local concentrations of pollutants. This technology along with greatly increased vehicle use and urbanization has created a problem of chronic, lower-level exposures often tens or even hundreds of miles away from pollutant sources. Thus scientists are now identifying many ways by which air pollution can contribute to ecosystem injury or death. Needle damage to ponderosa pines was first observed in the 1950s in California's San

Bernardino Mountains. Growth reduction in this tree and white fir, California black oak, and Jeffrey pine was also noted. The cause was eventually identified as ozone, produced mainly by automobile and truck exhausts, from the nearby Los Angeles Basin 75 miles to the west. Ozone-injured trees were also found to be more susceptible than healthy ones to insect attack and root rot.[79]

In the late 1980s several major studies confirmed what had long been suspected: Acid rain was turning many lakes and streams, especially in the northeastern United States and eastern Canada, so acidic that aquatic life could not survive in them. The pollution was also affecting forests in the Northeast, notably the red spruce, which accounts for a quarter of all the trees on the topmost slopes of eastern mountains.* A decline in growth of red spruce, sugar maple, and Fraser fir has been evident since the early 1960s. Frequent bathings in acid mist, clouds, and fog have made these trees more susceptible to the stresses of winter and the drying effect of winds. In a five-year study done at Cornell University, scientists found that red spruce needles are damaged by acidic clouds and mist, impairing photosynthesis. The research also found that the acid in the mist upsets the tree's nutritional balance. "At high concentrations it kills the needles outright," said Jay S. Jacobson, director of the experiment.[80] Even though the red spruce is the most severely affected, the Fraser fir is also succumbing in large numbers.

High Signs

In the high-elevation environments of the Appalachian Mountains where red spruce and Fraser fir are declining, average ozone levels are twice those at nearby lower elevations. At some high-elevation sites, forests have been declining for more than twenty-five years; at others the damage is more recent. What they have in common are higher concentrations of ozone, which forms when hydrocarbons and nitrogen oxides (produced from burning fossil fuels) undergo photochemical changes in the air. Acid deposition is also high in mountainous sites, especially in the eastern United States, where forests may be covered in fogs and clouds with the pH of vinegar for more than one hundred days a year.

Mixed Signals

In Southern pine forests, which are one of the nation's most valuable commercial stands, the federal Forest Response Program found that ozone reduced growth and accelerated the aging of needles in some types of pine seedlings. Subsequent research revealed that ozone penetrates the pores of leaves and needles, damaging the membranes of cells that contain chlorophyll, leading to reduced levels of photosynthesis and thus inhibiting growth.

* Acid rain is a catchall term for rain, snow, sleet, mist, fog, and clouds containing sulphur dioxide and nitrogen oxides produced mainly by the burning of coal, oil, and gasoline. Natural causes such as prolonged drought also produce symptoms similar to those of acid-rain damage. For an in-depth examination of the causes and effects of acid precipitation, see *Design for a Livable Planet*, pages 96-111.

Studies at Cornell indicate that ozone interferes with winter hardening of red spruce, making them more vulnerable to extreme cold.[81] It is also believed that ozone combined with drought, root diseases, and other natural stresses may have retarded growth of pines in Georgia, North Carolina, and South Carolina.

Perhaps even more of a threat to forests than direct damage to foliage is the *indirect* effect of acid deposition on the soil. As you can see in the accompanying illustration, it can: leach vital nutrients out of the soil, replacing them with hydrogen ions and acidifying them in the process; release aluminum ions from minerals in the soil, damaging the fine roots of trees and blocking the uptake of magnesium and calcium; produce high levels of aluminum that impede the flow of water within the tree, increasing its sensitivity to drought; through its nitric-acid component, overload forest ecosystems with nitrogen. Excessive nitrogen can also make a tree more liable to freeze in winter or dry out in summer.[82]

AIR POLLUTION ATTACKS A TREE

Based on an illustration by Gardiner Morse, with permission.

Deadly Invaders

As we emphasize throughout this book, the balance of forces that create and sustain an ecosystem are often precarious. A slight change in annual rainfall may lead to the destruction of a tree species that has dominated a woodland for hundreds of years. The disappearance of a certain species of bird can allow harmful insects to devastate a forest. In the pages above we have witnessed the devastation brought about by ax blades, chainsaws, combustion of fossil fuels, and other forms of commercial activity. We now look at what can (and did) happen when human action *unwittingly* lets loose a deadly natural predator. At the turn of the century there were millions of robust, towering American chestnut trees in the United States and Canada, providing ample shade in summer, excellent timber, and tasty nuts. In 1904 groundkeepers at New York City's Bronx Botanical Garden observed that the chestnut trees looked unhealthy. Within nine months the trees had died. Four years later most of the chestnut trees in New York State were gone, as the disease spread westward. The cause was eventually identified as a fungus imported accidentally from the Far East by a ship docking at the Port of New York. By the 1930s the deadly blight had killed virtually all of the mature chestnut trees in the United States. Similar devastation occurred when Dutch elm disease, carried from tree to tree by elm bark beetles, wiped out most of the elms in North America. In another case, that of the gypsy moth, the insect was imported *intentionally* by an entomologist in Massachusetts who wanted to crossbreed it with a silkworm to create a new silk-producing species. But after a few of them were blown out of their cage by a gust of wind, they bred prolifically and are still defoliating millions of trees in the Northeast more than a century after their escape. In this age of frequent flyers, more and more importations of "exotic" predators such as the Mediterranean fruit fly and the zebra mussel are to be expected.*

El Yunque Rain Forest

Located in the Caribbean National Forest in the northeastern corner of Puerto Rico, El Yunque is decidedly damp. It rains 350 days out of 365, with a total annual rainfall of 200 inches. When you climb its steep terrain, as I did a few years ago, you pass through three distinct plant communities. The forest that extends from the hot agricultural lowlands up to about 2,000 feet is the tropical rain forest proper; it is known as the **tabanuco** forest because it is dominated by the white-trunked tree of that name, a member of the family group that includes the trees of biblical fame that produce myrrh and frankincense, the elephant tree of southern Arizona and Baja California, and the gumbo-limbo of south Florida. Between 2,000 and 2,500 feet it gets noticeably cooler as you climb through a heavy foglike cloud in what botanists call a **montane** thicket, characterized by **palo colorado** trees and occasional stands of **sierra palms**. Then, on the highest ridges, above 2,500 feet, you come upon a misty, windswept landscape crowded with thickly growing vegetation and gnarled, dwarfed trees, known as the elfin forest.[83]

* For a report on the invasion of the zebra mussel, see chapter 1, pages 36-37.

The trees in El Yunque's lower forest form a dense canopy, heavily shading the sparse undergrowth and forcing vines to grow rapidly upward as they seek the light of the sun they need to produce leaf, flower, and fruit. The tabanuco trees, which form a canopy 80 to 90 feet above the forest floor, have slender trunks with broad, buttresslike bases. Their root systems develop mostly over the ground and are usually covered by fallen leaves, which are rapidly broken down by microorganisms and fungi. These nutrients, and others released by decomposing bedrock, are absorbed by feeder roots in the soil and are replenished when leaves and other parts of plants fall. In this way, vegetation in the tropical rain forest thrives, even when soil fertility is low.

Perhaps the most striking feature of El Yunque is the great diversity of its trees. In addition to the dominant tabanuco, more than 165 other species have been recorded, with 33 different kinds of trees appearing in any given acre. This variety is unmatched by any forest in the mainland United States.[84] Sadly, El Yunque is all that remains of a much larger rain forest that covered much of Puerto Rico's mountains one hundred years ago. Today agriculture and heavy logging have reduced it to a 10,000-acre preserve that makes up a third of the Caribbean National Forest.

What You Can Do to Help Save the Forests

1. Write to the U.S. President and tell him that you want his administration to work more actively to protect the nation's forests rather than the interests of timber-related industries.

2. Encourage your congressional representatives to work for legislation that will strengthen existing laws to protect the forest systems, such as the Forest Biodiversity and Clearcutting Prohibition Act (H.R. 1969), and to press for the removal of federal subsidies to the timber-related industries.

3. Urge your members of Congress to oppose appropriations for Forest Service road building. Urge them to support wilderness designation for all remaining roadless areas (some 60 million acres) and support legislation outlawing the timber "Purchaser Credit Program."

4. Persuade your legislators to work more aggressively to pass measures limiting the combustion of fossil fuels in order to reduce acid rain and global warming.

5. Get a copy of the *Citizen Action Guide* published by Save America's Forests, 4 Library Court, SE, Washington, DC, 20003; (202) 544-9219. It shows how you can work effectively for forest protection by educating yourself, shaping law through citizen action, and working with groups to build coalitions; it also provides a wealth of information on forest-saving skills, legislative summaries, and other valuable resources.

6. Support one or more of the following:

— *The Adirondack Council*, Church Street, P.O. Box D-2, Elizabethtown, NY 12932; (518) 873-2240. Its member organizations include the Association for the Protection of the Adirondacks, NRDC, the Wilderness Society, and the National Parks and Conservation Association.

— *The Association of Forest Service Employees for Environmental Ethics*, P.O. Box 11615, Eugene, OR 97440; (503) 484-2692. Their quarterly *Inner Voice* is

one of the best publications on forest preservation and management.

— *Forest Reform Network*, 5934 Royal Lane, Suite 223, Dallas, TX 75230; (214) 352-8370.

— *Friends of the Ancient Forest*, P.O. Box 995, Three Rivers, CA 93271-0995. Their "Adopt-a-Sequoia" campaign is designed to raise funds for defense of the sequoia tree.

— *Greenpeace Canada National Forest Campaign*, 1726 Commercial Drive, Vancouver, BC V5N 4A3; (604) 253-7701.

— *The Native Forest Council*, P.O. Box 2171, Eugene, OR 97402; (503) 688-2600.

— *Northern Forest Lands Council*, 54 Portsmouth Street, Concord, NH 03301; (603) 224-6590. Established under a congressional mandate in 1991 by the governors of Maine, New Hampshire, Vermont, and New York with representation from the U.S. Forest Service, landowners, and environmental organizations.

— *Siskiyou Regional Educational Project*, P.O. Box 220, Cave Junction, OR 97523; (503) 592-4459. Opposes U.S. Forest Service plans to log the largest intact coastal forest in the U.S. outside of Alaska.

7. Support the following national organizations (for details see the Directory, pages 331-355): American Forestry Association; American Wildlands; Earth Island Institute; Environmental Defense Fund; Greenpeace U.S.A.; Lighthawk; National Audubon Society; National Wildlife Federation; Natural Resources Defense Council (ask about its Rescue the American Rainforests campaign); the Nature Conservancy; Save-the-Redwoods League; Rainforest Action Network; Rainforest Alliance; Sierra Club; and the Wilderness Society.

8. At home and work, vigorously support the use of recycled paper. This can save enormous numbers of trees from being cut down.

9. Wherever possible conserve energy and avoid the use of fossil fuels to cut down on air pollution, acid rain, and global warming.

10. Get involved in tree-growing projects. For sixteen ways you can do this, see *Design for a Livable Planet*, pages 128-130.

11. Beg, borrow, steal, or preferably buy Lou Gold's videotape (see page 186 for details).

RESOURCES

1. U.S. Government

The leading government departments and agencies dealing with forests are: the Forest Service, which comes under the jurisdiction of the Department of Agriculture; the Bureau of Land Management; the Fish and Wildlife Service; and the National Park Service, all of which come under the Department of the Interior. For addresses, phone numbers, and other details, see the Directory, pages 352-354.

2. Environmental and Conservation Organizations

See the Directory.

3. Further Reading

— *Ancient Forests of the Pacific Northwest*, Elliot A. Norse. Washington, D.C.: Island Press, 1990.

— *California Redwood Parks and Preserves*, John B. DeWitt. San Francisco: Save-the Redwoods League, 1985.

— *A Conservation Strategy for the Northern Spotted Owl*, Jack Ward Thomas et al. Portland, Ore.: U.S. Department of Agriculture, 1990.
— *The Earth First! Reader*, edited by John Davis. Salt Lake City: Peregrine Smith, 1991.
— *Eastern Forests*, Ann Sutton and Myron Sutton. Audubon Nature Guides. New York: Knopf, 1988.

— *A Field Guide to Trees and Shrubs*, George A. Petrides. Peterson Field Guide series. Boston: Houghton Mifflin, 1976.
— *Forest*, Jake Page. Alexandria, Va.: Time-Life Books, 1983.
— *The Forest and the Sea*, Marston Bates. New York: Vintage, 1960.
— *Forest Ecology*, Stephen H. Spurr and Burton V. Barnes. New York: John Wiley, 1980.
— *Forest Primeval: The Natural History of an Ancient Forest*, Chris Maser. San Francisco: Sierra Club, 1989.
— *Fragile Majesty: The Battle for North America's Last Great Forest*, Keith Ervin. Seattle: The Mountaineers, 1989.
— *The Fragmented Forest*, Larry D. Harris. Chicago: University of Chicago Press, 1984.
— "Green Giants," Doug Stewart. *Discover*, April 1990.
— *Last Stand: Logging, Journalism, and the Case for Humility*, Richard Manning. Salt Lake City: Peregrine Smith, 1991.
— *The Life of the Forest*, Jack McCormick. New York: McGraw-Hill, 1966.
— *Mountain Treasures at Risk*. Washington, D.C.: Wilderness Society, 1989.
— *New Directions for the Forest Service: Wilderness Society Recommendations*. Washington, D.C., February 1989.
— *Northern Forest Lands Study*, Stephen C. Harper, Laura L. Falk, and Edward W. Rankin. Rutland, Vt.: U.S. Department of Agriculture, 1990.
— *The Redesigned Forest*, Chris Maser. San Pedro, Calif.: R & E Miles, 1988.
— *The Redwoods*, Kramer Adams. New York: Popular Library, 1966.
— *Redwood National Park*, Edwin C. Bears. Washington, D.C.: National Park Service, 1969.
— *Secrets of the Old-Growth Forest*, David Kelly and Gary Braasch. Layton, Utah: Peregrine Smith, 1987.
— *This Well-Wooded Land: Americans and Their Forests from Colonial Times to the Present*, Thomas R. Cox. Lincoln: University of Nebraska Press, 1985.
— *The Tongass: Alaska's Vanishing Rain Forest*, Robert Glenn Ketchum and Carey D. Ketchum. New York: Farrar, Straus and Giroux, 1987.
— *The Wasting of the Forest*, Wilderness Society report. Washington, D.C., undated.
— *Western Forests*, Stephen Whitney. Audubon Society Nature Guides series. New York: Knopf, 1985.

NOTES

1. This concept is inspired by chapter 5 of Paul Colinvaux's *Why Big Fierce Animals Are Rare* (Princeton, N.J.: Princeton University, 1978).
2. Cited in Jake Page, *Forest* (Alexandria, Va.: Time-Life Books, 1983), p. 6.
3. Colinvaux, *Big Fierce Animals*, p. 46.
4. Ibid., p. 48.
5. Robert Leo Smith, *Elements of Ecology and Field Biology* (New York: Harper & Row, 1977), p. 183.
6. Colinvaux, *Big Fierce Animals*, p. 119.
7. Lee Green, "They've Been Raping the Giant Sequoias," *Audubon*, May 1990.
8. Page, *Forest*, p. 105.
9. Smith, *Elements of Ecology*, p. 445.
10. Ibid., p. 447.
11. Catherine Caufield, "The Ancient Forest," *The New Yorker*, May 14, 1990.
12. Ibid.
13. Based on Larry Stone, "Message from the Mountain," *Des Moines Sunday Register*, April 28, 1991; David Foster, "Perspectives," *Columbia Daily Tribune*, September 15, 1991; Marie Reeder, "Siskiyou Merit Park Protection," *Outdoors West*, Winter 1988, p. 4; and *Bald*

Mountain Bulletin, 1991/1992. See also Jon Naar, *Design for a Livable Planet* (New York: Harper & Row, 1990), p. 118.

14. Presentation at Land, Air, Water conference, Eugene, Oregon, March 13, 1992.

15. T. H. Watkins, "The Laughing Prophet of Bald Mountain," *Orion*, Winter 1990, pp. 16-18.

16. Personal communication, March 14, 1992.

17. "View from Bald Mountain," *Bald Mountain Bulletin*, 1991/1992, p. 4.

18. Dennis Farney, "Timber Firm Stirs Ire Felling Forests Faster Than They Regenerate," *Wall Street Journal*, June 18, 1990.

19. The devastation of vast acreages of timber in Montana by Plum Creek and Champion, for example, is graphically documented in Richard Manning's *Last Stand: Logging, Journalism, and the Case for Humility* (Salt Lake City: Peregrine Smith, 1991); see also Michael D. Lemonick, "Whose Woods Are These," *Time*, December 9, 1991; and "Our Nation Is Destroying Its Last Virgin Forests," *Forest Voice*, vol. II, no. 2 (1990).

20. Background material for this section is based on the Wilderness Society's "End of the Ancient Forests" (1988), Keith Ervin, *Fragile Majesty* (Seattle: The Mountaineers, 1989), and Caufield's "The Ancient Forest."

21. The definitive description of the old-growth ecosystem is Jerry F. Franklin's *Ecological Characteristics of Old-Growth Douglas-Fir Forests*, U.S. Department of Agriculture Forest Service, General Technical Report PNW-118, 1981.

22. Janine M. Benyus, *The Field Guide to Wildlife Habitats of the Western United States* (New York: Fireside, 1989), p. 237.

23. Green, "They've Been Raping the Sequoias."

24. Cited in Douglas H. Strong, *Dreamers and Defenders: American Conservationists* (Lincoln: University of Nebraska Press, 1988), p. 71.

25. Letter by Agriculture Secretary James Wilson dated February 1, 1905, cited in Dyan Zaslowsky, *These American Lands* (New York: Henry Holt, 1986), p. 65.

26. Ibid.

27. See especially *The Wasting of the Forest: A Wilderness Society Report*; Keith Ervin, *Fragile Majesty: The Battle for North America's Last Great Forest* (Seattle: The Mountaineers, 1989); John Daniel, "The Long Dance of the Trees," *Wilderness*, Spring 1988; *Forest Voice*, vol. II, no. 2 (1990); and Timothy Egan, "Dissidents Say Forest Service Shifts Its Role," *New York Times*, March 4, 1990.

28. Caufield, "The Ancient Forest," p. 56.

29. Ervin, *Fragile Majesty*, p. 86.

30. *The End of the Ancient Forests* (Washington, D.C.: The Wilderness Society, 1988), p. i.

31. *The Wasting of the Forest*, p. 4.

32. Caufield, "The Ancient Forest," p. 56.

33. Ibid.

34. "Forest Murder: Ours and Theirs," *New York Times*, September 20, 1989.

35. George Laycock, "Trashing the Tongass," *Audubon*, November 1987, p. 124.

36. Joel Connelly, "The Big Cut," *Sierra*, May/June 1991.

37. Egan, "Struggles Over the Ancient Trees."

38. Timothy Egan, "Struggles Over the Ancient Trees Shift to British Columbia," *New York Times*, April 15, 1990.

39. Connelly, "The Big Cut."

40. Ibid.

41. Ervin, *Fragile Majesty*, p. 20.

42. Timothy Egan, "Energy Project Imperils a Rain Forest," *New York Times*, January 26, 1990.

43. Ibid.

44. John Nielsen, "Living on the Edge," *National Wildlife*, October/November 1989.

45. Edward Goldsmith et al., *Imperiled Planet* (Cambridge, Mass.: MIT Press, 1990), pp. 38-39.

46. "The Decadent Forest," *Audubon*, March 1986.

47. Much of the information in this section is based on information given in Mark Bonnett and Kurt Zimmerman's comprehensive "Politics and Preservation: The Endangered Species Act and the Northern Spotted Owl," *Ecology Law Quarterly*, vol. 18, no. 1 (1991).

48. Ibid., pp. 113-14.

49. Ibid., p. 117.
50. U.S. Fish & Wildlife Service, *The Northern Spotted Owl Status Review Supplement 2.4 (1989)*, cited in Bonnett and Zimmerman, p. 118.
51. 16 USC Section 1532(6) (1988).
52. Ibid., p. 145.
53. David Kelly and Gary Braasch, *Secrets of the Old-Growth Forest* (Layton, Utah: Peregrine Smith, 1987). See also Keith Ervin, "The Shrinking Forest of the Primeval," *Sierra*, July/August 1987, as well as Ervin's *Fragile Majesty*, passim.
54. Kelly and Braasch, *Secrets of the Old-Growth Forest.*
55. Ibid.
56. Jay Heinrichs, "There's More to Forests Than Trees," *National Wildlife*, February/March 1988.
57. Ibid.
58. Charles McCoy, "Cut Down," *Wall Street Journal*, January 6, 1992.
59. Ibid.
60. Ibid.
61. Bonnett and Zimmerman, "Politics and Preservation: The Endangered Species Act and the Northern Spotted Owl," p. 150.
62. Ibid.
63. Ibid. See also Philip Shabecoff, "Science, Politics, and Survival," *New York Times*, April 30, 1989.
64. *Ecology USA*, January 12, 1992, p. 2.
65. Timothy Egan, "U.S. Declares Owl to Be Threatened by Heavy Logging," *New York Times*, June 23, 1990.
66. Timothy Egan, "Politics Reign at Spotted Owl Hearing," *New York Times*, January 9, 1992.
67. Ibid.
68. "Lujan Creates Yet Another Panel to Study Owl-Protection/Jobs," *Ecology USA*, vol. 21, no. 4 (February 24, 1992). See also Keith Schneider, "U.S. to Push for Logging in Owl's Forest," *New York Times*, February 20, 1992.
69. Jon R. Luoma, "New Logic Approach Tries to Mimic Nature," *New York Times*, June 12, 1990.
70. Ibid.
71. Ibid.
72. Ann Sutton and Myron Sutton, *Eastern Forests*, in the Audubon Society Nature Guides series (New York: Knopf, 1988), p. 65.
73. Ibid., p. 67.
74. Dave Foreman and Howie Wolke, *The Big Outside* (Tucson: Ned Ludd Books, 1989), p. 393.
75. An invaluable source of information on the Adirondacks is *The Adirondack Park in the Twenty-First Century* (State of New York, 1990). See also "Focus on the Adirondacks," a special issue of *Sierra Atlantic*, Winter 1990.
76. Cited in an appeal letter from the Adirondack Council, April 1991.
77. Sam Howe Verhovek, "Panel Urges Huge State Purchases to Save Adirondack Wilderness," *New York Times*, May 5, 1990.
78. Sam Howe Verhovek, "'Wish List' Site for Sale But Albany Can't Buy It," *New York Times*, April 10, 1991. See also Verhovek, "Cuomo Plans to Scale Back Proposal for Adirondacks," *New York Times*, August 14, 1991.
79. James J. MacKenzie and Mohamed T. El-Ashry, *Ill Winds* (Washington, D.C.: World Resources Institute, 1988), pp. 13–14.
80. William K. Stevens, "Researchers Find Acid Rain Imperils Forests Over Time," *New York Times*, December 31, 1989.
81. James J. MacKenzie and Mohamed T. El-Ashry, "Ill Winds," *Technology Review*, April 1989, p. 68.
82. Ibid.
83. See Robert H. Mohlenbrock, "Elfin Forest, Puerto Rico," *Natural History*, December 1987, pp. 21–22.
84. Robert H. Mohlenbrock, "El Yunque Rain Forest," *Natural History*, February 1987, p. 79.

7
DESERTS

"Despite its clarity and simplicity, however, the desert wears at the same time, paradoxically, a veil of mystery. Motionless and silent, it evokes in us an elusive hint of something unknown, unknowable, about to be revealed."

— Edward Abbey, *Desert Solitaire*

How do you perceive a desert?[1] A *New Yorker* cartoon of a wasteland under a harsh sun with a bedraggled person crawling over the sand toward a shimmering mirage of water? The *Lawrence of Arabia* movie version: photogenic sand dunes with camels on the skyline and white-robed Hollywood "Bedouins"? Or now, courtesy CNN, TV images of intrepid pilots zapping Iraqi tanks into oblivion? The reality is something else. In the United States deserts cover 25 million acres or about 8 percent of the country's landmass. Located in a 500,000-square-mile area running from southeast Oregon into Mexico and surrounded by the Rocky Mountains on the east and the Sierra Nevada on the west, they are divided generally into two distinct types — cold and hot deserts.

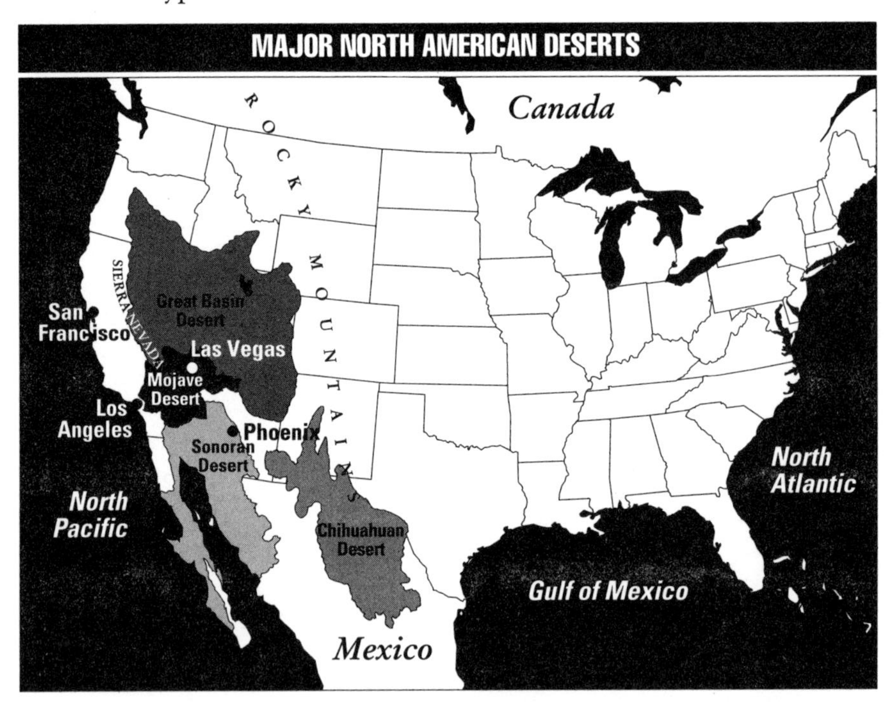

Deserts are indeed dry, defined by geographers as places where the evaporation of water is greater than the rainfall, where precipitation is less than 10 inches a year. The dryness comes from interaction of sunshine and wind, extreme temperatures, short, intense rainfalls, soil permeability, and other geologic factors. Movement of air masses over the land and sea is an important force in the formation of deserts. The zone of high pressure off the coast of California and Mexico deflects rainstorms heading south from Alaska to the east, preventing moisture from reaching the Southwest. Mountain ranges are another force. By creating a "rain shadow," the Cascade Mountains and the High Sierra intercept rain from the Pacific and help produce the dry conditions of the Great Basin Desert, which lies between the Sierra Nevada–Cascades axis on the west and the Rocky Mountains on the east. It is primarily the altitude of this region that accounts for its cool climate.*

* Many of the mountains in the Great Basin have peaks higher than 10,000 feet.

The low humidity of the desert allows as much as 90 percent of the sun's radiation to get through the atmosphere and heat the ground. At night the accumulated heat of the day quickly returns to the atmosphere with few clouds to impede it. This makes for the extreme daily temperature range common to deserts. Although they don't come often, desert rainstorms are usually very heavy, but much of the precipitation runs off the hard-baked soil. Because the water does not penetrate deep and because there is less vegetation than in temperate ecosystems, the thin desert soils lack the nutrients provided by decaying organic material in other ecosystems.

After violent storms the unprotected soil of the desert erodes easily and is further carried away by the wind. In those few places where the rain is able to penetrate, the desert may become rich in mineral nutrients carried back to the surface by evaporation of surface moisture. The permeability of the soil is determined by the nature of the subsurface rock, the amount of weathering and erosion that has occurred, and the topography of the surface itself. When the desert soil is rocky or pebbly, its coarse texture lets water flow down into plant roots, although it tends to dissipate fast. Soil composed of fine particles of sand and dust retains water long enough for plant roots to absorb it. However, when the soil surface rapidly gets wet from an intense summer thunderstorm, it quickly becomes saturated. The water doesn't penetrate into lower layers of the soil but flows along the surface, creating the typical flash flood of certain desert environments. Regardless of soil conditions, desert vegetation tends to be sparse, except at certain times of the year in the southwestern deserts, where you may come across brilliant displays of cliffrose, owl's clover, yucca, and other dormant plants that blossom extensively for a few weeks after a late-winter rain and before the onset of summer's desiccating heat.

Cold and Hot

The northern desert region — the Great Basin — is a cold desert, covering 160,000 square miles in Nevada, Oregon, Idaho, Wyoming, Arizona, and California, often referred to as the Intermountain West. The name Great Basin is misleading because the region is not one large basin but is composed of 150 basins alternating with some 160 separate mountain ranges. If you fly over it, you'll see why it is more accurately called the Basin and Range Province. The second major desert region is located in the southwestern U.S. and Mexico and is called hot, because that's what it usually is. It takes in the Mojave Desert (in southern Nevada, the southwest corner of Utah, and part of California), the Sonoran Desert* (in Arizona and California), and the Chihuahuan Desert (in southern New Mexico, western Texas, and extending into northern Mexico).

Plant and Animal Life

Because water is the key to life in the desert, many plants and animals have evolved through an ability to grow and reproduce during brief periods of rain or even of dew formation. Due to its cooler climate, the Great Basin Desert contains fewer plant species and life forms than the hot deserts to the

* Also called Colorado Desert. See page 225.

south. In the cold desert, some animal species are similar to those found in temperate grasslands — pronghorn antelope, coyote, jackrabbit, and the kangaroo rat, for example. Because of the general lack of water in the desert, the recycling of dead plant and animal matter occurs at a slower pace than in warmer biomes. Microbial decomposition is limited to short periods when moisture is available and takes place mainly through the action of termites and other scavenger insects, and of microbial recyclers.

Sagebrush Country

Whereas in the Mojave, the Sonoran, and the Chihuahuan deserts there are many types of cactus, agave, and yucca,* in the Great Basin most of the dominant plants are sagebrushes (*Artemisia*) or saltbushes (*Atriplex*). Big sagebrush takes up virtually all the water in the area covered by its roots, which often extend up to 90 feet in diameter or nine times the spread of its crown. Its leaves provide forage for a variety of wildlife species, while the crown offers good cover for them. As the name indicates, the sage grouse is particularly partial to sagebrush, getting 70 percent of its food from its leaves and buds. Mule deer, pronghorn antelope, and other small- or medium-sized mammals also eat sagebrush, but cattle don't like its taste. In general, the presence of large grazing animals increases the dominance of sagebrush, because they ignore it and feed on other plants, especially grasses, thus eliminating some potential sagebrush competitors.[3]

> *"This vast, empty quarter lost between the great dividing ranges is one of the wildest and least populated areas in the temperate Northern Hemisphere."*
> — Foreman and Wolke, *The Big Outside*[2]

When the uneaten sagebrush starts to spread, the shrubs grow thick, making it hard for sheep and cattle to squeeze through in their search for forage. Ranchers consider this kind of range "depleted," because it does not support their livestock. With financial aid from the government, private landowners and federal land managers try to "rehabilitate" the range by burning, cutting, or poisoning sagebrush, then reseeding with wheatgrasses from Russia and other countries.[4]

Traveling into the northern portions of the Great Basin, you begin to see grasses, defining the boundary between the arid deserts of Utah and Nevada and the moister, steppelike vegetation of Idaho, Oregon, and Washington.[5] At higher elevations the landscape, becoming wetter, gives way to pinyon/juniper woodland and aspen forest.

Huge Nevada, Foreman and Wolke write, is a "lean, hungry land, with little fat to make men rich," producing fewer pounds of beef than tiny Vermont, and it has more roadless areas of 100,000 acres or more than any other state except California. Despite its sparseness or perhaps because of it, the ecological Big Basin is exposed to a wide range of threats. Oil companies have begun exploring the region for oil and gas. Motorized tricycles,

* See pages 224-225.

The Great Basin Spadefoot

The spadefoot toad is a classic inhabitant of the sagebrush, perfectly adapted to escape the heat of the day and take advantage of whatever moisture it can find. It hides from the sun by digging itself underground to a level where the soil keeps its thin skin cool and moist. It shovels itself in, back-end first, by rotating its "spade," a crescent-shaped projection on the side of its hind foot. At nightfall it reverses the process, coming to the surface where it catches insects with its long, sticky tongue.[6] With the onset of the dry season adult spadefoot toads remain dormant for eight to nine months. When they detect a summer downpour, they come up to the surface, mate, and lay their eggs in puddles left by the rain. Within two weeks tadpoles appear as if by magic, providing the basis for the Navajo folklore that toads and frogs fall from the sky.

(Sources: *Field Guide to Wildlife Habitats of the Western United States; Arid Lands*)

skimobiles, and other off-road vehicles (ORVs) probe deeper into the backcountry. As in other parts of the West, the greatest threats to the Great Basin come from a combination of mining and livestock interests and from federal agencies "effectively in thrall to this rustic gentry." Destruction of wilderness in Nevada is less spectacular than that in the old-growth forests of Oregon or the grizzly habitat of Montana, state Foreman and Wolke, "but it is a slow, steady gnawing at the fabric of natural integrity like sheep grazing at a mountain meadow in July."[7] In northwestern Nevada, the Black Rock Desert's 640,000 acres, with elevations ranging from 4,000 to 8,600 feet, are under BLM jurisdiction. Inhabited by mountain lions, pronghorns, and wild horses, the desert is also the habitat of two rare plants — winged milkvetch and Barneby wild cabbage. It is threatened by ORVs, overgrazing, and some oil and gas leases.

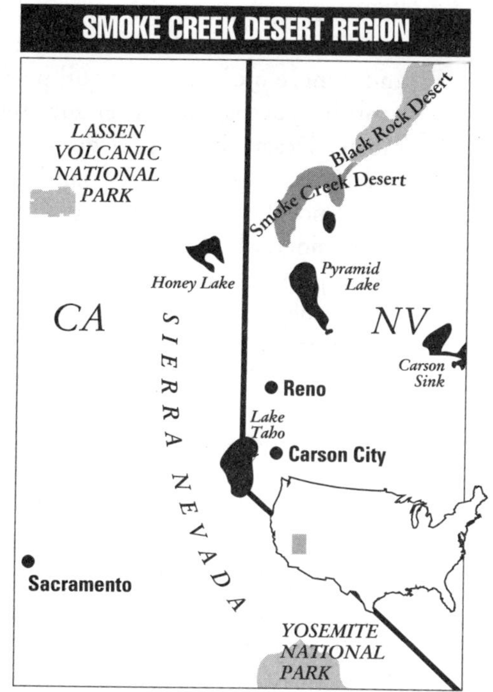

Flat Is Beautiful

Also in the northwestern part of Nevada, the Smoke Creek Desert (148,000 acres, BLM-administered) is a large, dry lake bed that serves as the basin for Smoke Creek (part of which is in California) and numerous other drainages from the Buffalo Hills and the Fox Range.The main threats to this largely untouched wild area come from dirt roads and jeep trails used by ranchers and mining prospectors and from low-level military flights.[8] As Foreman and Wolke note with a touch of irony, conservation groups have "obligingly neglected" this desert, while the BLM, "in keeping with their assumption that Wilderness

can't be flat or open, refused to even study this reservoir of tranquility for possible Wilderness recommendation." The authors make the point that wilderness is not necessarily limited to regions of spectacular grandeur and that true wilderness can often be found "in the most godforsaken zones of the planet." In this case, "flat is beautiful."[9]

The Hot Deserts

South of Lake Tahoe, Nevada shares several large roadless areas with California. They represent a gradient from the cool Great Basin to the hot Mojave Desert. By an accident of history this is a region shared also by the 1,350-square-mile Nevada Test Site, a federal bombing range and nuclear-test facility, which has provided an (inadvertent) opportunity to study the effects of radiation on the local flora and fauna. If you were allowed to wander around the test site, you would notice, along with an incredible vista of bomb craters reminiscent of a moonscape, the transition from sagebrush to creosote bush-dominated ecosystems. An indicator of this transition is the change in annual rainfall. The sagebrush communities occur when the annual rate goes above 7¼ inches. When it falls below that level, creosote bush dominates. When rainfall is just at or very close to 7¼ inches, there is a truly transitional community, a mixture of plant and animal species from northern and southern desert ecosystems sometimes found within a few yards of each other.[10]

Characteristic hot deserts are dominated by creosote and burro bushes and, where moisture and soil permit, by tall growths of acacia, saguaro, ocotillo, yuccas, and other succulents rising above the shrub level. In the Mojave Desert the sites are quite variable in composition, due to the presence of *bajadas* — deposits of sand and gravel skirting the foot of hillsides. Differences between plant communities at the bottom and top of a bajada can be considerable. Creosote bushes, with their pervasive, resinous odor, are well-equipped for survival in the desert. They are drought "tolerators" — that is, they avoid the harsh consequences of drying out by frugal living and stretching out their water savings. They have roots that extract moisture from the soil and leaves whose growth is slowed during periods of drought. Because their extensive root systems crowd out less aggressive competitors such as grass, they are, like sagebrush, considered a pest by ranchers whose cattle must range far and wide for sparse forage. However, along with another common desert plant, jojoba, creosote has excellent pharmaceutical properties as a preservative, lubricant, and, experimentally at least, as a means of controlling the growth of cancer cells.[11]

The high rocky hillsides of the southwestern deserts, where the shallow soil drains fast, are the domain of the ocotillo, or coachwhip. The plant bears this name because it consists of a cluster of slender thorny stems up to 15 feet in height that sway in the wind. For most of the year the ocotillo is bare, but it responds to spring and summer rains with a sudden bristling of tiny leaves. Sharing the same shallow soil as the ocotillo is the agave, known as the century plant because it takes many years — up to fifty but not one hundred —

before a single stalk emerges from its leaves, grows as high as 15 feet tall, and produces striking clusters of light-colored flowers. It fits well into the higher, drier altitudes of the desert, where its slow growth process allows it to be patient about rain.[12]

Cacti are another plant family that adapt well to desert conditions. They have evolved shallow roots that spread out laterally over a wide area, absorbing rainfall before the water runs off, evaporates, or sinks out of reach. In place of leaves, they have spines with succulent tissues that store large amounts of water. Unlike leaves, spines permit heat to be given off, require little energy to maintain, and expose a minimal surface area to drying winds. Giant saguaro cacti, some of which achieve a height of 50 feet and a diameter of 10 feet, are said to store 6 to 8 *tons* of water in their stems.[13] However, the claim that many desert travelers have survived for days on the water and juice sucked out from these plants appears to be more myth than reality.

Desert Survival

For desert animals survival strategies require more than simply holding out until the rain comes. They have to seek out food, avoid predators, and deal with competition from other desert creatures. Of course they must also be well adapted to conserving moisture. To save energy, many animals confine their activity to the cool desert nights. Like the spadefoot toad, some are burrowers. Scorpions, for example, have enlarged claws for digging. (If you plan to sleep out in the desert, watch out that you don't get one in your shoes. They can give you a very nasty bite!) Others — certain lizards and snakes — dive headfirst through the sand with nostrils upturned or equipped with protective valves to seek refuge from the scorching surface of the desert sand in insulated burrows. Even on the hottest day in the desert the temperature 12 inches below the surface is cool enough for many creatures to survive in comfortably. Reptiles rely on extremely tough outer coverings to reduce surface evaporation as a protection against the scorching heat. During the day, birds are the most active creatures in the desert because they can most easily avoid the hot desert surface as they search for food. They nest during the cooler rainy season when food is more abundant for their young.

The Sonoran Desert

The Sonoran Desert, which includes the extreme southern part of California, covers much of Sonora, Mexico, the peninsula of Baja California, and Arizona south of the Mogollon highlands. Although its lowest elevation is 250 feet below sea level (at Salton Sea, California), it ranges generally between 1,000 and 4,000 feet above sea level. Its southerly location and biseasonal rainy periods produce more diverse vegetation and support more cacti and other succulents than do any other North American desert ecosystems. The part of the Sonoran Desert in western Arizona and California is known as the Colorado Desert because it is located near the Colorado River, occupying its drainage and that of the Gila River as well as all of the Salton Sea basin. Receiving a scant 2 to 5 inches of rainfall throughout the year, it has less luxuriant plant life than the rest of the Sonoran. Its typical plants are smoke and elephant trees, desert iron wood, California fan palms, mesquite, and teddy-bear cholla.

The Mojave

The Mojave covers southern portions of California and Nevada, part of western Arizona, and the extreme southwestern corner of Utah. Its climate and vegetation are intermediate between the Great Basin and the Sonoran. With some elevations as high as 7,000 feet, the Mojave generally lies below 3,000 feet, and in Death Valley National Monument it has the lowest elevation in North America, 282 feet below sea level. Temperatures are extreme, ranging from 120 degrees Fahrenheit to below freezing. In the Mojave precipitation averages 8 to 12 inches a year, falling mainly in winter, with Death Valley averaging less than 2 inches annually and in some years getting none at all. Where the Mojave abuts the Great Basin, it gets several feet of winter snow.

The Chihuahuan Desert

The Chihuahuan takes up almost one-third of North America's total desert area, covering much of the Mexican Plateau and ranging from the state of San Luis Potosi in the south across the Rio Grande into the Trans-Pecos region of Texas. If you visualize this desert as a hand, three fingers protrude into New Mexico, with the middle one including the gypsum dunes of the White Sands National Monument and the tip of the thumb edging into southeastern Arizona. The Chihuahuan Desert's lowest elevation is 1,800 feet (along the Rio Grande), with shrub desert extending to 3,600 feet, above which the sotol-grassland complex goes up to 5,000 or more feet. The Chihuahuan, like the Great Basin, is basically a shrub desert, but it also supports several larger species of yucca and agaves and has more grasses in its uplands than does the Sonoran Desert.

Desert Not So Solitaire

"The desert is a vast world, as deep in its way and complex and various as the sea." — **Edward Abbey,** *Desert Solitaire.*[14] Abbey wrote these words in 1967 at Nelson's Marine Bar in Hoboken, New Jersey, accounting perhaps for the ocean analogy. He also warned that most of what he had written about was already gone or going fast. "This is not a travel guide," he added, "but an elegy. A memorial. You're holding a tombstone in your hands. A bloody rock. Don't drop it on your foot. Throw it at something big and glassy. What do you have to lose?" Sometimes viewed as 25 million acres of sand, dirt, and cactus, the California deserts contain 90 mountain ranges (including some with summit coniferous forests), more than 40 different plant communities, and 2,500 plant and animal species, many of which are rare. Although they have something in common with oceans, desert ecosystems take much longer to recover from environmental damage. As we have seen in North Africa, tracks of Rommel's tanks are still clearly etched into the landscape some fifty years after World War II's Battle of El Alamein. Closer to home, it is disquieting to discover that in the California desert there are nine U.S. military bases, 36,000 miles of roads and dirt tracks,* 12,000 miles of pipeline, and 3,500 miles of high-capacity powerline.[15] Then there is the impact of mining, cattle grazing, ORVs, 2.5 million resident people, and 20 million visitors a year.

* Compared with the *entire* U.S. Highway System's 44,000 miles of highways.

Desert soils are not very rich in organic matter. They often have lower layers composed of clays, salts, and very fine sand that are virtually cemented together by calcium carbonate, silica, or even iron compounds. The layers form hardpans that are impervious to water and to penetration by plant roots, contributing to their extreme fragility. Compared with other soils, the damage done by human activities in a desert ecosystem is usually much greater. For example, a single motorcycle traveling 1 mile over the desert can displace three-quarters of a ton of topsoil, and a larger vehicle displaces considerably more.[16] Twenty-five years ago Abbey described places where vehicles had removed all the topsoil over many acres, gouging out gullies 6 or more feet deep. The loss of this topsoil diminishes vital plant life needed by desert creatures for food and cover. Bearing in mind that it takes about a thousand years for 1 *inch* of top soil to accumulate, the prospect of restoring the ravaged terrain in the arid environments of the Southwest is remote indeed.[17]

Destroying the Tortoise and Other Desert Species

As desert habitats are increasingly fragmented by humans, they become less able to support the larger and wider-ranging species — grizzly bears, jaguars, and California condors (all extinct in deserts today), and more recently, desert tortoises — that cannot adapt rapidly to the changes, intrusions, and imbalances being created within the desert ecosystems. The case of the desert tortoise (*Gopherus agassizii*) is sadly but one example in a long process of attrition now under way. Up until the 1940s healthy desert tortoise populations were widespread in the Southwest. Today the species is in decline, concentrated in two main areas of the Mojave and Sonoran deserts that are separated geographically and genetically by the Colorado River. In 1980 a small population of the tortoise in southwestern Utah was declared threatened by the U.S. Fish and Wildlife Service (FWS), and five years later the Service declared the species threatened or endangered over its entire range but added that it was "precluded by other, higher priorities," which meant that it received very little attention from the FWS or other federal agencies. Only after a petition from the NRDC, EDF, and Defenders of Wildlife did the FWS in 1989 grant endangered species status to all wild desert tortoises north and west of the Colorado River. But even this was done on a temporary basis; the tortoises' endangered status expired in April 1990.[19]

> *"The health of the desert tortoise reflects the health of the ecosystem it lives in. Without sufficient undisturbed habitat, a constant food source, and protection from predators, including man, the tortoise will not survive."*
>
> — The Wilderness Society[18]

Time and the Tortoise

In 1988 the first traces of upper-respiratory syndrome among wild tortoises were detected at the Desert Tortoise Natural Area, a 38-square-mile preserve near California City. The disease is thought by the Bureau of Land

A Desert at the Crossroads*

"Burrobushes, creosote bushes, and other desert plants had been run over, their branches broken, crowns crushed, and roots damaged. Some of the smaller plants had been completely uprooted. All around us we could see how the bikes had ripped open the ground as the riders executed sharp turns and circled around bushes, in the process pulverizing the fragile crust that protects the desert's soil from being blown away by the wind. The burrows of many kangaroo rats, pocket mice, and other small animals had been crushed, and we found a fresh tire track within inches of a horned lark nest that still, amazingly, contained four young birds."

* This description of a visit to the western Mojave Desert is excerpted with permission from the article "A Desert at the Crossroads" by Robert C. Stebbins appearing in California Academy of Science's *Pacific Discovery*, Winter 1990, the whole issue of which is devoted to the California desert.

Management to be transmitted by pet tortoises released accidentally into the desert. There is at present no known cure for it. Another, and as yet unidentified, disease struck tortoises in the Colorado Desert at Chuckwalla Bench, an area of critical BLM concern where their numbers have dropped by 60 percent since 1982, partly due to the disease.[20] In the Mojave's Desert Study Natural Area, BLM scientists in 1990 found only 200 tortoises, compared with 481 four years earlier. Aside from disease brought in from the outside, tortoise populations that have tracked across the deserts for millions of years are severely imperiled by the cumulative effects of off-road vehicles, cattle grazing, mining, road and transmission line construction, and other habitat-disturbing activities. As people develop desert lands to put up houses, shopping malls, motels, and office buildings, they disturb and destroy the desert's natural environment. One byproduct of this development, especially evident in southern California and in the Las Vegas, Phoenix, and Tucson areas, is the escalation of camping, hiking, hunting, biking, driving ORVs, and other outdoor recreational activities.

ORVs, many of them sanctioned by the BLM, fragment tortoise (and other species') habitat, crush burrows, and kill individual animals. In 1979 a report by the Council on Environmental Quality noted that up to a million acres of the California Desert had been badly scarred by ORVs and other destructive activities in the previous five years, compared with the 150 years it took strip-mining to desecrate 2 million acres in the entire United States.[21]

Cattle Grazing

For more than one hundred years large segments of the desert have been used as rangeland, and today some 5 million acres of California desert are leased by ranchers from the BLM for cattle and sheep grazing. The desert also supports populations of feral (nondomesticated) animals introduced by explorers, prospectors, and miners. These animals trample tortoises and other small creatures and their burrows and destroy vegetative shelter and food sources.

Even on BLM lands cattle are allowed to compete for food with bighorn sheep, to whom they apparently transmit disease. Inevitably the native species — birds, small mammals, ungulates, amphibians, and reptiles, all of which are smaller than the exotic animals — lose the struggle for water, food, and space to the larger herds, which consume as much as 10,000 pounds of plant matter per animal each year.

Grazing by livestock and feral animals also causes shifts in the composition of plant communities. In the Mojave River drainage, for example, grazing is thought to be largely responsible for the spread of "aggressive" salt cedars that are replacing native vegetation. In Eureka Valley, California, grazing has promoted the spread of Russian thistle and other competitive, exotic weeds that threaten the dune community and its rare species.[22]

The disappearance of the bighorn from the Mojave and Sonoran deserts and the sharp decrease in tortoise population are believed to be classic symptoms of a dramatic vegetation decline that first affects larger species (which consume a greater amount of plants per capita) and then spreads to smaller and more numerous species.

The Tamarisk Invasion[23]

Tamarix sp., called deciduous tamarisk or salt cedar, is a virulent pest in desert riparian areas because it withdraws and transpires water from the ground at a high rate, displacing native vegetation. Originally planted by immigrants from southern Europe as ornamental shrubs or to create windbreaks, the tamarisk dispersed widely along watercourses. Wild tamarisk growth was first noticed along the upper Gila River in Arizona after a flood in 1916, when it was extensively found in central Utah and in the Rio Grande and Pecos valleys of New Mexico and Texas. By the early 1960s tamarisk occupied an estimated 1,400 square miles of floodplain land in the western U.S.

On the California desert, tamarisk is so well established along parts of the Colorado and Mojave rivers and around the Salton Sea that eradication is well nigh impossible. However, at Eagle Borax Spring, a large marsh in Death Valley National Monument, the invader was successfully removed by a combination of burning, chainsawing, and herbicides carried out in a ten-year program by the National Park Service. Once the tamarisk had gone, the marsh made a rapid recovery: "The surface water has returned, to be used by migratory birds; the grasses and reeds are flourishing; and the grove of mesquite trees is again healthy."[24] The Bureau of Land Management and the Anza-Borrego Desert State Park are separately initiating programs of tamarisk control at other important areas of the California Desert. (For further information, write Riparian, P.O. Box 193, Lucerne Valley, CA 92356.)

Mining

Although mining for gold, silver, borates, and other minerals has disastrous impact on the slow-healing desert environment, it is allowed to continue in many wilderness study areas (WSAs). Under provisions of the Federal

Land Policy and Management Act of 1976, existing mining was allowed to go on in the same way as before that date and new mining is permitted as long as it doesn't impair the WSA's suitability for wilderness designation.[25] Unfortunately, even when these activities *do* degrade wilderness values, new mining operations are routinely approved by the Bureau of Land Management.[26] In addition to attracting people, vehicles, and heavy equipment, mining destroys vegetation, opens up large areas of the soil, creates pits where animals get trapped, and leaches cyanide, heavy metals, and other pollutants into water sources on which many species depend.

Quoth the Raven

In the desert areas of California, Nevada, Utah, and New Mexico vast flocks of ravens are attracted to the ever-growing number of town dumps, landfills, and sewage sites that are being built. In addition to preying on young desert tortoises, the ravens devour rodents, snakes, and birds important to the desert ecosystems.

Desert Storm in a Teacup?

By the mid-1970s the pressures on the desert from mining, grazing, urban development, recreation, and vandalism were beginning to take on a political complexion. Essentially, the ball was in the court of the BLM, which administers half of the desert's 25 million acres, while the Department of Defense administers 3.2 million, the National Park Service 2.5 million, and the state of California 1 million. The remaining 6 million acres are privately owned. In 1976 Congress directed the BLM to come up with a plan to deal with the crisis in the desert, especially as it related to the protection of Native American relics and habitats of unusual significance for wildlife. Four years later the BLM released its plan, which was based on user demands rather than on the capacity of the desert to survive them. For example, it proposed closing a mere 17 percent of the area to vehicle use. Guidelines for mining in regions with rare or sensitive species were just about the same as those applying anywhere else in the desert, and almost no restraints were placed on grazing. In 1989 the General Accounting Office (GAO), the investigative arm of Congress, found the BLM plan defective in many respects, especially when it came to wildlife preservation. Most importantly, the BLM was not monitoring wildlife and habitat, a procedure considered necessary in virtually every habitat management plan. Ignoring one of its own (1985) reports warning that mining, grazing, and ORVs were destroying the habitat of the Mojave ground-squirrel, the BLM had collected no further data on the species. The failure, said the GAO, came from the BLM's favoring the interests of mining, grazing, and recreation, including establishing off-road-vehicle free-play areas in desert-tortoise habitat. The BLM has not demonstrated the willingness to take actions necessary to protect wildlife interests, the GAO concluded bluntly.[27]

Defending the Desert Ecosystems

Conservationists agree that the impact of human disturbance and habitat destruction in desert ecosystems is best countered by protecting large and continuous areas of wildlands. The Wilderness Society, National Audubon, the Sierra Club, and many other environmental groups believe there would be a better chance of protecting these ecosystems if responsibility for them is transferred from the BLM to the National Park Service.*

> *"The best means to protect the full range of biological diversity in the California Desert — remnant populations of species isolated across mountaintops, riparian communities and other relict habitats; plants and animals specifically adapted for survival in the hot, dry environment; and the wide-ranging species dependent on large chunks of intact habitat — is preservation of continuous tracts of the diverse desert habitat matrix."*
>
> — Dennis D. Murphy and Kathy M. Rehm,
> Center for Conservation Biology, Stanford University[28]

Rewriting the Map

In 1986 bills were introduced in Congress that would in effect rewrite the map of southeastern California, radically changing access to the desert and how it can be used. In November 1991 the California Desert Protection Act (H.R. 2929), introduced by Congressman Mel Levine (D-California), was passed by the House of Representatives. If a similar bill is passed by the Senate, the law would designate 4.1 million acres of BLM wilderness in seventy-three separate areas, create the 1.5-million-acre Mojave National Monument, and add another 1.5 million acres to Death Valley and Joshua Tree national monuments and redesignate them as national parks. ORVs would be barred in some of these areas; legitimate mining claims would be honored unless the claims were condemned; cattle grazing would continue in the BLM wilderness areas but be phased out in the new national parks.

Supporters of the act see it as a big step toward saving the severely threatened wildlife of the desert, because some 60 percent of the populations of extremely rare plants and animals are located on lands that the act would set aside and because people tend to become more respectful of the environment when they are in national parks. Excluding vehicles from wilderness areas, they believe, would make it easier for the BLM to patrol and stop the vandalism, shooting, and habitat destruction that presently destroy so much wildlife.[29]

The Senate bill, sponsored by former Senator Alan Cranston (D-California), was tougher than H.R. 2929 in that it would protect 8.5 million acres of wilderness. It faces a hard fight because it is strongly opposed by

* For more on the BLM and the Park Service, see chapter 8, pages 240-245.

mining, ranching, and military interests in the region.* Even important as the legislation may be, many conservationists and environmental groups such as Defenders of Wildlife, EDF, and NRDC agree with Judy Anderson of the California Desert Protection League that no amount of laws will take the place of a better funded, better run, and more environmentally sensitive BLM.[30]

What You Can Do to Help Save the Desert's Wildlife

1. To find out the status of the California Desert Protection Act (CDPA), contact the Senate Documents Room, B-04, Hart Senate Office Building, Washington, DC 20510; (202) 224-7860; and the House Documents Room, H-226, U.S. Capitol, Washington, DC 20515; (202) 225-3456.
2. Write your support for implementation of CDPA to: the chairperson of the Senate Subcommittee on Public Lands, 306 Dirksen Senate Office Building, Washington, DC 20510; and of the House Subcommittee on Public Lands and National Parks, 812 House Annex 1, Washington, DC 20515.
3. Urge your own senators and representatives to support expanded wilderness protection in the California desert and let them know you oppose hunting on the proposed East Mojave National Monument.
4. Contact the Governor of California, State Capitol, Sacramento, CA 95814; (916) 445-2851, urging your support for CDPA and other necessary measures to save the deserts.
5. Support the efforts of the Sierra Club Desert Task Force, 730 Polk Street, San Francisco, CA 94109; (415) 776-2211; and of the Wilderness Society's San Francisco office, 116 New Montgomery, Suite 526, San Francisco, CA 94105; (415) 541-9144. These were the Desert Protection Act's prime backers among national environmental groups. Also support the National Audubon Society's Western Regional Office, 555 Audubon Place, Sacramento, CA 95825; (916) 481-5332; the Desert Protective Council, and the Desert Tortoise Council (see details below).
6. To find out more about the encroaching military operations in the West, contact Nevada Outdoor Recreation Association, P.O. Box 1245, Carson City, NV 89702.
7. Do not buy cactus or other desert plants, tortoises, or other pet animals that have been imported illegally from the Southwest, Mexico, and other desert regions.

* For an update on this bill, see "Pending Legislation," page 361.

RESOURCES

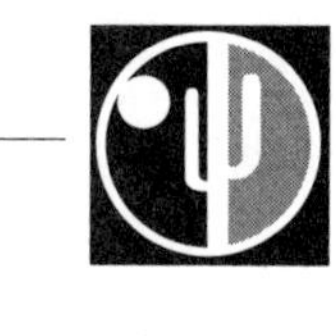

1. U.S. Government

— *Great Basin National Park*, Baker, NV 89311; (702) 234-7331.
— *National Park Service National Monuments:*
Craters of the Moon, P.O. Box 29, Arco, ID 82313.
Death Valley, Death Valley, CA 92328; (619) 786-2331.
Joshua Tree, 74485 National Monument Drive, Twenty-nine Palms, CA 92277.
Organ Pipe Cactus, Route 1, Box 100, Ajo, AZ 85321.
Saguaro, 36933 Old Spanish Trail, Tucson, AZ 85730.
— *General Accounting Office*, P.O. Box 6015, Gaithersburg, MD 20877; (202) 275-6241. The GAO's review of the BLM's wildlife management in the California Desert Conservation Area provides a historical overview of the area and the issue (see "Further Reading" below).

2. Environmental and Conservation Organizations

— *Audubon Desert Task Force*, P.O. Box 260, Porterville, CA 93258; (209) 784-4477.
— *California Desert Protection League* can be reached at (818) 248-0402. Its members include the California Native Plant Society, (916) 447-2677; the California Wilderness Coalition, (916) 758-0380; Citizens for Mojave National Park; the Desert Protective Council; Desert Survivors, (213) 465-7010; the National Audubon Society; the Sierra Club; and the Wilderness Society.
— *Citizens for Mojave National Park*, P.O. Box 106, Barstow, CA 92312; (619) 256-9561. Devoted to the creation of a national park in what is now designated as the East Mojave National Scenic Area.
— *Desert Bighorn Council*, 1500 North Decatur, Las Vegas, NV 89108 (no telephone). A clearinghouse for information on the desert bighorn.
— *Desert Protective Council*, P.O. Box 4294, Palm Springs, CA 92263; (805) 942-3662. The DPC was founded by a group of botanists and concerned citizens to protect Joshua Tree National Monument against prospecting and mining. Today the DPC is involved in a wide range of conservation efforts to protect desert plants and animals, cliffs and canyons, dry lakes and sand dunes, and historic and cultural sites. It promotes the enjoyment and wise use of desert resources and supports stronger legislation for desert protection. The DPC publishes an impressive list of educational bulletins by leading experts, which may be ordered from DPC Publications, 3750 El Canto Drive, Spring Valley, CA 92077.
— *Desert Tortoise Council*, P.O. Box 1738, Palm Desert, CA 92261; (619) 431-8449.
— *The Wilderness Society*, Southwest Region, 234 North Central Avenue, Phoenix, AZ 85004; (602) 256-7921.

3. Research Organizations and Museums

— *Arizona-Sonora Desert Museum*, 2021 North Kinney Road, Tucson, AZ 85743; (602) 883-1380. Located 14 miles west of Tucson, the museum has a live collection of some 200 desert animals and 400 plants.
— *Chihuahuan Desert Research Institute*, P.O. Box 1334, Alpine, TX 79831; (915) 837-8370. Dedicated to understanding this desert region, the Institute operates a visitors' center 3 miles south of Fort Davis, Texas.
— *High Desert Museum*, 59800 South Highway 97, Bend, OR 97702; (503) 382-4754. Its 150 acres of desert, sagebrush flatlands, forest, and mountains are surrounded by the Deschutes National Forest.

4. Further Reading

— *Adventuring the California Desert*, Lynne Foster. San Francisco: Sierra Club, 1987.
— *The American Southwest: A Vanishing Heritage, Report 2 — Ecological Values of the California Desert*. The Wilderness Society, August 1989.
— *Arid Lands*, Jake Page. Alexandria, Va.: Time-Life Books, 1984.
— *California Desert: Planned Wildlife Protection and Enhancement Objectives Not Received.*

U.S. General Accounting Office GAO/RCED-89-171, June 1989.
— *The California Deserts*, Edmund C. Jaeger. Stanford, Calif.: Stanford University Press, 1965.
— *Decision for the Desert: The California Desert Protection Act.* Prepared by the Sierra Club, the Wilderness Society, and the National Audubon Society for the California Desert Protection League, July 1989.
— *Desert: The American Southwest*, Ruth Kirk. Boston: Houghton Mifflin, 1973.
— "A Desert at the Crossroads," Robert C. Stebbins. *Pacific Discovery*, Winter 1990.
— "Desert Folly, Desert Hopes," Peter Steinhart. *Defenders*, January/February 1990.
— *The Desert Reader*, edited by Peter Wild. Salt Lake City: University of Utah Press, 1991.
— *Deserts*, James A. MacMahon. Audubon Society Nature Guides series, New York: Knopf, 1985.
— *Deserts of America*, Peggy Larson. Englewood Cliffs, N.J.: Prentice-Hall, 1970.
— *Desert Solitaire: A Season in the Wilderness*, Edward Abbey. New York: Ballantine Books, 1985.
— "Desert Tortoise Gets Partial Protection," Jasper Carlton. *Earth First! Journal*, September 22, 1989.
— *The Desert World*, David F. Costello. New York: Thomas Y. Crowell, 1972.
— *Failure in the Desert.* Wilderness Society report. Washington, D.C., 1989.
— *The Hidden Life of the Desert*, Thomas Wiewandt. New York: Crown, 1990.
— *Human Causes of Accelerated Wind Erosion in California's Deserts*, Howard G. Wilshire. Palm Springs: Desert Protective Council Publications, 1992.
— *The Land of Little Rain*, Mary Austin. New York: Penguin Books, 1988. A nature-writing classic, first published in 1903.
— "The Lands No One Knew." *Wilderness*, Spring 1989.
— "Living in a Land of Extremes," Elizabeth Pennisi. *National Wildlife*, April/May 1989. A report on Organ Pipe Cactus National Monument.
— "Off-Road Vehicles and the Fragile Desert," Robert C. Stebbins. *American Biology Teacher*, vol. 36, nos. 4 and 5 (April and May 1974).
— "Rambo's Racers," Elena Jarvis. *Defenders*, January/February 1990.
— "Sky Islands of the Great Basin: Nevada's New Wilderness Areas," Doug McMillan. *Motorland*, March/April 1991.

NOTES

1. Aside from Jon Naar's spending nine months in Egypt's Western Desert and separate visits by him and Alex Naar to arid lands of the U.S. Southwest, the Great Basin, and Mexico, this section derives much valuable material from James A. MacMahon, *Deserts*, in the Audubon Society Nature Guides series (New York: Knopf, 1988); and from Janine M. Benyus, *Field Guide to Wildlife Habitats of the Western United States* (New York: Fireside, 1989).
2. We are indebted to Dave Foreman and Howie Wolke, *The Big Outside* (Tucson: Ned Ludd Books, 1989), for much of the material in this section.
3. MacMahon, *Deserts*, p. 37.
4. Benyus, *Field Guide to Wildlife Habitats*, p. 167.
5. MacMahon, *Deserts*, pp. 36-37.
6. For more on the fascinating lifestyle of the Great Basin spadefoot, see Benyus, *Field Guide to Wildlife Habitat*, pp. 174-75.
7. Foreman and Wolke, *The Big Outside*, p. 206.
8. See Leon Czolgosz, "The Afghanization of the American West," *The Earth First! Reader*, pp. 50-55, for a report on the U.S. Air Force and Navy turning large parts of several western states into military operation areas (MOAs) or supersonic operations areas (SOAs).
9. Foreman and Wolke, *The Big Outside*, p. 206.
10. MacMahon, *Deserts*, p. 49.
11. Jake Page, *Arid Lands* (Alexandria, Va.: Time-Life Books, 1984), p. 154.
12. Ibid.
13. Ibid., p. 98.
14. Edward Abbey, *Desert Solitaire* (New York: Ballantine Books, 1968), p. xii.
15. Peter Steinhart, "Desert Folly, Desert Hopes," *Defenders*, January/February 1990.

16. *Failure in the Desert*, Wilderness Society report (Washington, D.C., February 1989), pp. 24-25.
17. Gary Paul Nabhan, "Replenishing Desert Agriculture with Native Plants and Their Symbionts," in *Meeting the Expectations of the Land,* edited by Wes Jackson, Wendell Berry, and Bruce Colman (San Francisco: North Point Press, 1984), pp. 172-82.
18. Wilderness Society, *Failure in the Desert*, p. 28.
19. Blake Edgar, "Suffering Symbol of the Mojave," *Pacific Discovery*, California Academy of Sciences, Winter 1990, pp. 18-19.
20. Ibid., p. 19.
21. Wilderness Society, *Failure in the Desert*, pp. 24-25.
22. Ibid., p. 15.
23. Based on William M. Neill, "The Tamarisk Invasion of Desert Riparian Areas," Bulletin 83-84, Desert Protective Council (Spring Valley, CA 92077).
24. Ibid.
25. Federal Land Policy and Management Act, Section 603.
26. Wilderness Society, *Failure in the Desert*, p. 27.
27. Kim Heacox, "The Mojave Desert: Who Will Rule It?," *Audubon*, May 1990, p. 75.
28. *Ecological Values of the California Desert*, Wilderness Society, August 1989.
29. Steinhart, "Desert Folly," p. 29.
30. Ibid., pp. 28-30.

Part 3
THE PEOPLE

YOSEMITE NATIONAL PARK

8 PUBLIC LANDS[1]

"The newly formed United States was gifted with more wild land than any other modern nation in the world. The land was wasted at an appalling rate, accelerated further by such improved axes, saws, and firearms as the colonists could acquire."

— Dyan Zaslowsky, *These American Lands*[2]

Today 29 percent of the United States' land area is owned by the federal government in a vast mosaic of grasslands, forests, mountains, wetlands, lakes, rivers, seashores, islands, and deserts. According to figures supplied by the U.S. Office of Governmentwide Real Property Relations, the federally owned lands in the fifty states account for 662.2 million acres (1.03 million square miles) out of a total U.S. land area of 2,271.4 million acres (3.55 million square miles). Looked at another way, the land owned by the U.S. government adds up to about the same area as that occupied by the three largest states — Alaska, Texas, and California — with Massachusetts, Rhode Island, and Delaware thrown in for good measure.[3]

Who Runs Our Public Lands?

Here's how the U.S. public lands are distributed among different government agencies and jurisdictions:

U.S. Department of the Interior:

— *Bureau of Land Management* (BLM) — *266.3 million acres* (416,000 square miles). The BLM administers National Resource Lands, which consist mostly of arid and mountainous countryside that *didn't* get grabbed up in the billion-acre General Land Office giveaway of the nineteenth century. Sometimes called "the lands no one wanted," the BLM's vast holdings located throughout ten western states and Alaska are of unique natural, archaeological, and historic importance.

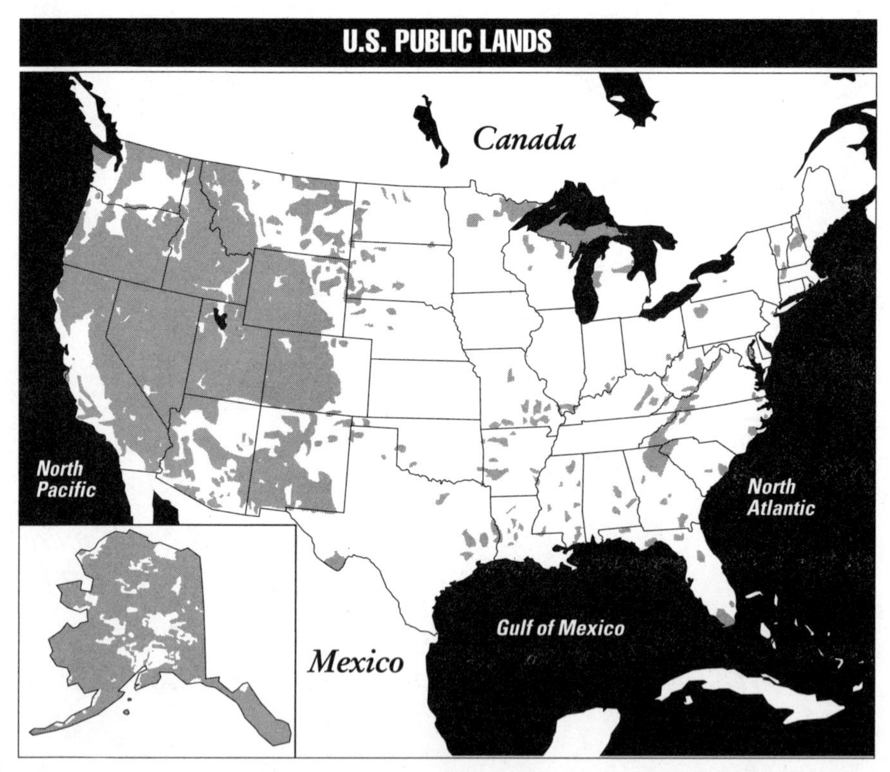

The Federal Public Lands of the United States

- National parks — 74.2 million acres
- National forests and grasslands — 201.5 million acres
- Wildlife refuges — 83.5 million acres
- Bureau of Land Management lands — 266.3 million acres
- Bureau of Reclamation lands — 6 million acres
- Bureau of Indian Affairs lands — 2.75 million acres
- Department of Defense lands — 26 million acres
- Other lands — 2 million acres

Total — 662.25 million acres

- National Trails — 21,620 miles
- Wild and Scenic Rivers — 10,500 miles

— *Fish and Wildlife Service* (FWS) — *83.5 million acres* (130,000 square miles). FWS's prime responsibility is to administer the National Wildlife Refuge System.*

— *National Park Service* (NPS) — *74.2 million acres* (116,000 square miles). The NPS administers the National Park System under a broad congressional mandate to conserve† and protect U.S. geologic, biologic, scenic, and historic resources in perpetuity, while making them available for public enjoyment and education. In addition to national parks, the system includes a mixed bag of monuments, battlefields and other historic sites, memorials, preserves, seashores and lakeshores, riverways, scenic trails, and national parkways. In general (but not always!), units of the NPS get better protection than lands under jurisdiction of other agencies.

— *Bureau of Reclamation* — *6 million acres* (9,375 square miles) in water project developments in the West, mostly for irrigated agriculture in California.

— *Bureau of Indian Affairs* — *2.75 million acres* (4,300 square miles), excluding reservations which are not federal property.

Department of Agriculture

— *U.S. Forest Service* (FS) — *201.5 million acres* (315,000 square miles). The FS administers the national forests, primarily for "sustained-yield" productivity of timber, fish and wildlife habitat, and a wide variety of recreational interests. Trying to keep all these balls in the air at the same time is a tricky juggling act called "multiple use."‡

Department of Defense (DOD)

— *26 million acres* (41,000 square miles), including: U.S. Army, 10.4 million acres; Air Force, 8.1 million acres; Navy, 2 million acres; and Army Corps of Engineers, 5.5 million acres.

---

* See pages 260-266.

† It is important to distinguish between conservation, which is a continuous process, and preservation, which means keeping things the way they were — i.e., stopping the clock. For more on this, see "The Divided Crusade," page 258.

‡ For details on how the FS handles these often conflicting interests, see chapter 6, pages 191-193.

Other agencies

— Including the Department of Energy (nuclear weapons plants) and the Tennessee Valley Authority — total approximately *2 million acres* (3,000 square miles).

Note: *91 million acres (142,000 square miles) of the forests, parks, refuges, and resource lands listed above are designated by Congress as "wilderness areas" and are administered under the National Wilderness Preservation System. By law, a wilderness is defined as an area "where the earth and its community of life are untrammeled by man, where man himself is a visitor who does not remain." Such areas are off-limits to vehicles, engines, or buildings of any kind.*

These lands, *our* lands, represent a million years of evolving landscapes and ecosystems, containing timber, minerals, and other tangible resources as well as priceless intangibles: wilderness, wildlife, pristine watersheds, wild rivers, scenic beauty, recreational opportunities, and a virtually untapped storehouse of scientific information. Migratory birds by the tens of millions use the public lands as breeding grounds and stopover resting places. The lands are also the permanent habitat of untold numbers of elk, caribou, wolves, bears, antelopes, and other species, and the visiting place of some 100 million people who each year use national parks, forests, wildlife refuges, and other federal lands for hiking, camping, hunting, fishing, boating, and other nature-related activities.

The BLM — Mission Well Nigh Impossible

With more acreage than the Park and Forest services combined and a fraction of the budget, the BLM is expected to keep its lands healthy while they are overgrazed by livestock, stripped by miners, pulverized by dirt bikes and ORVs, dive-bombed by the air force, nuked by the Department of Energy, and hiked and camped on by an assortment of birdwatchers, conservationists, and nature lovers, all ready to sue for their conflicting rights at the drop of a Stetson. Long thought of as a stepchild agency, the BLM was created in 1946 when the General Land Office and the federal Grazing Service were merged. However, it didn't have a charter until the Federal Land Policy and Management Act of 1976. All public lands, the law mandates, are to be retained for the long-term use of the American people unless "it is determined that disposal of a particular parcel will serve the public interest." The BLM was given multiple-use, sustained-yield* goals, somewhat similar to those of the Forest Service, with land-use planning as a cornerstone of its management of the lands now identified as natural resource lands. Also, the United States must receive "fair market value of the use of the public lands and their resources," but with "areas of critical concern" designated to protect historic, cultural, and natural values."[4] How to reconcile these conflicting goals and interests has been at the heart of the BLM's dilemma since its inception.

The Sagebrush Rebellion

In June 1979 the Nevada legislature proclaimed all public lands in the state "not previously appropriated" to become state property. The declaration,

* "Sustained yield" means cutting no more trees than are grown.

symptomatic of a long antifederation tradition in the West, was echoed in the legislatures of most other western states, giving rise to what became known as the Sagebrush Rebellion. "Count me as a rebel," shouted presidential candidate Ronald Reagan to a Utah campaign crowd in 1980. When elected, one of his first moves was to appoint as secretary of the interior (and thus the BLM's boss) James Watt, a man from Wyoming who made clear his contempt for federal conservation programs. Before his appointment Watt had, as attorney for the prodevelopment Mountain States Legal Foundation, opposed Interior Department rulings on more stringent strip-mining controls and had sued to speed up oil and mineral exploitation in wilderness regions.[5] Although the Sagebrush rebels failed in their quest to have the national resource lands transferred to the western states, a Middle East energy crisis in 1980, as in 1991, gave the administration all the ammunition it needed to press for the accelerated exploitation of oil, gas, and mineral resources on the public lands of the United States.

Even since Watt's departure from the DOI in 1984, the BLM's planning process and land-management practices have not been consistent with the bureau's mandate. In 1988 the General Accounting Office (GAO) reported that 60 percent of the public rangeland was in unsatisfactory condition, and little or nothing was being done to correct overgrazing. Stating that the BLM staff was demoralized, the GAO cited the case of a BLM area manager who caught a rancher cutting trees without authorization in a sensitive riparian zone and ordered him to stop. Word was sent down through political channels from Washington that the manager was to apologize to the rancher — and deliver the wood to his house.[6] In 1989 Cy Jamison was appointed director of the BLM; he had worked since 1980 as an adviser to a Western congressman well known for his opposition to environmental causes, especially wilderness protection. Despite this unpromising background, Jamison has won the (qualified) support of many conservationists for his efforts to move the BLM toward a more balanced approach to the environment.

> *"What a beautiful and thrilling specimen for America to preserve and hold up to the view of her refined citizens and the world, in future ages! A* nation's park, *containing man and beast, in all the wild{ness} and freshness of their nature's beauty!"*
>
> — George Catlin, painter of Native Americans, 1832[8]

The National Park System

Catlin was one of the first to call for preservation of the wilderness he saw as "destined to fall before the deadly axe and desolating hands of cultivating men."[7] Others joined in advocating reserving wild areas, including Henry David Thoreau, who proposed creating parks of primitive forests of 500 or a thousand acres in his own and surrounding Massachusetts townships. They should be publicly owned and made sacrosanct, he urged. In 1832 the Arkansas Hot Springs became the

first place to be set aside as a natural reservation, and in 1864 the federal government granted Yosemite Valley to California "for public use, resort, and recreation." Although the valley covered only 10 square miles and Yosemite did not become a national park until 1890, the legal preservation of public domain for scenic and recreational values established an important precedent in American history. The first national park, Yellowstone, was established in 1872 when President Ulysses S. Grant signed an act designating more than 2 million acres of northwestern Wyoming and southwestern Montana "as a public park or pleasuring ground for the benefit and enjoyment of the people."[9]

In the last decades of the nineteenth century Congress gave away huge chunks of the country to railroad companies, timber interests, and settlers. The Homestead and Desert Land acts offered every American family its share (160 acres) of the public domain. At the same time the federal government set aside a number of spectacular sites — including Sequoia and General Grant (later incorporated into Sequoia) forests in California, Mount Rainier in Washington, and Crater Lake in Oregon — as national parks. But their administration of these places was cavalier in spirit and usually in fact. Until 1916 Yellowstone was actually run by the U.S. Cavalry! In response to gross vandalism and theft of Anasazi relics at Mesa Verde, Colorado, Congress in 1906 gave the President the power to designate as national monuments sites containing historic, scientific, or scenic treasures. By 1916, when President Wilson signed the National Park Service Act, twenty such monuments had been declared. The pressing need for a truly protective agency for the parks had been demonstrated in the bitter battle over a proposed dam (to provide water for the city of San Francisco) that would submerge Hetch Hetchy, a valley in the northwest corner of Yosemite, whose beauty the great preservationist John Muir considered second only to that of the Yosemite Valley itself. In 1913 Congress voted approval for the dam, and a few years later Hetch Hetchy was lost forever. Under its first director, Stephen Mather, who was appointed in 1916, the National Park System expanded to include Lassen National Volcanic Park in California, Hawaii National Park, and Mount McKinley in Alaska, Zion in Utah, Grand Canyon in Arizona, and Acadia (originally called Lafayette) in Maine. Three years later there were nineteen parks. The expansion was accompanied by a rapid influx of tourists, many of them transported by automobiles. This represented a policy favored by Mather. He had been instrumental in establishing the National Park-to-Park Association, whose members would see the construction of roads so that all the parks would be accessible. At the time only a few people foresaw that the automobile, convenient as it might be, would become the system's greatest problem.

Today national parks are administered by the National Park Service. Encompassing 74.2 million acres throughout the United States and its territories, the system incorporates 354 separate units — national parks, monuments, preserves, lakeshores, rivers (including wild and scenic rivers), seashores, historic sites, memorials, military parks, battlefields, historic parks, recreation areas, parkways, scenic trails, parks "without national designation," the National Capital Parks, the White House, and the National Mall. Thirty-five million acres of NPS lands, 90 percent of them in Alaska, are designated as wilderness under the Wilderness Act of 1964.* They are managed to retain their pristine character, and motor vehicles are prohibited within their boundaries. Other roadless areas still await wilderness designation by Congress.

Rivers and Trails

Under NPS jurisdiction are the National Wild and Scenic River System† and the National Trails System. The thousands of miles of trails include the Pacific Crest Trail, running from the Canadian border through Washington, Oregon, and California to the Mexican border, and the Appalachian Trail, running through twelve states from Springer Mountain in Georgia to Mount Katahdin in Maine. In the view of many conservationists, these trails and rivers systems are neglected stepchildren lacking national support and accorded low priority and inadequate funding. As Zaslowsky writes in *These American Lands*, "part of the reason behind this fitful and retarded growth lies in the inadequacies and confusions of the law that created the [rivers] system, but an equally significant obstacle has been the absence of any clear, organized constituency for its continued health and growth."[10]

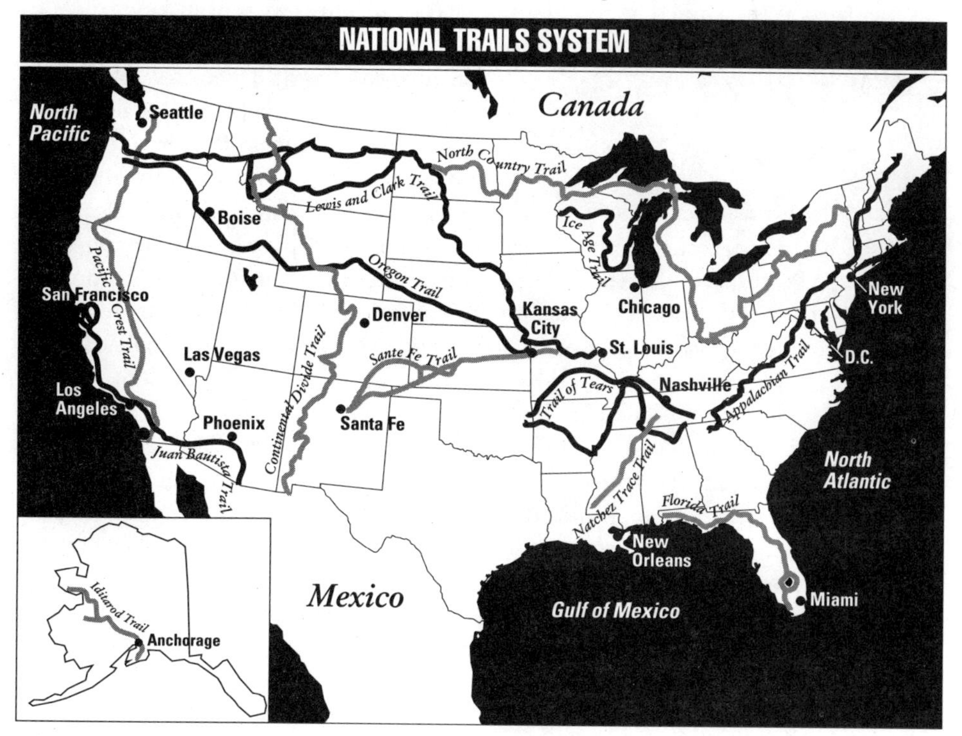

The National Trails System Act of 1968 created three categories of trails: *National Scenic Trails* are continuous extended routes of outdoor recreation within protected corridors; the first of these were the Appalachian and the Pacific Crest trails. *National Recreation Trails* are existing trails recognized by the feder-

* For more on this act, see chapter 9, pages 283-284.
† For more on Wild and Scenic Rivers, see pages 42-43 and pages 274-275.

al government as part of the national system. *Side and Connecting Trails* provide additional access to and between components of the system. In 1978 a fourth category was added — *National Historic Trails* — which recognizes past routes of exploration, migration, and military action.

Today there are seventeen national scenic and historic trails, crisscrossing the U.S. from coast to coast and border to border. Twelve of them are administered by the Park Service, four by the Forest Service, and one by the BLM. **Mileages of Scenic Trails:** Appalachian Trail: 2,144; Pacific Crest Trail: 2,638; Continental Divide Trail: 3,200; North Country Trail: 3,200; Ice Age Trail: 1,000; Potomac Heritage Trail: 700; Natchez Trace Trail: 110; Florida Trail: 1,300. **Mileages of Historic Trails:** Iditarod Trail: 2,450; Oregon Trail: 2,000; Lewis and Clark Trail: 3,700; Mormon Pioneer Trail: 1,300; Overmountain Victory Trail: 300; Cherokee Trail of Tears: 2,052; Nez Perce Trail: 1,170; Santa Fe Trail: 1,200; Juan Bautista de Anza: 1,200 miles.

The Appalachian Trail

Stretching across 2,100 miles of wooded, pastoral, and wild lands of Appalachian Mountain ridgelines from Maine to Georgia, the Appalachian Trail (AT) was designated as the United States' first public footpath. It is used primarily for walks and day hikes by all manner of nature enthusiasts from birders to wildflower photographers. Backpackers make weekend escapes to it from Boston, New York City, Philadelphia, Washington, D.C., and other heavily urbanized areas. Along the AT backbone, arctic plants — paper birch and black spruce, native to Quebec's Gaspé Peninsula — penetrate deeply into ecosystems as far south as the northern mountains of Georgia. Protected for more than 96 percent of its way by federal or state ownership of the land or by rights-of-way, it was created in the 1920s and '30s by volunteer hiking clubs joined together by the Appalachian Trail Conference (ATC). The U.S. Forest Service and individual states acquired and now administer 850 and 420 trail miles respectively, while the Park Service retains overall responsibility. In 1968 the National Trails Act made the AT, on paper at least, a linear national park and authorized funds to surround the entire route with public lands, protected from incompatible use. In practice, the number of trails built, reconstructed, or maintained by the Forest Service has steadily declined from a peak of 144,000 miles in the 1940s to less than 100,000 today.

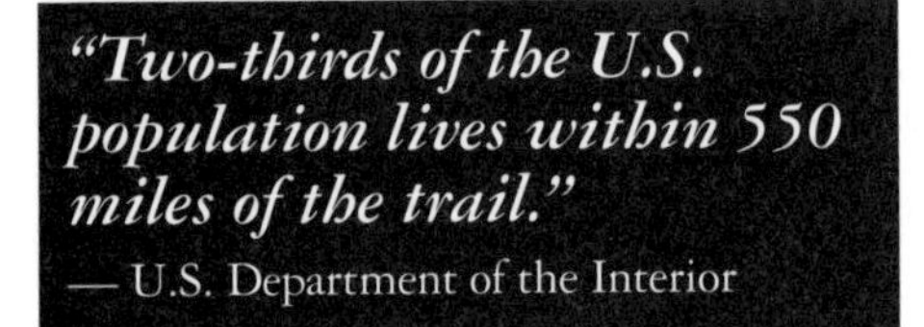

Part of the problem stems from the Forest Service's viewing regions as isolated planning units rather than as integrated environments and ecosystems that cut across public and private lands and political boundaries. As the Wilderness Society proposes, new designated wilderness could protect the wilderness character of major segments of the Appalachian Trail and help ensure the integrity of natural corridors through the headwaters country of the Chattahoochee, Nantahala, and Tallulah rivers and nearby creeks and smaller streams. As an example, the Society says that designating 9,400 acres at Blood Mountain would protect country on both sides of 11 spectacular miles of the trail as it winds through the mountains north of Dahlonega, Georgia.[11]

(For more on national trails, contact the National Park Service, National Trails Branch, P.O. Box 37127, Washington, DC 20013-7127; (202) 343-3780.)

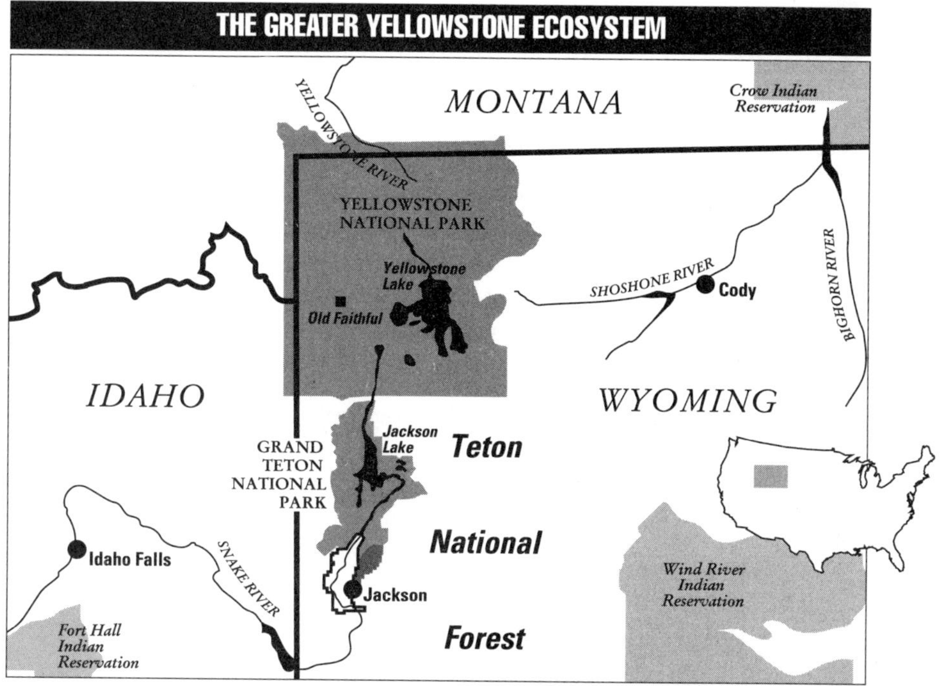

In its discussion of ecological fragments and systems the Wilderness Society selects three regions of the United States where coherence should be — and can be — restored and preserved: the Southern Appalachian Highlands of the Southeast, the pond-and-forest country of the three northernmost New England states, and, by far the largest of the three, the Yellowstone region of the Rocky Mountains known as the Greater Yellowstone Ecosystem (GYE).[12] This vast complex of designated and de facto wilderness lands stretches across the northwest corner of Wyoming and takes in large chunks of Montana and Idaho. The GYE is a unified ecological region defined by mountain ranges, rivers, vegetation, wildlife, and other natural features. "It encompasses the rugged Teton Mountains, Wind River Mountains, Absaroka Mountains, alpine meadows, deep canyons, blasting geysers, forests, lakes, crystal-clear rivers, and one of the most dazzling arrays of megafauna left on Earth."[13]

At its heart is Yellowstone National Park, the symbol of wild America since 1872. Surrounding it are more than 12 million acres including Teton National Park, administered by the Park Service; portions of seven national forests (composing roughly three-quarters of the GYE), managed by the Forest Service; three national wildlife refuges, run by the FWS; the Centennial Mountains, coming under BLM jurisdiction; and other federal, state, and private lands.

The Yellowstone landscape is shaped by the world's largest undisturbed geothermal region and characterized by a rich diversity of wildlife

that includes the largest herds of elk and one of the few free-roaming bison herds in the world. The region provides critical habitat for grizzly bears, bald eagles, whooping cranes, peregrine falcons, and other endangered or rare species. Its watershed takes in the headwaters of three major river systems — the Yellowstone-Missouri, the Snake-Columbia, and the Green-Colorado. The predominant vegetation consists of spruce forests and lodgepole pine. Because it is a comparatively untouched wilderness, the GYE provides an unparalleled environment to study and understand how natural systems operate.[14]

> *"Large, coherent natural areas defined by river drainages, vegetative types, and the home ranges of wildlife are utterly essential to the maintenance of a healthy resource of wilderness — and the diversity of life it nurtures. But too many areas are threatened with fragmentation by roadbuilding, mining, logging, and other uses, and by administrative confusion and conflicts between public and private ownership."*
> — Wilderness in America

An ecosystem the size of Yellowstone is not easily defined by administrative boundary lines of national parks and other lands. Just as river systems, geothermal aquifers, and forests extend beyond these confines, so do many animal species range well beyond them to meet their seasonal habitat needs. When Congress initially established the park boundaries, it was preoccupied with preserving Yellowstone's geothermal curiosities. Today with the advantage of better understanding of the ecological connections between the parks and surrounding lands, conservationists and others are calling for an ecosystem-based approach to the management of our public lands.[15] For years the GYE has suffered from an administrative case of "too many cooks," added to which are the often conflicting interests of tourism, real estate, oil and gas exploration, logging, and wilderness and wildlife preservation. According to recent studies by the Forest Service and the Wilderness Society, the highest *economic* as well as ecological gains lie not in commercial exploitation but in managing Yellowstone's seven national forests and adjacent federal lands for recreation, stream protection, wildlife habitat, and wilderness purposes.[16] Yet, incredibly, the Forest Service still wants to increase logging to the detriment of recreation and watershed protection — and is reluctant to recommend *any* significant additional wilderness protection.

Because the Forest Service controls so much of the GYE, its multiple-use policy of allowing commercial activities is of great concern to environmentalists. By definition, wilderness is an area that operates solely according to the dictates of natural process and thus must remain free of commercial or industrial enterprise. Although there are about a million acres of established wilderness in Greater Yellowstone, they consist mainly of high-elevation terrain considered to be of little commercial value. Most conservationists believe that at least twice that amount is needed to provide for the type of large, coherent *system* of protected wild country vital to the integrity of the whole complex. This would include 200,000 acres just northwest of Yellowstone

National Park in the Gallatin Range, constituting critical habitat for resident grizzly bears, wolverines, bighorn sheep, wintering elk, and the increasingly threatened cutthroat trout. In the Centennial Mountains, another important wildlife habitat, where gray wolves have been sighted, at least 92,000 acres should also be designated, according to the Wilderness Society, which says that the same rationale should be applied to 16,000 acres of grizzly-bear habitat in the Lionhead enclave that straddles the Continental Divide between Montana and Idaho. In all, the Society proposes, there are fourteen such crucial areas whose preservation would go a long way toward ensuring the future health of Yellowstone and the lands around it.

Commercial timbering, especially in the Targhee and Gallatin national forests, has been extremely destructive, while livestock are allowed to graze both in the "protected" wilderness areas and all of the seven national forests of the GYE.

In 1990 the joint working committee of the various administrations of Greater Yellowstone drafted "Vision for the Future," a proposed coherent-management plan for the Ecosystem. If implemented, the plan would make large amounts of land unavailable for timber, livestock, and other multiple-use ventures. This prompted a vigorous protest from highly organized industrial groups such as the Yellowstone Regional Citizens' Coalition, which claimed that the plan would lead to "antidevelopment use" and would hamper citizens' rights to "snowmobiling...off-highway vehicle use...mining, grazing, and timber harvesting."[17] According to an editorial update in *Wild Earth*, pressure from John Sununu, then President Bush's chief of staff, and from several western congressmen apparently resulted in "Vision" being completely rewritten to uphold the interests of the resource extraction industries. The magazine also believes that the firing of Lorraine Minzmyer, regional director of the National Park Service, was in part due to her work on the "Vision" draft.[18]

In their article "Last Chance to Save the Yellowstone Ecosystem," Bill Willers, professor of biology at the University of Wisconsin, and Jasper Carlton, director of the Biodiversity Legal Foundation, propose that as a first step all national forests within the GYE be removed from the land base of the Forest Service because of its betrayal of public trust. As a constructive solution they propose the establishment of a Yellowstone World Biological Preserve that would become a model for other preserves, with a future determined strictly by biologists free of industrial and political pressures. "If we lose Yellowstone, with its name so much part of the American scene, it is doubtful that big wilderness anywhere can be preserved," they conclude.[19]

The Great Yellowstone Fire

Summer 1988 was the time of the greatest fire in Yellowstone since before the Civil War. You probably saw the dramatic reports on TV. A prolonged drought had transformed the landscape into a tinderbox. Under normal conditions the grasslands would have burned more slowly than heavily forested areas, but the drought had deprived the grass and sage of most of its moisture. Wind-driven flames raced across broad expanses of open ground, traveling as much as 10 miles a day, incinerating vast meadows and consuming all organic

material in its path. Fire temperatures reached 1,000 degrees Fahrenheit, evaporating the moisture that remained in the vegetation. Firestorms created hurricane-force winds, fanning the flames, lifting huge columns of smoke tens of thousands of feet in the air and carrying hot embers, which ignited new fires in front of the fire line.

The fire, which began on Forest Service land, soon spread into Park Service territory, where in line with its policy of suppressing all fires not started by lightning or volcanic action, the Service tried to suppress it. Unfortunately, the 60-mile-an-hour wind and the dry conditions made it impossible to control. The army and the marines were summoned to support hundreds of civilian firefighters already in the field fighting the spreading inferno. They could do little more than save Old Faithful Inn and a few other log buildings. According to the postfire assessment, the largest firefighting effort in U.S. history "probably did not significantly reduce the acreage burned."[21] Nature took its course. As the long, hot, dry summer drew to a close, the weather cooled, bringing rain that tamed the fires. Many of them burned themselves out. Then a brief quarter-inch snowfall blanketed the region, achieving in days what the $125 million human effort could not do in weeks. Once the fires of Yellowstone were out, the question arose: How much destruction had they actually done?

> ***"Fire is a stimulant and as important to the ecosystem as sunshine and rain."***
> — Robert Barbee, superintendent, Yellowstone National Park[20]

Assessing the Damage

Immediately after the fire Yellowstone looked like a scene out of Dante's *Inferno*. Close to a million acres of trees and plants had been reduced to a charred, smoldering wasteland. The earth was scorched and blackened, arousing fears that the intense heat had sterilized the life-giving soil. Yet, when scientists began collecting data before the arrival of the heavy winter snows, they found encouraging signs of life less than an inch below the charred top layer of the soil. In virtually all of Yellowstone's burned areas, living matter — plant root crowns and rhizomes — had survived. Grasses, fireweed, heartleaf arnica, and other deep-rooted plants began to respond to the increased amounts of nitrogen and other nutrients released by the fires and to the increased sunlight now absorbed by the charred black soil.

Considering the extent and intensity of the fire, the number of large animals immediately lost was relatively small: 246 elk, 9 bison, 4 deer, 2 moose, and 2 grizzly bears. Most of these died from smoke inhalation. However, because 30 to 50 percent of the winter feeding ranges had been burned, there was increased competition for dwindling food resources, resulting in losses of more than 600 bison and perhaps a third of the park's elk herds. Although this winter kill was larger than in previous years, it represented a healthy culling out of unnaturally large herds by eliminating many of the old and infirm animals. When the bears came out of their den in the spring of 1989, they found a larger-than-usual number of winter-killed animals to feed on. Yellowstone was vibrant with impressive new growth. One of the first plants to rise out of the fire-blackened landscape was the dogtooth violet or glacier lily. Then came the grasses, growing more than 2 feet tall and carpeting over the ravaged open terrain. Even the trees came back to life in a process scientists call **relay floristics**.

Lodgepole pines, which are predominant in Yellowstone, adapt to and are even dependent on fire. As explained by Don Despain, a research and plant biologist at the park, the heat of the fire melts the resinous glue in the lodgepoles' serotinous cones, enabling them to drop their seeds, which are then dispersed by the winds. In early spring the seed would germinate and start the process of creating a new forest.[22] Asked about reports that the full extent of the fire's destruction might take hundreds of years to heal, Despain said that the totally burned areas were mostly old forests of spruce, fir, and lodgepole. "It might take three centuries for a given patch to go from black ground to mature lodgepole, then climax forest. But from an ecological point of view, there is no damage, there's still a forest ecosystem out there, nothing this park was established to preserve is lost."[23] The lodgepole seedlings at Yellowstone in the spring of 1989 have laid the foundations for a new generation of healthy and productive forests that will take many human life spans to complete. The basic lesson of the Yellowstone experience is that fire is a naturally regenerating process, not the destructive force many people portray it to be. It is arrogant to assume nature doesn't need fire simply because *we* don't want it to intrude on our convenience. Nature has had to live with fire much longer than we have and has developed its ecological responses in accord with it.

Relay Floristics

Relay floristics is a change in apparent composition of a species based on differential growth and maturation rates. In this process, species start out at the same time but, because some species grow more slowly than others, their expression as a dominant is limited until enough time has passed to allow their growth forms to dominate the community. Over time the seedlings of a lodgepole pine, for example, grow taller than the shrubs around them and replace them in dominance. However, since all the species were essentially present from the outset, it is a temporal factor, not a change in species composition, that underlies the process at Yellowstone described here.

What You Can Do to Save the GYE

1. Join the Greater Yellowstone Coalition (GYC). It was founded in 1983 by a group of regional citizens who recognized that an ecosystem whose land forms, watercourses, plants, and animals are woven together across political and administrative boundaries must be managed similarly — across boundaries and as a whole.

From a few dozen ecosystem-minded conservationists, GYC has grown into one of North America's most powerful environmental coalitions, incorporating nearly a hundred member organizations and 5,000 individual members throughout the region. Among its members are the Alliance for the Wild Rockies, American Rivers, American Wildlands, Bear Creek Council, Friends of the Teton

Wilderness, the Medicine Wheel Alliance, the Montana Wildlife Federation, the National Audubon Society, the National Parks & Conservation Association, the Sierra Club Legal Defense Fund, the Wilderness Society, the Wolf Fund, and the Yellowstone Grizzly Foundation.

GYC organizes local and regional initiatives for participation in resource planning and protection for the GYE. Its on-the-ground activities include analyzing such environmental threats as proposed clearcuts, gold mining in grizzly bear habitat, or dam proposals on blue-ribbon trout streams; it also critiques federal and state fire policies and wildlife management decisions. When necessary, GYC takes direct action to challenge agency decisions that inadequately protect ecosystem resources. GYC's long-term planning program includes the Greater Yellowstone Tomorrow Project, designed to provide an ecologically sound blueprint for ecosystem management, and the Snake River Watershed Protection Project, emphasizing comprehensive protection of an entire river basin.

2. Order a copy of GYC's "Greater Yellowstone: An Environmental Profile." (For details on how to do this and on GYC and its membership, see the Directory, page 336.)

3. Contact members of the congressional delegations from Idaho, Montana, and Wyoming, as well as the governors of those states, and urge them to do all they can to preserve the forests, waters, wildlife, and wild places of Greater Yellowstone for the generations that follow.

4. Read George Wuerthner's "Greater Yellowstone Ecosystem Marshal Plan" in *Wild Earth*, vol. 1, no. 4 (Spring 1991). For more on this important proposal, see "What We Can Do to Save the Public Lands," pages 266-267.

5. Read *The Greater Yellowstone Ecosystem: Redefining America's Wilderness Heritage*, edited by Robert B. Keiter and Mark S. Boyce. Published in late 1991, this book provides a comprehensive review of the issues involved.

National Parks — Endangered Species

> *"America's national parks are screaming to us, from sea to shining sea, from rim to rim, to do something to save them."*
> — Paul Pritchard, president, National Parks & Conservation Association

Nineteen ninety-one was the seventy-fifth anniversary of the National Park Service and a good time to reflect on their present and future conditions. As we have seen, there is much to be proud of in terms of their existence, vast areas set aside for wilderness, wildlife, and human recreation. In terms of public attendance the system is a roaring success. In fact, there's the rub — too few resources to handle too many people. In 1991 the number of visitor days to the National Park System totaled 300 million. The actual number of visitors was 57.4 million, an increase of 6 percent over 1990. By the year 2,000 the numbers are expected to double. Pressure of crowds and their automobiles is, unfortunately, only one of the critical problems faced by the NPS and, by extension, all of us. As we shall see in the following section, urban encroachment, air pollution, acid rain, clear-cutting, overgrazing, too many elk and deer, and polluted and insufficient water are among the problems that are making our national parks an endangered species.

THE TEN MOST ENDANGERED NATIONAL PARKS

1. *The Everglades* is considered the most endangered park in the system, having lost 90 percent of its marine birds since the 1930s. Less than thirty panthers remain where there used to be hundreds. Main problem: The park's water has been diverted for agricultural use and polluted by residential development. (For more on the Everglades, see chapter 2, pages 64-69.)

2. *Shenandoah National Park*, which extends 80 miles along the crest of Virginia's Blue Ridge Mountains, is on the edge of disaster, says its superintendent, J. William Wade.[24] Main problem: Air pollution, most of it blown by winds that carry a high proportion of acids produced from factories and power plants in the Midwest, is the worst in any of the national parks, according to the NPS. The pollution not only obscures Shenandoah's panoramic views but is slowly poisoning its trout streams and killing its ferns, flowers, and trees. The NPS contends that building nineteen proposed new coal-burning power plants in Virginia would worsen air pollution in the park.[25]

3. *Acadia National Park* on the coast of Maine has some of the worst air quality in the park system, blowing in from the Midwest and up from Boston. One of the smallest national parks (35,000 acres), Acadia is visited by more than four million people a year, causing intense overcrowding and traffic gridlock. After Great Smoky Mountains National Park (8.7 million visitors), Acadia receives the greatest crowds annually — more than the Grand Canyon (3.9 million), Yosemite (3.5 million), and Yellowstone (3 million).

4. *Yellowstone* in Wyoming is crowded with too many human visitors and a large surplus of bison and elk. Efforts by conservationists and some congresspeople to reintroduce the wolf, a natural predator of bison and elk, have been

blocked by cattle raisers and the secretary of the interior. At Yellowstone, bison, grizzly bears, king salmon, and many other species need much more space than what is inside park boundaries to stay alive. For them the greatest threats are urban encroachment and simply too many tourists.

5. *Grand Canyon* in Arizona, one of the world's most breathtaking natural wonders, is in danger of fading from view because of air pollution. On the worst days, smog is so bad that you can't see across the 10-mile divide from one rim to another. Much of this pollution comes from the 2,250-megawatt coal-burning Navajo Generating Station near Page, Arizona, 60 miles to the northwest. Navajo is owned in part by the Bureau of Reclamation, which long resisted EPA efforts to get the power station to install "scrubbers"* to cut down on the 300 tons of sulfur dioxide gas (SO_2) that rise into the air from its stacks every day.† Another source of Grand Canyon pollution comes from Los Angeles, particularly from its aerospace industry, some 400 miles to the west.[26] Studies by meteorological experts and scientists show that the Grand Canyon pollution affects all the national parks, monuments, and recreation areas in the "Golden Circle."

6. *Zion National Park* in southern Utah is threatened by a project that would deprive it of the centerpiece of its spectacular Zion Canyon — the damming of the east fork of the Virgin River (flowing through the canyon) in order to divert water for the fast growing community of Saint George 45 miles north of the park. Zion is also threatened by the proposed building of a giant wraparound movie screen, a 350-seat theater, a motel, and a parking lot at Springdale, backing directly on the park's main campground and abutting the entrance station. Strong opposition is voiced by environmental groups and the National Parks and Conservation Association, an important watchdog group with more than 300,000 members nationwide.[27]

7. *Yosemite* in California, a four-hour drive from San Francisco, is the epitome of an incomparable natural wonder being loved to death by overuse, development, and pollution. Visited by some 3.5 million people a year in more than one million cars, the 750,000-acre park is fittingly described as "the embattled wilderness" in a 1990 book of that title.[28] On Memorial Day 1985 traffic gridlock became so great that the NPS instituted a quota system for controlling automobiles inside the valley where John Muir, Theodore Roosevelt, and the great naturalist photographer Ansel Adams each found their greatest source of inspiration.

Aside from massive overcrowding, the main problem is commercial development. In 1980 the NPS issued a general development plan for Yosemite that would, in principle, begin the process of restoring the park to its pristine character. According to William Penn Mott, Jr., director of the Service from 1984 to 1988, the plan was faulty in several ways: Funds were not available except for building some new housing and starting a

* These are mechanical control devices that wash (most of) the gases before they escape from the plant.
† When *Gemini 111* astronauts circled the earth in 1965 the only sign of life they could detect was the smoke coming out of Navajo's chimney stacks.

maintenance complex at El Portal and bringing utilities throughout the park up to code; there is not enough land at El Portal to accommodate needed buildings for housing without massive grading; many of the structures in the valley are on the National Register of Historic Buildings and cannot be razed; the Merced River Corridor from Yosemite to beyond El Portal has been designated a wild and scenic river, prohibiting the construction of buildings as suggested in the plan; and there is not enough water at Wawana to provide for the proposed increased housing. In late 1991 the park completed an eighteen-month survey that should enable Yosemite to develop a much improved plan intended to eliminate the congestion and improve the environmental quality of the area. The park has already removed a number of campsites along the Tuolumne River and more will be removed so that the riparian growth can develop and give the river its appropriate setting.[29] In December 1990 a

The Colorado Plateau and the Golden Circle

The Grand Canyon is an integral part of a much larger region and ecosystem — the Colorado Plateau, first delineated by John Wesley Powell in the late nineteenth century as one of the major physiographic provinces of the United States. The region's plateaus, mesas, and canyons are deeply carved by the Green, San Juan, Escalante, Dolores, Gunnison, and Colorado rivers and their many tributaries. Its rugged landscapes embrace an astonishing ecological diversity representing nearly all the life zones on earth from arid and sparse deserts to the largest ponderosa pine forests in the world. Ownership and management of Colorado Plateau are predominantly in government hands. Some 55 percent of the region's total land area (100 million acres) is federal lands, including twenty-six units of the National Park System, seventeen national forests, and 31.4 million acres of BLM lands. Native American reservations occupy nearly 25 percent of the land or 25 million acres. Coming within this vast publicly owned area is the Golden Circle of national parks — Grand Canyon, Zion, Bryce, Capitol Reef, Canyonlands, Arches, and Mesa Verde — and many less-known but significant gems such as Walnut Canyon, Natural Bridges, and Wupatki national monuments. The Plateau is sparsely populated, with about seven persons per square mile, which is one-tenth the national average. Although only five of its cities number more than twenty thousand residents, its periphery is ringed by some of the fastest growing metropolitan areas in the U.S. — Phoenix, Denver, Las Vegas, Salt Lake City, and Albuquerque.

In August 1991, after intense negotiations between the Grand Canyon Trust,* the Environmental Defense Fund, and the owners of the Navajo power plant, the EPA announced a negotiated settlement requiring a 90 percent reduction in SO_2 emissions. The settlement is expected to have significant effect throughout the Colorado Plateau by implementing the Clean Air Act, which calls for "the greatest degree" of protection for all federal park land.[30]

* For more on the work of the Grand Canyon Trust, see the Directory, page 336.

Japanese takeover of the park's lucrative concessions from MCA, a California-based entertainment conglomerate, stirred debate over who should reap profits from America's natural treasures. According to an Interior Department report, the concessioners reaped revenues of half a billion dollars in 1988, but paid the government only $12.5 million in franchise fees.[31] Critics, who say that Yosemite has been turned into a "theme park," challenged the takeover by proposing a new conservationist-controlled company, the Yosemite Restoration Company, to compete for the concessions contract that comes up for renewal in 1993.[32]

8. *Sequoia and Kings Canyon*, 865,000 acres in two neighboring parks 50 miles east of Fresno, California, with two million visitors a year, are among the most polluted in the NPS system, with airborne wastes from millions of acres of factory farms and smog from Los Angeles and other urban areas creating some of the most turbid air in North America rising up the mountainsides and choking the great trees.

9. *Olympic National Park*, 30 miles west of Seattle, Washington, and 30 miles south of Victoria, British Columbia, is plagued with major problems whose causes originate outside of its own jurisdiction. For example, clear-cutting has so destroyed the wooded hillsides next to Olympic and Rainier national parks (30 miles southeast of Seattle) that the resulting soil erosion has filled streams with silt, severely inhibiting fish runs. At Olympic, the NPS wants to have two hydroelectric dams on the Elwa River torn down. Neither dam provides for passage of fish and thus would prevent salmon and trout from reaching the park. The loss of the fish affects the region's entire ecosystem, we were told by the park's superintendent, Maureen Finnerty.[33] Mountain goats, intentionally introduced in the 1920s (before the establishment of the park), are not native to the Olympic Peninsula and have no natural predators. Now prolific, they are tearing up natural plant communities and eroding the fragile alpine and subalpine soils. After a scheme to airlift them out of the park failed, the NPS attempted to shoot them but with little success.[34]

10. *Glacier National Park*, 1 million acres in northwestern Montana and the adjoining *Waterton Lakes National Park* in southern Alberta, is unparalleled glaciated mountain country with spectacular lakes, glaciers, waterfalls, flower-strewn meadows, and thick, diverse, coniferous forests. Its problems are many. Subdivisions outside the park threaten wildlife, which includes grizzly and black bears, moose, and gray wolves. The wolves, recently reintroduced, are the only active pack in the West. They are trapped and hunted in Canada. Moreover, at the insistence of local ranchers, the U.S. Department of Agriculture has set up a $40,000 program to kill the wolves, while the U.S. Fish and Wildlife Service makes efforts to protect them! Also from across the border, coal mining and oil and gas drilling threaten the north fork of the Flathead River; air pollution from the Columbia Falls smelter reaches the park. In the northern areas of Glacier, clearcuts have penetrated every major drainage beneath Canada's Great Divide up to about 5,000 feet.

What You Can Do to Help Save the National Parks

1. Urge your congressional representatives to support increased funding for national parks and legislation to have them managed as ecosystems that cut across political and administrative boundaries.

2. Join the National Parks and Conservation Association (NPCA), the leading citizen-funded organization dedicated to protecting, promoting, and improving the national parks system. Since 1919 NPCA has fought on behalf of the parks, blocking pressure from the National Rifle Association to open parks for hunting and trapping, preventing a nuclear waste dump being located next to Canyonlands National Park in Utah, winning a lawsuit that rid Mammoth Cave National Park in Kentucky of a source of serious pollution, and protecting the mountain bear in New Mexico and the grizzly bear in Wyoming and Montana.

3. Subscribe to *National Parks*, NPCA's excellent bimonthly magazine. You get it automatically when you join the association.

4. Find out about the NPCA's long-range plan, which is based on: establishing the National Park Service as an agency responsible to the President and Congress rather than the Interior Department; settling park boundaries along ecological lines; placing science, historic preservation, and conservation programs high among NPS priorities; and improving resource management, visitor facilities, education, and interpretive programs for the national parks. (For more information on NPCA, see the Directory, page 338.)

5. If you plan to visit one of the more crowded national parks, try to get there by using public transportation or car pooling; and visit midweek. Especially avoid holiday weekends. When you are there, find out from the visitors' center which areas may be less crowded.

6. Visit one of the less popular parks. These include the following ones, which had fewer than 225,000 visitors in 1991: Isle Royale, Michigan, the largest island in Lake Superior, which drew fewer than 25,000 visitors; Voyageurs, Minnesota; Guadalupe Mountains, Texas; Channel Islands, California; and Great Basin, east of Ely, Nevada. A good source of information is "The National Parks: Lesser-Known Areas," available for $1.50 from the U.S. Government Printing Office, Washington, DC, 20402. Order GPO Stock No. 024-005-009-11-6.

7. Read John Muir's classic book *Our National Parks*. First published in 1901, it was reissued in 1991 as a paperback by Sierra Club Books, San Francisco. His message is still relevant today: Appreciate the grandeur of the parks; set aside additional lands to protect wildlife from the hunter's slaughter; and preserve forests from the lumberjack's ax.

The National Forest System—"Land of Many Uses"

"In the administration of the forest reserves. . . all land is to be devoted to its most productive use for the permanent good of the whole people and not for the temporary benefit of individuals or companies."
— Gifford Pinchot[35]

These were Pinchot's guidelines for the newly formed U.S. Forest Service (FS). In 1905, as its first chief, he took control of 86 million acres of forestland transferred from the Department of the Interior to the Department of Agriculture. One of America's first conservationists, Pinchot advocated a forestry program tied into a nationally unified policy of resource management including efficient use of minerals, control of water power development, and regulation of grazing. Unfortunately, when President Taft succeeded Theodore Roosevelt in 1909, Pinchot lost support in the White House and in Congress for the concept of the government's acting as steward of the land. At this time a feud was developing between the Forest Service and the National Park Service. It was partly a matter of intergovernmental rivalry, because the creation of new national parks meant the transfer of some forestlands from the Forest Service to the NPS. However, there was a deeper dimension to the conflict, echoing a basic split within the conservation movement itself.

The Divided Crusade

By 1900 a national movement was taking shape around the idea that the transformation of the wilderness into "civilization" — farms, rangelands, dammed rivers, railroads, highways, and urban areas — might not be a universal panacea. The crusade to save what was left was divided into two major camps, one led by John Muir, founder of the Sierra Club and the prototypical *preservationist*, the other by Gifford Pinchot, the consummate *conservationist*. Both Muir and Pinchot wanted to save the forests but they differed on how to do it. Their opposing views came to a head over the issue of Hetch Hetchy.* Muir believed the valley must be protected from any human intrusion whatsoever because it was located within a national park. Pinchot, whose position prevailed, felt that the valley was there to serve what he considered "the public good" — in this case to be dammed so that it could provide water for the city of San Francisco.

Recognizing Wilderness

The principle of "usefulness" was incorporated into the multiple-use policy of the Forest Service† and remains even today a thorny issue within and without the conservation movement. Although mandated by the National Forest Management Act of 1976 to maintain a balance between the forest's many

* For details on the Hetch Hetchy case, see page 244.

† For details on the Forest Service's multiple-use policy, see chapter 6, page 191.

users, the Forest Service is perceived as favoring commercial interests over considerations of wilderness, wildlife, watershed, or recreation. As we have seen above, the FS pricing policy results in large amounts of national — that is, public — forest timber being sold to private companies below cost. In June 1984 the federal General Accounting Office analyzed more than three thousand timber sales in the western regions and found that in 1981, 27 percent, and in 1982, 42 percent of the sales did not cover FS costs, accounting for a $156 million loss to the taxpayers in just those two years. On the basis of simple economics, the GAO concluded, "some national forests should not be managed for timber production."[36]

Integrity of the Forests

The loss of hundreds of millions of dollars each year in below-cost timber sales became an issue of great concern not only to environmentalists but to many others appalled by the hidden use of public money to subsidize private interests. In 1985 the Wilderness Society and other environmental groups filed lawsuits challenging the Service's plans for the San Juan and Grand Mesa national forests in Colorado. Part of their argument was that the Wilderness Act of 1964* established wilderness in a wide array of resource values as a legitimate use and land-management category of the national forests.[37] Outdoor recreation, the society says, is only one of many reasons for maintaining wilderness. More significant is the unsurpassed value of wilderness as a means of maintaining the integrity and quality of watersheds — half of all the water in the West flows out of the national forests, for example. Then there is the vital need for biological diversity and to preserve a wide variety of habitats to enrich the great gene pool that contains the future of all life. As *Wilderness in America* explains, we need more wilderness simply because the living quality of our society demands it. "Wilderness is a museum, a biological fountain, a scientific resource from which we can learn — and to which we can turn for some of our most basic needs."[38]

The National Wilderness Preservation System

"When we see land as a community to which we belong, we may begin to use it with love and respect."
— Aldo Leopold

The idea of setting aside an administratively designated wilderness was given form in 1924 when Aldo Leopold, one of the founders of the Wilderness Society, persuaded the Forest Service to so designate a 540,000-acre roadless area in New Mexico's Gila National Forest. Leopold's "land ethic" was institutionalized forty years later in the Wilderness Act, which described wilderness as "an area where the earth and

* For details on this landmark legislation, see chapter 9, pages 283-284.

its community of life are untrammeled by man, where man himself is a visitor who does not remain." Although 9 million acres of national forest were immediately designated by the Forest Service, the Wilderness Preservation System remained a mere skeleton of what the act intended. It took twenty-five years for the system to reach a significant dimension, totaling in 1989 some 90 million acres of public lands stretching from 7 million acres in Alaska's Gates of the Arctic National Park to 36 acres on Florida's Passage Key National Wildlife Refuge. As the Wilderness Society said at that time, the United States must address the future of at least 90 million more acres of wildlands that remain unprotected.[39]

Where's the Wilderness?

- Nationwide there are 474 wilderness areas designated as national wildlife refuges (NWRs) — forty-three in Alaska and the remainder in the lower forty-eight states and Hawaii, except for Connecticut, Rhode Island, Delaware, Maryland, Kansas, and Iowa, which have none.
- These areas represent every major biome species in North America, containing thousands of plant, animal, and insect species including 600 bird, 220 mammal, and 63 federally listed endangered species.
- The nation's largest wilderness (8.7 million acres) is in Wrangell-St. Elias National Park; the smallest (6 acres) is the Pelican Island National Wildlife Refuge in Florida, which was the first national wildlife refuge in the U.S., established in 1903.
- The Northeast has the smallest amount of wilderness, with less than 200,000 acres (two-tenths of 1 percent of the total U.S. wilderness) in the eleven states from Maine to Maryland, which have one-quarter of the country's population.
- Fifty-six and a half million acres of wilderness — 62.3 percent of the total — are in Alaska, with most of the rest (one-third) located in the West.
- Outside of Alaska, California has the most wilderness (almost 6 million acres), with Washington and Idaho each having roughly 4 million acres each.

(Sources: The Wilderness Society and the Sierra Club)[40]

The National Wildlife Refuge System

In 1988 the Wilderness Society reported that "many of our 445 national wildlife refuges are refuges in name only."[41] A few months later the U.S. General Accounting Office followed with *National Wildlife Refuges: Continuing Problems With Incompatible Uses Call for Bold Action*. The eighty-four-page report confirmed that more than 90 percent of the refuges hosted harmful uses unrelated to wildlife, including mining, off-road joyriding, powerboating, waterfowl hunting, and military air exercises. Heavy demands for such uses were diverting the refuge staff's attention from wildlife and habitat management, the report warned.[42]

Systemwide Problems

The Fish and Wildlife Service has long acknowledged that all was not well in the system. Its 1982 draft report to Congress identified 6,956 threats to the natural resources of individual refuges that "will continue to degrade certain fish and wildlife resources until such time as mitigation measures are implemented."[43] But when the final version of the report was published these references to threats were taken out. As the Wilderness Society explained, Reagan appointees at the Interior Department didn't want to heighten concern about the health of the refuges because their goal was not to protect wilderness but to expand "economic" operations such as cattle grazing, farming, timber cutting, hydroelectric generation, tourist concessions, commercial hunting and fishing, and oil and gas drilling. The problems were, and still are, serious, and endemic throughout the system. One of the greatest threats is the development and exploitation of refuge resources for profit. In addition to inappropriate tourist development at some locations, you can find oil and gas rigs at more than fifty refuges, livestock grazing on nearly one hundred, and military bombing and strafing at others. There are widespread habitat losses affecting the migration of millions of birds, degradation of water, destruction of wetlands and other ecosystems, massive fish die-offs, and the widespread killing of endangered animal species. Some idea of the extent of the threats may be seen in the following review of the ten most endangered national wildlife refuges on the Wilderness Society list:[44]

The Ten Most Endangered Wildlife Refuges

1. *Arctic National Wildlife Refuge* (19.3 million acres) stretches across northeast Alaska on the Canadian border. Together with adjacent protected lands in Canada, it forms North America's largest wildlife refuge. Bisected by the peaks and glaciers of the eastern Brooks Range, the refuge is best known for its northern portion, the coastal plain running along the Arctic Ocean. Designed to embrace the range of the great Porcupine caribou herd, ANWR is home to free-roaming herds of musk-oxen, Dall sheep, packs of wolves, and to wolverines, polar and grizzly bears, and other solitary species.*

2. *Key Deer Refuge* (7,500 acres), located on fifteen of the western Florida Keys, was created to prevent extinction of the tiny (27-inch-high) Key deer, which was one of the first additions to the Endangered Species List. The refuge is also an important haven for the survival of manatees, peregrine falcons, bald eagles, and other endangered creatures, as well as the endangered Key tree cactus. Every year as much as one-fifth of the Key deer population is killed by cars on U.S. 1 and state roads, while poorly planned residential construction gobbles up land vital to a healthy deer population and critical to the ecological integrity of the refuge. Additional problems include water pollution from inadequately treated domestic waste and unmanaged boating threatening wading birds and other wildlife.

3. *Stillwater Refuge* (24,000 acres), 75 miles east of Reno, Nevada, forms part of the 165,000-acre Stillwater Wildlife Management Area, consisting of

* The problems of the ANWR are examined at greater length on pages 160-165 and pages 284-285.

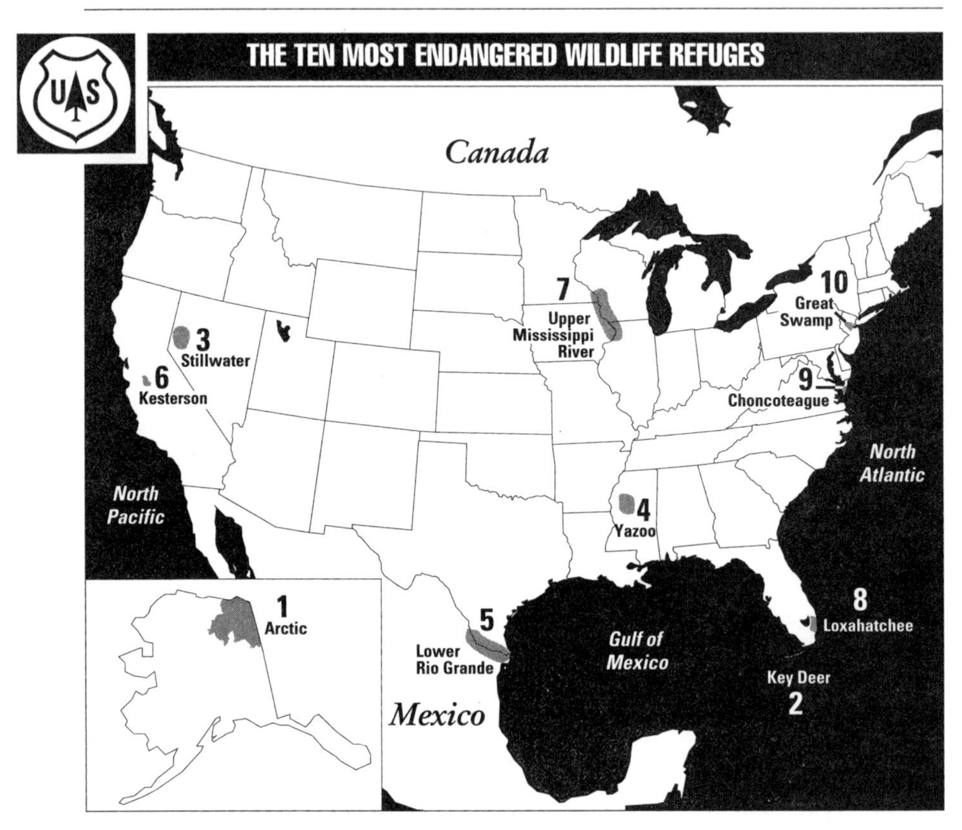

desert, marshes, ponds, and open salt water. Its prolific wildlife includes more than 160 bird species, especially waterfowl (an estimated quarter of a million stop over to refuel) and the endangered bald eagle. As one of North America's most important inland staging areas for migrating birds, Stillwater was designated a unit of the Western Hemisphere Shorebird Reserve network. Located at the terminus of the Carson and Truckee rivers, this area receives all of its water from a U.S. Bureau of Reclamation irrigation project and is the dumping ground for a huge amount of heavy metal- and pesticide-laden agricultural runoff that has killed hundreds of thousands of birds and fish. Because of dams and water diversion, less than 5 percent of the management area's marshes (including those at the nearby Fallon National Wildlife Refuge) remain, seriously threatening two endangered fish: the giant Lahonton cutthroat trout and a sucker called by its Native American name, *cui-cui*.[45] Other afflictions visited upon Stillwater and Fallon are dogfights and bombing exercises from the Fallon Naval Air Station, ORVs, unmanaged boat traffic, and poaching.

4. *Yazoo Refuge* (12,500 acres), 30 miles south of Greenville, Mississippi, is part of the Yazoo Delta in the northwest of the state. Its lakes, marshes, bottomland forests, and fields provide winter habitat for migratory wildfowl and other wildlife. The refuge is seriously threatened by a series of U.S. Army Corps of Engineers operations that comprise the largest drainage project in American history and that, if continued, will drain half of Mississippi's remaining 4 million acres of delta hardwood bottomlands, at a projected cost of $2 billion.

5. *Lower Rio Grande Refuge* (26,500 acres) runs along 200 miles of the Rio Grande River, from Boca Chica Beach to Falcon Dam, Texas. Situated at the intersection of four climatic and geographical environments (river valley, Gulf coastal plains, subtropical uplands, and Chihuahuan Desert), the refuge is known as an international biological crossroads. It also lies at the intersection of two major bird migration routes and links ten distinctive vegetative communities. Biologists say that the region's incredible plant and wildlife diversity is rivaled in the U.S. only by the southern tip of Florida. It provides habitat for 525 wildlife species, including the endangered ocelot, jaguarundi, brown pelican, and peregrine falcon. Ninety percent of the native brushland has been wiped out for agriculture and other uses, breaking up the wildlife habitats the refuge is intended to preserve. Because it is fragmented into many small tracts, it is particularly susceptible to agricultural runoff and the effects of more than one hundred pesticides regularly applied to the surrounding cropland. Another threat comes from building projects, especially a huge proposed resort, Playa del Rio, that may be built near the mouth of the Rio Grande and that, according to the FWS, would have an unacceptably adverse impact on wetlands, shellfish beds, fisheries, wildlife habitat, and recreation areas. There is also danger of wildlands flooding from the building of dams and reservoirs near Brownsville.

6. *Kesterson Refuge* (6,000 acres), 18 miles north of Los Banos, California, contains scattered remnants of wildlife that include migrating waterfowl and more than two hundred other bird species as well as the endangered San Joaquin kit fox. Created to mitigate the loss of wetlands caused by a Bureau of Reclamation project, the refuge incorporates native grasslands, shallow marshlands, and unique vernal pools. Toxic agricultural runoff was so great and so many birds were killed that by 1984 the FWS had to use shotguns and explosive devices to scare away wildlife heading for Kesterson's tainted pools lest they too became victims. The refuge is now officially closed. (For more on Kesterson and its impact on San Francisco Bay, see chapter 3, page 91.)

7. *Upper Mississippi River Wildlife and Fish Refuge* (90,000 acres) consists of riverside parcels along 284 miles of the river from Rock Island, Illinois, to Lake City, Minnesota. They include marshes, sandbars, and wooded islands used by waterfowl traveling the Mississippi Flyway and other wildlife. With more than three million visitors annually, it is the most heavily used refuge in the entire NWR system. "Each year we see an increase in recreational use, particularly boating craft on the river, and we're concerned about how this may affect our heron rookeries, eagle nesting and other wildlife species," said refuge manager James Lennartson.[46] Several other outdoor activities — ORV and snowmobile driving, especially — are clearly incompatible with the refuge's mission. A major problem is soil erosion from poor agricultural practices and badly managed development. This causes widespread sedimentation, which is often dumped on the refuge, covering wildlife habitat, changing drainage patterns, and depositing concentrated pollutants on the land.

8. *Loxahatchee Wildlife Refuge* (145,000 acres), about 10 miles west of Boynton Beach, Florida, is the only substantial remnant of the northern Everglades natural system and the only protected portion of that region.

Located at the southern end of the Atlantic and Mississippi flyways, it hosts some of North America's rarest plants and animals. The area is a mixture of tree islands, saw grass, wet prairies, and sloughs. The wildlife includes wood storks and snail kites (both endangered), alligators, Fulvous whistling ducks, white ibis, herons, and egrets. Loxahatchee's problems are mostly water-related. Construction of canals and levees has dramatically changed the natural flow of water and the alternation of wet and dry seasons characteristic of the Everglades. Development of private airports, subdivisions, roads, and other facilities within the watershed has created increased dependency on the refuge for flood control, so that excessive amounts of water are sometimes sent there. The abrupt changes in water flow and levels damage nesting and reduce food supplies for many wading birds. For the wood stork, for example, a 2-inch increase in water level can be a margin of survival because it disperses the fish that the stork depends on.* Excessive nitrogen and phosphorous from agricultural runoff are changing the vegetation at an alarming rate. Along the edges of the refuge, prime saw grass is being replaced by cattails and other undesirable plants, destroying vital breeding and nesting habitats. Contaminants in the runoff have led to high levels of toxins in fish and birds. A proposed solid-waste disposal plant on several hundred acres just east of the refuge would in all probability leak pollutants into the groundwater.

9. *Chincoteague Wildlife Refuge* (9,500 acres) on the south end of Assateague, a 45-mile barrier island straddling the Maryland-Virginia border, was created to protect snow geese and other migratory birds. It is well known for its wild horses, descended, according to legend, from mustangs from a sixteenth-century Spanish ship wrecked offshore. Lying along the Atlantic Flyway, it features salt marshes, ponds, and stands of loblolly pine and oak. Its wildlife includes piping plover (threatened), Delmarva Peninsula fox squirrel, and peregrine falcon (both endangered). Chincoteague is home for 70 percent of all piping plovers that nest in national wildlife refuges. A major threat comes from increasing numbers of summer visitors. Beachgoers disturb wildlife nesting. ORVs driving along the southern end of the refuge's beach destroy sand dunes, cause erosion, and disrupt wildlife. Land is appropriated for parking lots, bathhouses, and other tourist facilities. To its credit, the FWS closed one section of beachfront during plover nesting season; but a master plan restricting public beach use elsewhere remains to be implemented. FWS clear-cutting of more than 40 acres in the southwest of the island resulted in habitat loss for the Delmarva squirrel and many bird species.

10. *Great Swamp Wildlife Refuge* (6,936 acres), in Basking Ridge, New Jersey, 26 miles west of New York City, consists mainly of cattail marsh, hardwood ridges, swamp woodland, and grassland with many large old oak and beech trees, as well as stands of mountain laurel and rhododendron. Wildlife includes wood duck, red and gray fox, and white-tailed deer plus blue-spotted salamander and bog turtle (both endangered). More than three hundred species of creatures have been recorded there. Great Swamp is at the

* See chapter 2, page 71.

receiving end of polluted runoff from many sources including two sewage plants, road salts in wintertime, and fertilizer and pesticides from farms, golf courses, and suburban lawns. The impact spreads beyond the refuge, which acts as a filter for the Passaic River, the source of domestic water for downstream communities. Within the refuge are a 5-acre asbestos dump (toxic enough to be a Superfund candidate) and two landfills suspected of containing hazardous substances. Urban development in the watershed causes greater and faster water runoff. The resulting sharp changes in water flows into the refuge imperil waterfowl and other marsh life. Growing traffic volume through Great Swamp has led to more roadkills, dust, noise, and pollution in the refuge. The FWS proposes to cut down red maples

In Land We Trust

In an era of shrinking government funds there has been a sizeable growth of land trusts, which are usually privately-run, nonprofit groups dedicated to acquiring and preserving open spaces that would otherwise be developed. Many of these spaces are in critical areas on coastlines or bordering on national and state parks. Today there are 900 land trusts in the United States, twice as many as a decade ago, with holdings of about 2.7 million acres. Increasingly when land goes on the market, public funds are not available to buy it. One of the larger trusts is the San Francisco-based Trust for Public Land (TPL), which works to protect land for people to enjoy as parks and recreation areas. The Jackson Hole Land Trust, for example, was founded with TPL assistance to protect ranch lands at the gateway to Grand Teton National Park in Wyoming. Similarly, TPL assisted the Adirondack Land Trust in upstate New York and the Marin Agricultural Land Trust near Point Reyes National Seashore in California.

Many of the land trusts are located in the Northeast, where there is greater pressure to develop remaining areas of public land and wilderness; there are 116 in Massachusetts, 114 in Connecticut, and 64 in New York, according to the Land Trust Alliance, a national organization providing services and programs for regional and local land trusts. The trusts, which have attracted wide public support, vary widely in their approaches. They get most of their funds from tax-deductible, private contributions. Some of them own the land they manage. Others oversee conservation easements, which means that someone else holds title to the property but has agreed to leave it in its natural state. Larger land trusts such as TPL operate nationally in the U.S. and in Canada; others are based in individual towns and communities.

To find out more about land trusts, contact:

— *The Land Trust Alliance*, 900 Seventeenth Street, NW, Suite 410, Washington, DC 20006; (202) 785-1410. Executive director, Jean W. Hocker.

— *The Trust for Public Land*, 116 New Montgomery Street, 4th floor, San Francisco, CA 94105; (415) 495-4014. Regional offices in Boston, New York, Seattle, Santa Fe, New Mexico, and Tallahassee, Florida. President, Martin J. Rosen.

and other trees and replace them with shrubs and other vegetation that, they say, would attract woodcocks. But it would also cause habitat loss for migratory songbirds and other birds that need old-growth wooded areas, which are scarce in this highly urbanized region.

What We Can Do to Save the Public Lands

The public lands of the United States — the forests, the seashores, the parks, rivers, refuges, and wilderness — are a priceless heritage of which we are both the beneficiaries and the stewards. If they are to survive in an ecologically intact way, we must be prepared to take radical action to save them and to pay the price it will cost. Such action can best be carried out in light of our understanding both the causes of the threats and what is at risk if we allow them to go unchecked. Certainly, such ecologically harmful practices as clear-cutting, mining, excessive cattle grazing, oil and gas exploitation, and indiscriminate use of ORVs on public lands must be more effectively controlled than they are at present.

Central to any long-term solution must be increased awareness by the managers of our public lands of the need to preserve and conserve the ecosystems to which those lands belong. Most of those ecosystems are not confined to the boundaries of one public agency or another, nor are they necessarily limited to publicly owned lands. Once this is understood — that, for example, air pollution from Los Angeles can seriously affect the Grand Canyon — we must as concerned and responsible citizens press our government to take the legislative, judicial, and executive action necessary to safeguard the future of our public lands *and* pay the cost of the investment.

As we shall see in the following chapter, an impressive body of laws designed to protect the environment has been built up over a long period of time. Many of these laws mandate the responsibilities of those agencies of government that have jurisdiction over the public lands. When an agency such as the U.S. Forest Service loses sight of its prime responsibility to the land of which it is the steward, it must be corrected by the full power of our governmental system guided by an understanding of ecological principles and consequences. Administrative boundaries between one type of public land and another or between public and private lands must be replaced by "resource boundaries" that respect natural ecosystems. The Greater Yellowstone Complex, for instance, could be a forerunner of this future type of land management. A North American Wilderness Recovery Strategy as proposed by *Wild Earth* outlines how this can be done: The strategy would be based on identifying large core wildernesses, surrounding them with buffer zones where human activity is limited, connecting the cores together with biological corridors, and assisting nature in restoring native ecological conditions complete with large predators.[47] Such a strategy presupposes moving government agencies (and other decision makers) away from traditional "enclave" management to a more ecologically-balanced system of resource management. A great many people, interest groups, and public agencies have a lot of investment in the traditional boundary lines, which define their turf.

This explains why legislation intended to preserve public lands but that would cut across traditional boundaries often languishes in Congress. Yet, paradoxically, Congress has moved forward with important environmental laws such as the Clean Water Act and the Wild and Scenic Rivers Act that are "almost, by definition, ecosystem-based and resource-oriented."[48] As we have indicated earlier, some public agencies seem to be more receptive than others to an ecosystems approach to solving the problems they confront. A case in point is the National Park Service. "As we gained knowledge through research on the migration of plants, we have learned better to manage the parks. We are constantly looking for the possibility of defining the boundaries of parks topographically as well as ecologically," says William Penn Mott, Jr. Although expanding park boundaries is often restricted by skyrocketing land costs, lack of congressional funding, and other factors, Mott explains, "We should still be able in the limited areas we control to manage plant and animal communities so as to preserve the species and their gene pools in a healthy condition."[49]* Where the awareness has been conspicuously absent since 1980 is in the presidency and the administration. Without leadership and support from the White House it is unlikely that true progress can be made in saving our public lands. If we want to protect them from further deterioration, we must vigorously press the president (and, of course, Congress) to become stewards of our public lands on our behalf in the broadest ecological context possible.

RESOURCES

1. U.S. Government

The leading federal departments and agencies dealing with public lands are:

— *The Department of Agriculture*, with jurisdiction over the Forest Service and the Soil Conservation Service.

— *The Department of the Interior*, with jurisdiction over the Bureau of Indian Affairs, the Bureau of Land Management, the Bureau of Mines, the Bureau of Reclamation, the Geological Survey, the National Park Service, the Office of Surface Mining Reclamation and Enforcement, and the U.S. Fish and Wildlife Service.

For addresses, phone numbers, and other details, see the Directory, pages 352-353.

2. Environmental and Conservation Organizations

Virtually every environmental and conservation group is concerned with public land issues. Except where provided in this chapter, addresses, phone numbers, and other details on these groups are listed in the Directory.

3. Further Reading

— *America 200: The Legacy of Our Lands*, Virginia S. Hart, editor. Washington, D.C.: U.S. Department of the Interior, 1976.

— *Beyond the Hundredth Meridian: John Wesley Powell and the Second Opening of the West*, Wallace Stegner. Lincoln: University of Nebraska Press, 1982.

— *The Birth of the National Park Service*, Horace M. Albright. Salt Lake City: Howe Brothers, 1985.

* The need to save habitat in order to save species is examined from the perspective of a leading conservation biologist in chapter 10, pages 306-307.

— *The Bureau of Land Management*, Marion Clawson. New York: Praeger, 1971.
— *Dreamers and Defenders : American Conservationists*, Douglas H. Strong. Lincoln: University of Nebraska Press, 1988.
— *The Greater Yellowstone Ecosystem: Redefining America's Wilderness Heritage*, Robert B. Keiter and Mark S. Boyce, editors. New Haven: Yale University Press, 1991.
— *How Not to be Cowed: Livestock Grazing on the Public Lands*, An Owner's Manual, Johanna Wald, Ken Rait, Rose Strickland, and Joe Feller. San Francisco: Natural Resources Defense Council, 1991.
— *The Lands No One Knows*, T. H. Watkins and Charles S. Watson, Jr. San Francisco: Sierra Club Books, 1975.
— *Mountains Without Handrails: Reflections on the National Parks*, Joseph L. Sax. Ann Arbor: University of Michigan Press, 1980.
— *National Parks: The American Experience*, A. Runte. Lincoln: University of Nebraska Press, 1987.
— *Our Common Lands: Defending the National Parks*, David J. Simon, editor. Washington, D.C.: Island Press, 1988.
— *Our Landed Heritage*, Roy M. Robins. Princeton, N.J.: Princeton University Press, 1976.
— *Our National Park Policy: A Critical History*, John Ise. Baltimore: Johns Hopkins University Press, 1961.
— *Playing God in Yellowstone: The Destruction of America's First National Park*, A. Chase. Boston: Atlantic Monthly Press, 1986.
— *These American Lands*, Dyan Zaslowsky. New York: Henry Holt, 1986.
— *The U.S. Forest Service: A History*, Harold K. Steen. Seattle: University of Washington Press, 1976.
— *The Waste of the West: Public Lands Ranching*, Lynn Jacobs. Tucson, Arizona: Lynn Jacobs Publications, 1992.
— *Wilderness and the American Mind*, Roderick Nash. New Haven: Yale University Press, 1973.
— "Winning [and losing] the West: The Perplexing Saga of the Bureau of Land Management's Wilderness Review," James Baker, *Sierra*, May/June 1985.
— *Yosemite: The Embattled Wilderness*, Alfred Runte. Lincoln: University of Nebraska Press, 1990.

NOTES

1. We are especially indebted to William Penn Mott, Jr., and his colleagues at the National Park Service Western Region for the valuable help they provided for this chapter.
2. Dyan Zaslowsky, *These American Lands* (New York: Henry Holt, 1986), p. 4.
3. *Real Property Owned by the United States Throughout the World*, U.S. General Services Administration.
4. Cited in Zaslowsky, *These American Lands*, p. 139.
5. Ibid., p. 140.
6. Richard Conniff, "Treasuring 'the Lands No One Wanted.' Environmentalists Take Heart as the BLM Looks Past Mines and Ranches to a New Era," *Smithsonian Magazine*, September 1990, p. 34.
7. George Catlin, *North American Indians*, vol. 1 (Philadelphia, 1913), pp. 294–95, cited in Roderick Nash, *Wilderness and the American Mind* (New Haven: Yale University Press, 1973), p. 101.
8. Ibid.
9. Cited in Joseph Sax, *Mountains Without Handrails: Reflections on the National Parks* (Ann Arbor: University of Michigan Press, 1980), p. 6.
10. Zaslowsky, *These American Lands*, pp. 321–22.
11. *Wilderness in America: A Vision for the Future of America's Wildlands* (Washington, D.C.: The Wilderness Society, 1989), p. 18.
12. Ibid., p. 15.
13. Bill Willers and Jasper Carlton, "Last Chance to Save the Yellowstone Ecosystem," *Wild Earth*, Winter 1991/1992.
14. One of the most recent and comprehensive sources is *The Greater Yellowstone Ecosystem*, edited by Robert B. Keiter and Mark S. Boyce (New Haven: Yale University Press, 1991).

15. Ibid.
16. Ibid., p. 15.
17. Ibid.
18. Editorial update in Bill Wllers and Jasper Carlton, "Last Chance to Save the Yellowstone Ecosystem," *Wild Earth*, Winter 1991-1992, p. 47.

19. Ibid., p. 48.
20. Cited in Peter Matthiessen, "Yellowstone Burning," *New York Times Magazine*, September 11, 1988, p. 128.
21. Cited in Thomas Hackett, "Fire," *The New Yorker*, October 2, 1989, p. 73.
22. Matthiessen, "Yellowstone Burning," p. 123. See also "Yellowstone's Burning Question," *Nova* TV Show #1619, December 5, 1989, a transcript of which may be ordered from Journal Graphics, New York.
23. Matthiessen, "Yellowstone Burning," p. 128.
24. R. Drummond Ayres, Jr., "Pollution Shrouds Shenandoah Park," *New York Times*, May 2, 1991.
25. Ibid.
26. Report from National Public Radio, June 30, 1991, on chemical tracer studies done at the Grand Canyon and the University of California at Irvine. See also Seth Mydans, "Grand Canyon's Air Is Being Polluted on 2 Fronts," *New York Times*, October 15, 1989.
27. Robert Reinhold, "Aisle Seats to a National Park? Well, Maybe," *New York Times*, February 18, 1991.
28. Alfred Runte, *Yosemite: The Embattled Wilderness* (Lincoln: University of Nebraska Press, 1990).
29. Personal communication, December 6, 1991.
30. For a more detailed examination of this issue, see Tom McNichol, "The Great Divide," *San Francisco Chronicle Image*, June 2, 1991.
31. See Jeanne McDowell, "Fighting for Yosemite's Future," *Time*, January 14, 1991, p. 46.
32. Robert Reinhold, "MCA Agrees to Sell Interest in Yosemite to End Dispute," *New York Times*, January 9, 1991.
33. Personal communication, August 13, 1990.
34. Ibid.
35. Cited in Douglas H. Strong, *Dreamers and Defenders: American Conservationists* (Lincoln: University of Nebraska Press, 1988), p. 71.
36. GAO report cited in Zaslowsky, *These American Lands*, p. 98.
37. The Wilderness Society position on this subject is outlined in *Wilderness in America*, passim.
38. Ibid., p. 8.
39. Ibid., p. 6.
40. See especially "The National Wilderness Preservation System," pamphlet (San Francisco: Sierra Club, 1988).
41. "Ten Most Endangered Wildlife Refuges," Wilderness Society, Washington, D.C., 1988, p. 1. Referred to hereafter as "Endangered."
42. GAO/RCED-89-196, pp. 16–30.
43. Wilderness Society, "Endangered," p. 2.
44. Based on material from John Mitchell, "You Call This a Refuge," *Wildlife Conservation*, March/April 1991, pp. 70-93; GAO/RCED Report-89-196, especially Appendix IV; and from the Wilderness Society, National Audubon Society, Sierra Club, and other sources.
45. George Laycock, "What Water for Stillwater?," *Audubon*, November 1988, pp. 12-25.
46. Elizabeth Dawn Amato, "Are National Refuges Safe for Wildlife?" *Outdoor America*, Spring 1991, p. 15.
47. *Wild Earth*, Spring 1991, has ecological foundations for big wilderness as its issue theme. Its articles include: "Greater Yellowstone Ecosystem Marshall Plan" by George Wuerthner; "Dreaming Big Wilderness" by Dave Foreman; and "A Native Ecosystems Act (Concept Paper)" by Reed Noss. See also *Wild Earth*, vol. 1, no. 4.
48. Joseph L. Sax, Foreword to *Our Common Lands: Defending the National Parks*, edited by David J. Simon (Washington, D.C.: Island Press, 1988), p. xiv.
49. Personal communication, December 6, 1991.

LAWS OF THE LAND

"Public land law is at the core of the history of national economic development."

— George Cameron Coggins and Charles F. Wilkerson, *Federal Public Land and Resources Law*

As we have noted throughout this book, there is a considerable body of law dealing with issues affecting ecosystems — ocean dumping, damming of rivers, the use of national forests, and endangered species, for example. In this chapter we provide an overview of natural resources law, which is designed primarily to preserve and protect public resources — water, rangeland, minerals, forest, and wildlife. We also include a short section on key environmental legislation, which concerns itself with the consequences of the use of these resources.*

The doctrine that certain resources are held by the state for the benefit of all the people as a public trust has roots in the Roman concept of *res communes* (the common thing) and the English idea of the commons. It is also central to the beliefs of the original inhabitants of North America, for whom land, natural resources, animals, and people are part of a great whole.[1]

Initially limited to rights of access to and the use of waterways, the concept of national public domain was first invoked in 1778 when the small, land-poor state of Maryland protested against Virginia, Massachusetts, and other "landed" states that were reluctant to give up vast lands in the West granted to them under the royal charter. All lands, Maryland claimed, had been fought for by "common blood and treasure" of the thirteen states and must be commonly owned.[2] Two years later, the Continental Congress agreed to the claim, making it a condition of membership in the federal union. Congress also resolved that any unappropriated lands should be disposed of "for the common benefit of the United States" and be settled and formed into new states that would become members of the Union, enjoying equal rights with the others. By 1802 the lands were transferred, and the new government became owner of more than 233 million acres of land. The area doubled one year later when President Thomas Jefferson agreed to pay France $27 million for its Louisiana Territory. By 1854, as a result of further purchases, diplomacy, and war, the federal sovereignty extended to more than 1.4 billion acres. The Northwest Ordinance of 1785 defined how the public lands were to be surveyed and sold by auction to settlers moving west, but there was a basic philosophical difference as to who should settle the land. Jefferson wanted to put the land into the hands of yeoman farmers, which meant selling it cheap, while Alexander Hamilton, believing that land sales represented the only sure source of income for the government, wanted it sold at the highest price possible. However, before the debate could be resolved, speculators and squatters moved in.

By 1828, for example, two-thirds of the residents of the new state of Illinois lived on land belonging to the federal government. Accepting the inevitable, Congress, with the *Preemption Act of 1841*,† legitimized this de facto occupation by allowing squatters to stake a claim to unsurveyed land at $1.25 an acre. It also

* A more detailed examination of environmental law is given in our companion book, *Design for a Livable Planet*, on pages 228-257.

† Preemption, the preferential right of settlers on public lands to buy their claims at a modest price, had been generally conceded, except in New England, to persons at the edge of the frontier by the time of the American Revolution.

began giving land away to promote settlement and as gifts to wagon-road companies (3 million acres), railroads (94 million acres), land-grant colleges (77 million acres), and swamp reclamation projects to provide land for farmers (65 million acres). The final blow to the Hamiltonians was the *Homestead Act of 1862*, the impact of which we shall examine below (page 279).

The Water Resource

We begin with water because all other resource uses are dependent on its availability. Mining, farming, and livestock raising all require great quantities. Timber production depends on water for growing trees, and rivers are used for transporting logs to mills as well as for navigation, recreation, and, in many places, for generating electricity. And as we have seen in the preceding chapters, water is an essential part of the ecosystems we need to preserve for future generations.

According to traditional English common law, if you owned land adjacent to a waterway, you acquired riparian rights in it as part of your estate in real property and you were entitled to the full "natural flow" of the stream. This meant that no one could use the water for other than "natural uses" such as for the household or farm. In-stream uses for mills were allowed insofar as they preserved or did not diminish the natural flow.

In the United States this riparian doctrine was modified in response to the demands of fast-growing commerce, industry, and agriculture, leading to a rule of "reasonable use" based on a balancing test to resolve competing riparian rights. A second doctrine — that of prior appropriation — resulted from the government policy of encouraging water development for economic purposes in the water-scarce West. An appropriator, whether or not the owner of riparian land, obtained a vested property right superior to all later users, if water was diverted out of the stream for "beneficial" purposes — a term that generally included domestic and economic uses such as mining, manufacturing, agriculture, and livestock watering. "First in time, first in right" became the watchword, even if the first user took all the water in the stream.[3]

Up until the early part of this century, states developed and enforced their own systems of water law, ignoring for all intents and purposes the role of federal law. But in 1908 the U.S. Supreme Court recognized the doctrine of federal reserved water rights.[4] The full implications of this doctrine did not become apparent until 1963, when Arizona filed an action against California for an allocation of the Colorado River's water.[5] Although the basic dispute concerned the amount of water each state had the right to use, alleged federal rights were also involved. In its ruling the Supreme Court supported the principle of reservation of water rights by "federal establishments such as National Recreation Areas and National Forests." Although it rejected the claim of the United States that it was entitled to use the water without charge, it agreed that "all uses of mainstream water within a State are to be charged against the State's apportionment, which of course includes uses by the United States."[6]

The following is a summary of the major legislation relating to the water resources of the United States.

Clean Water Act

The Clean Water Act of 1977 (CWA)[7] is the culmination of many pieces of legislation on water pollution control including the important *Water Pollution Control Act of 1972* (WPCA),[8] which optimistically envisioned fishable, swimmable waters by 1983 and the elimination of the discharge of pollutants into the waterways two years later. To achieve these goals WPCA initiated a $5-billion-a-year federal grants program to finance construction of local sewage treatment systems; it also required all municipal and industrial wastewater to be treated before being discharged into waterways. The Environmental Protection Agency was required to set federal limits on the amounts of specific pollutants that could be released by municipal and industrial facilities based on the level of cleanup achievable through existing technologies and taking into account costs to the regulated community.

Although industrial discharges have been generally brought under control, many estuaries, lakes, and rivers, especially in heavily populated areas, still suffer from degradation. In seeking to correct these stubborn problems Congress, over a presidential veto, passed the *Water Quality Act* of 1987.[9] The goal of this amendment to the CWA was "to restore and maintain the chemical, physical, and biological integrity of the Nation's waters." Enhancing the EPA's power to prosecute water polluters, the act directs the EPA and state officials to supplement existing technology-based standards with a water-quality approach to control persistent pollution problems.

Addressing the many "hot spots" where concentrations of toxic pollutants were unacceptably high, the 1987 law requires states to identify such spots, determine their specific sources, and develop individualized control strategies to eliminate the problem within three years. However, the EPA's narrow interpretation of a hot spot has brought its implementation of this provision under attack. The amendment attempted to fill one gigantic gap in the CWA. Roughly half of all water pollution does not come from specific discharges but from nonpoint sources — runoff from agricultural and mining operations, road salting, construction sites, and city streets, for example. Unfortunately the permit program does not apply to these sources, and for all intents and purposes they remain unregulated.

Section 404 of the CWA requires a permit to place any dredged or fill material into "waters of the United States." "Waters" are defined broadly and include wetlands. Permitting authority lies with the U.S. Army Corps of Engineers. A permit can be vetoed by the EPA, but the agency has been generally reluctant to use this veto power. The key requirement for the grant of a permit is a "showing" or demonstration that no practicable alternative is available that would not involve filling in waters. The Section 404 permit program has been the source of many pitched battles over permits for subdivisions, shopping malls, marinas, and other developments in swamps and coastal wetlands.

Wild and Scenic Rivers[10]

In 1968 the *National Wild and Scenic Rivers Act* instituted a legislative program to study and protect free-flowing river segments by making them part of the National Wild and Scenic Rivers System. To qualify for inclusion a river must

be undammed and have at least one outstanding resource attribute — recreation, scenery, wildlife habitat, history, geology, or a related feature. Rivers may be added to the system either by act of Congress or by order of the secretary of the interior upon official request by a state.

According to Section 16(a) of the act, "river" means a flowing body of water or estuary or a section, portion, or tributary thereof, including rivers, streams, creeks, runs, kills, rills, and small lakes.[11] Section 2(b) defines the three main designations: *wild river areas* — rivers or sections of rivers free of impoundments and generally inaccessible except by trail, with watersheds and shorelines essentially primitive and water unpolluted. They represent vestiges of primitive America; *scenic river areas* — rivers or sections of rivers free of impoundments, with shorelines or watersheds still largely primitive and shorelines largely undeveloped, but accessible in places by roads; and *recreational river areas* — rivers or sections of rivers readily accessible by road or railroad that may have some development along their shorelines or may have undergone some impoundment or development in the past.

The act makes clear that *all* types of free-flowing rivers are to be brought into the system, including those that are very remote and those that flow through more developed areas, provided that they meet the required criteria. Wild river designation is not intended to "lock up" an area under a strict preservation plan, because Congress recognized that a river corridor can be conserved without depriving current residents of their futures, provided development occurs in ways that recognize the outstanding values of the river area. Congress also recognized the need for partnerships among landowners, local, state, and tribal governments, and federal agencies in determining the future of a river area.

National Rivers Inventory

In 1982 the National Park Service published the National Rivers Inventory (NRI), a listing of rivers in the lower forty-eight states (excepting Montana, which declined to take part) of potential candidates for inclusion in the National Wild and Scenic Rivers System. Although never fully completed, it found that only about 2 percent of the 3 million miles in the lower forty-eight states appeared to be eligible — that is, 1,524 river segments comprising 61,700 river miles. By the end of 1988, the system's twentieth anniversary, only 119 river segments covering 9,200 miles had been protected as wild and scenic. By early 1993 the totals were 153 segments covering 10,500 miles.

Marine Laws

The *Marine Mammal Protection Act* (1972)[12] is intended to protect, conserve, and encourage research on marine animals. The essential feature of the act is a moratorium on any "taking" (*take* is broadly defined to mean "to harass, hunt, capture, or kill") and importation of marine mammals.* This moratorium is subject to

* Marine mammals include whales, dolphins, seals, and manatees.

exceptions (for example, it does not apply to subsistence hunting by Eskimos), but the basic idea is that marine mammals are off-limits.

The *Marine Protection, Research, and Sanctuaries Act* (1972),[13] also known as the *Ocean Dumping Act*, regulates the dumping of material in U.S. waters to protect the marine environment. It forbids outright the ocean dumping of radiological, chemical, and biological warfare agents and high-level radioactive waste, and requires an EPA permit to dump any other material in the ocean. The EPA can issue such a permit only if the need for the dumping (taking into account possible alternative disposal methods) outweighs the harm to human health and welfare (including economic, esthetic, and recreational values) and to marine ecosystems.

In 1977 Congress amended the act to eliminate the then-common East Coast practice of dumping sewage sludge in the ocean. New York City and neighboring New York and New Jersey counties ignored the prohibition and continued to dump tons of sewage sludge each day. The EPA tried to stop this practice but lost in federal court because it was unable conclusively to show environmental harm. In 1988, over vociferous protest from New York City, Congress imposed a flat ban on ocean dumping of sewage sludge after 1991. In June 1992 the last bargeload of New York City sludge was dumped into the Atlantic.

The Coastal Zone Management Act (1972)[14] encourages states to develop, in cooperation with federal and local governments, water- and land-use programs for coastal waters and adjacent shorelines. A 1976 amendment requires the programs to plan for shorefront access, shoreline erosion, and energy facilities; it also provides for 90 percent federal funding of the programs and for 50 percent federal funding to states for acquisition of land for estuarine sanctuaries and public access to beaches. These provisions were updated and strengthened by the *Coastal Zone Management Improvement Act* (1980).

The Port and Tanker Safety Act (1978)[15] empowers the U.S. Coast Guard to supervise vessel and port operations and to set standards for the handling of dangerous substances. The act and the regulations cover the design, construction, alteration, repair, maintenance, operation, equipping, personnel, and manning of vessels and set minimum standards for ballast tanks, oil washing systems, and cargo protection systems. The act also mandates a national program for annual inspection of vessels.

The Act to Prevent Pollution from Ships (1980)[16] requires pollution reception facilities at ports and terminals to be "adequate" to receive the residues and mixtures containing oil or noxious liquid substances from seagoing ships. Akin to this act is the *Oil Pollution Act,*[17] making it unlawful to discharge oil or oily mixtures from tankers in prohibited zones except in cases of securing the safety of a ship, cargo, or saving life at sea.

The Intervention on the High Seas Act[18] authorizes the Coast Guard to take measures on the high seas to prevent, mitigate, or eliminate the danger of harm from any oil spill that poses "a grave and imminent danger to the coastline or related interests of the United States." The determination of whether such a danger exists must be based in part on consideration of threats to human health, to fish and other marine resources, and to wildlife.

The Outer Continental Shelf Lands Act (1978)[19] imposes liability for the consequences of oil pollution from offshore activities in outer-continental shelf waters upon the owners, operators, and guarantors of the sources of that pollution. The act established the Offshore Oil Pollution Fund, administered by the secretaries of the treasury and of transportation, and the Fishermen's Contingency Fund, administered by the secretary of commerce. These funds are designed to ensure that money is available to pay for prompt and effective removal of oil spilled or discharged as a result of activities on the outer continental shelf.

The Rivers and Harbors Appropriations Act (1899),[20] known also as the *Refuse Act*, prohibits the depositing of refuse from any vessel or shore establishment into the navigable waters of the U.S. It is one of the operating statutes for the Army Corps of Engineers, whose regulations govern the issuance of permits to obstruct these waters.

The Marine Plastic Pollution Research and Control Act (1987),[21] implementing Annex V of the International Convention for the Prevention of Pollution from Ships (known as MARPOL, for marine pollution), prohibits the dumping of plastics at sea, severely restricts the dumping of other ship-generated garbage in the open ocean or in the waters of the United States, and requires all ports to have adequate garbage-disposal facilities for incoming vessels. The act applies to all watercraft, including small recreational vessels.

Mineral Resources

During the headlong rush to exploit the West, the federal government, it seemed, was far more concerned with giving land away than protecting it. For example, starting with the discovery of gold in California in 1849, every major mining strike in the West took place on public land. Yet it took twenty-three years for Congress to create a federal law regulating the extraction of federal treasure from federal land.[22] Even then the legislation it enacted — the *General Mining Law of 1872* — was not very restrictive. Under its provisions, you could get a mining patent simply by making a "valid" mineral discovery, paying for a boundary survey, applying to a land office for the land included in the survey, buying it for either $2.50 an acre for surface mining or $5 for underground mining, and investing $100 a year in improvements for five consecutive years. This was the government's share in an estimated $20 billion in gold, silver, and other minerals extracted from the public lands during the busiest mining decades in the developing West.[23] Perhaps even more disconcerting is that this law is still active and being applied in the national forests and in the public lands of the Bureau of Land Management (BLM).

The General Mining Law was originally intended to "dispose" of public lands by allowing prospectors to claim and patent public lands at bargain prices. The extraction of gold, silver, uranium, and other metals, it declares, is the "highest and best use" of more than 400 million acres of

public lands in the western United States. By the same token it leaves public-lands managers and concerned citizens powerless to stop destructive mining operations on public lands, regardless of how great the natural or recreational values of the area.[24] In Grand Canyon National Park, for example, there are today some fifty thousand private mining claims scattered over 2 million acres of surrounding land. Ten uranium mine sites have been developed near the park in recent years, with one of them only 100 yards from the park boundary. Wildlife, scenery, and archaeological sites in and around the park are threatened by this continued development. In 1989 approximately $1.6 billion in gold was mined in Nevada, much of it on public lands, with no payment of royalties returned to the U.S. Treasury.[25]

The need for reforming the General Mining Law has become more urgent in recent years because of the widespread use of "heap-leach" mining in western states. This technology involves the use of cyanide, mercury, arsenic, and other toxic chemicals to leach gold from ore containing as little as the amount in one dental filling per ton. The leaching solution is collected in holding ponds that attract wildlife, killing tens of thousands of birds and seeping into the groundwater.

What You Can Do about the General Mining Law

- Urge your representatives in Congress to completely revise this outmoded law by supporting S. 433, the bill sponsored by Senator Dale Bumpers (D-Arkansas).* Such a reform, endorsed by the Sierra Club, National Audubon, and other environmental organizations, would give the public greater say in all decisions regarding public lands: It would take away mining's special privileges and put it on an equal level with other public land uses, including recreational activities and wildlife habitat. Mining operations would be held accountable for the environmental damage they cause, and mining interests would be required to pay royalties to the public for the use of public lands and for minerals taken from those lands.
- For further information contact: the Mineral Policy Center, Room 550, 1325 Massachusetts Avenue, NW, Washington, DC 20005; (202) 737-1872; or 1872 Mining Law Reform, Sierra Club, 408 C Street, NE, Washington, DC 20002.

Surface Mining Control and Reclamation Act (1977)

Under this act, operators of coal mines affecting more than 2 acres of ground are required to obtain permits from the U.S. Department of the Interior or from a state with a DOI-approved program.[26] The permit must contain a reclamation plan and general provisions requiring the operator to prevent damage to the environment and to restore the land to approximately its original contours and conditions. New surface coal mining is prohibited within 300

* For an update on this and other bills in Congress, see Appendix B, "Pending Legislation," pages 357-363.

feet of a public or community building, a public park, or within 300 feet of any occupied building without the consent of the owner. Title IV of the act set up an Abandoned Mine Reclamation Fund, financed by a reclamation fee on coal produced by surface or underground mining, or by 10 percent of the value of the coal at the mine, whichever is less.

The Laws of the Range

Perhaps no single law has changed the face of a country more than the *Homestead Act of 1862*. Under its provisions, any American citizen twenty-one years old or the head of a household could claim 160 acres of public land, live on it for six months, and then buy it for $1.25 an acre, or live on it for five years and cultivate it, then receive title for nothing more than a nominal filing fee.

"One hundred and sixty acres. If anything unifies the story of the American West — its past and its present, its successes and its dreadful mistakes — it is this mythical allotment of land."

— Marc Reisner, *Cadillac Desert*[27]

Between 1862 and 1882 more than a half million entries were filed in what soon became a mad scramble for the heartland of America and beyond. By 1900 some 80 million acres of public land were virtually given away, but many of the recipients were not the small family farmers the Homestead Act was intended to help. In the post-Civil War years rampant fraud, usually in the form of unchecked fake entries to the General Land Office, which was responsible for administering the act, enabled large mining, ranching, and farming enterprises to grow even larger through the gift of government land. As we have seen throughout this book, the consequences of this uncontrolled land acquisition — overgrazing, soil erosion, and destruction of ecosystems, to mention only the most outstanding — are still very much with us today.

The Taylor Grazing Act

Ever since the cattle boom of the early 1920s, the unregulated rangelands of the West became crowded with more and more animals. Overgrazing led to widespread soil erosion that with the advent of the Depression in the early 1930s turned into the famous Dust Bowl. In 1934, eighteen months after the election of Franklin Roosevelt, the *Taylor Grazing Act* became law. Hailed somewhat prematurely as "the Magna Carta of Conservation," it closed 80 million acres (later increased to 142 million) of public land to further settlement. The act created grazing districts, within which qualified local ranchers were to be issued annual grazing permits for an allotted number of animals. It authorized the Secretary of the Interior to set grazing fees, 25 percent of which were earmarked for range management and improvement. The original fee of 5¢ per animal per month was bitterly opposed by the cattlemen who resented paying even this trivial fraction of what grazing on private land would cost for what they considered to be a God-given right. Even today, as

we have noted, any attempt to increase grazing fees is greeted with the same hostility.* A second flaw in the Taylor Act was its establishment of advisory committees composed of local stockmen who were supposed to cooperate with the district managers. In practice, the committees were able to exercise absolute control over the severely understaffed managers, who had no incentive to antagonize their neighbors. The fee controversy came to a head in 1946 when a House subcommittee found that the Interior Department had been too easy on the stockmen. In the words of Representative Jed Johnson of Oklahoma, "they have made a joke out of the Grazing Service." In 1947 the Service's budget was cut from an already meager $1.7 million to about $500,000.[28]

The Federal Land Policy and Management Act of 1976 (FLPMA)[29]

More of our public land is managed by the Bureau of Land Management (BLM) than by any other agency, including the Forest Service. Because much of this land is adjacent to national parks, development activities permitted on BLM lands have major impacts on the parks. Up until the early 1970s these activities were carried out under an assortment of 2,500 statutes that virtually ignored their environmental consequences. The FLPMA was a landmark achievement because for the first time it provided in one law comprehensive authority and guidelines for managing and protecting vast tracts of federal lands and resources under jurisdiction of the BLM and defined in that law as the "public lands."[30] It mandated that public lands were to be retained in federal ownership for the benefit of the entire nation, unless it was in the public interest to dispose of a given parcel. It established that public lands are to be managed under the principles of "multiple use and sustained yield" and in a manner "that will protect the quantity of scientific, scenic, historical, ecological, environmental, air, atmospheric, water resource, and archaeological values; that, where appropriate, will preserve and protect certain public lands in their natural condition; that will provide food and habitat for fish and wildlife and domestic animals; and that will provide for outdoor recreation and human occupancy and use."[31]

For protecting national parks, one of the FLPMA's most important provisions is the inventory and planning requirements of Sections 201, 202, and 603, which give priority for designation and protection of "areas of critical environmental concern" and a mandate for review and recommendation of areas qualified for wilderness designation. Another key provision is the prohibition against impairing the suitability of areas under consideration by the secretary of the interior for designation as wilderness.[32]

As part of the inventory process, the secretary is charged with reporting to the President on the suitability of all roadless areas of 5,000 acres or more and roadless "islands" of the public lands having wilderness characteristics for inclusion in the National Wilderness Preservation System.† Within two years of each report, the President must make recommendations for final des-

* See chapter 5, pages 152-153.

† See also pages 259-260.

ignation by Congress. As well as providing the opportunity to "buffer" the national parks by designating additional wilderness areas, the FLPMA contains other provisions for managing adjacent BLM lands so that they will protect the parks. In the development and revision of land-use plans, the BLM is instructed to consider the relative scarcity of the values involved and the availability of alternative means, including recycling, and to weigh long-term benefits to the public against short-term gains. The law also requires the BLM to coordinate its land-use planning and management programs with those of other federal departments and agencies.[33]

Legislating the Forests

Section 24 of the *General Revision Act of 1891*, usually known as the *Forest Reserve Act*, authorized the President of the United States to withdraw from settlement or exploitation any forest area of the public lands that, in the opinion of the secretary of the interior, required protection and timber preservation. It was followed in 1897 by the *Forest Service Organic Act*, creating forest reserves to "improve and protect the forest within the boundaries for the purpose of securing favorable conditions of water flow, and to furnish a continuous supply of timber for the use and necessities of citizens of the United States." It placed these reserves under the administration of the General Land Office of the Department of the Interior. In 1905 the *Reorganization Act* transferred these reserves to the Department of Agriculture and created the U.S. Forest Service. In 1911 the *Weeks Act* appropriated $9 million to establish national forests in the eastern United States. In 1960 the *Multiple Use and Sustained Use Act* redefined the purposes of the national forests to give equal consideration to outdoor recreation, range, timber, watershed, and wildlife and fish habitat. However, the Forest Service chose clear-cutting as the prime way to carry out its mandate. By 1969, 61 percent of the harvest from western national forests and 50 percent from eastern forests were clear-cut. As soil erosion, water runoff, silting and warming up of streams, and other consequences of clear-cutting became apparent, opposition to the Forest Service practices intensified. Even the timber industry joined with conservationists to press Congress to reformulate the policy for the national forests.

Congress responded in 1976 with the *National Forest Management Act* (NFMA), the most far-reaching piece of legislation for the national forests since their creation some eighty years earlier. Although NFMA did not ban clear-cutting outright, as many conservationists had urged, it adopted guidelines severely limiting clear-cutting on federal lands. The act required the sale of timber from each forest to be limited to an amount equal to or less than that which the forest could replace on a sustained-yield basis, once all multiple-use aims were met. At the same time the Forest Service was instructed to "provide for diversity of plant and animal communities." This included growing a wide range of tree species, not simply those that were commercially saleable. It was left to the Department of Agriculture, parent agency of the

Forest Service, to write specific regulations to implement this and other NFMA goals. The regulations were adopted in 1982.*

NFMA also mandated that all uses of the forest were to be determined in detailed fifty-year plans, one for each of the 155 national forests. They were to be completed by 1985 and subject to public review.[34] When the first forest management plans were published in 1985, they were so widely criticized by conservationists and concerned citizens nationwide—lawsuits were filed by the Wilderness Society challenging the plans issued for San Juan and Grand Mesa national forests in Colorado, for example—that many of them were sent back to the drafting board.

The Recreation Principle

In 1872 the *Yellowstone Act* created Yellowstone "as a public park or pleasuring ground for the benefit and enjoyment of the people." In establishing the first national park in the U.S., the government reinforced the principle expressed in the Homestead Act ten years earlier that every American family was to have its share of the public domain free of monopolization by the rich. Yellowstone (and subsequently other spectacular sites) could be preserved for the average citizen by the government's holding it to be used and enjoyed by everyone. Despite this declaration of principle, Congress remained ambivalent about how to carry it out. On the one hand it had created Yellowstone as a national park; on the other hand it voted no money to operate it. For the next twenty-two years its superintendents had no legal authority to deal with the hordes of poachers and vandals who infiltrated the park. Nor did Yellowstone's enabling act provide for the protection of its wildlife, including the last few wild herds of bison left in the country. Only after poaching reduced their numbers to twenty-two did Congress act to save the bison by appropriating funds to buy domesticated specimens to breed with the remaining wild ones.[35]

To a great extent Congress supported the idea of national parks, because it shared the common belief that they were economically "worthless," and for this reason the California redwood forests and other resources that were not considered to have important commercial value were kept out of the system for more than fifty years.

Protecting the Parks

The *National Park Service Organic Act of 1916* represents an important step in the protection of park lands but still leaves a number of issues unresolved. In addition to creating the National Park Service, the act requires the secretary of the interior to administer the system to conserve scenery, natural and historic objects, and wildlife, *and* to provide for public enjoyment, while ensuring that the parks are "left unimpaired for the enjoyment of future generations."[36] Although the act presented the secretary with what sometimes were conflicting responsibilities, a 1978 amendment and subsequent court rulings determined that the secretary's first duty must be to assure that park resources are not

* The regulations dealing with the maintenance of "viable populations" of plants and animals are discussed on pages 289-290.

damaged or irretrievably lost. Simply put: Without protection of national parks, there could be no public enjoyment of them. Yet even this strong mandate is qualified by an exceptions clause indicating that the secretary cannot take action outside of this mandate "*except* as may have been or shall be directly and specifically provided by Congress."[37]

The lack of clear jurisdictional authority leaves the secretary of the interior in another potentially conflicting position, especially as she or he is also responsible for managing the lands of the Bureau of Land Management. While the secretary must respond to external threats to the parks, the courts have not yet extended this authority beyond park boundaries or against landowners whose activities threaten park resources. In the case of federal lands of the BLM or national forests, Congress has mandated that they be managed under multiple-use standards that sanction road building, logging, mining, oil and gas exploration, and other development operations that could prove harmful to an adjacent national park. Before a satisfactory solution is reached two questions must be addressed: Will Congress revise existing park legislation to give the secretary power in dealing with external threats so that the parks can be better protected? And will Congress put the national forests once again under the jurisdiction of the Department of the Interior?

National Trails Systems Act

The *National Trails Systems Act*, signed into law on October 1, 1968, the same day as the *Wild and Scenic Rivers Act*, authorized the inclusion of scenic and recreational trails across the United States into a national trails system. In addition to the 2,144-mile Appalachian Trail, which was included immediately, fourteen other trails were named for consideration by the secretaries of the interior and agriculture on lands under their jurisdiction. Other trail designations could be made through individual acts of Congress. In 1978 an amendment established historic trails as an additional category. (For more on the National Trails System, see chapter 8, pages 245-247.)

Legislating Wilderness

The *Wilderness Act* (1964)[38] established the National Wilderness Preservation System (NWPS), consisting of all national-forest areas that had been administratively designated as "wild," "wilderness," or "canoe areas," amounting to 9.1 million acres. The act directed the Forest Service to study all of its remaining "primitive" areas to determine which of these should be preserved as wilderness. It also directed the National Park Service and the Fish and Wildlife Service to study their lands for possible wilderness designation. In 1971 the Forest Service conducted a Roadless Area Review and Evaluation (RARE) of national-forest lands to assess their suitability for inclusion in NWPS. Using a stringent "purity" standard for wilderness eligibility, the agency held that for roadless areas to qualify they must be totally removed from even the "sights and sounds of civilization." Therefore, it declared, no roadless areas in the East qualified because eastern forests had previously

been impacted by road building and timber cutting. The RARE study was challenged in the courts, and Congress in 1973 passed the *Eastern Wilderness Act*, which added sixteen parcels of land in thirteen eastern states for a total of 207,000 acres.

Roadless Areas

In 1978 Congress further repudiated the Forest Service's "purity" standard by enacting the *Endangered American Wilderness Act*, which designated significant wilderness areas in the West and established that wilderness areas could no longer be exempted because cities or towns could be seen or heard from them. Prodded by growing public concern and a lawsuit from the Sierra Club forcing the Forest Service to comply with the environmental impact statements required by the *National Environmental Protection Act of 1969* before it developed any roadless areas, the service published its RARE II study in 1979. RARE II covered 62 million acres of roadless forestlands, of which, the Forest Service recommended, 15.4 million should be designated as wilderness, 10.4 million set aside for further study, and 36 million to be released for development. After a challenge by the state of California, a federal appeals court in 1982 held that the Forest Service study was legally inadequate because its environmental impact statement did not comply with the NEPA and that the agency was biased against wilderness.[39] Despite this ruling, RARE II served as a starting point for wilderness designations on a state-by-state basis. In 1984, 8.6 million acres of new wilderness were designated by Congress in twenty-one states, the largest amount given over to wilderness preservation since the Alaska National Interest Act of 1980. Many of the bills passed designated more land than the Forest Service had recommended.

Alaska National Interest Lands Conservation Act (ANILCA)

In 1980 ANILCA tripled the size of the wilderness system by designating a total of 56.5 million acres of wilderness — 5.4 million in national forests, 32.4 million in national parks, and 18.7 million in national wildlife refuges. It is a remarkable piece of legislation not simply because of its bulk (fifteen titles and 180 pages in the Statutes at Large) but because it represents the first attempt in history to design the land-use patterns for a region approaching the size of a subcontinent. Even though it was enacted only after much compromise by its supporters, it offers protection for about 80 percent of the acreage they wanted to include. There are, however, important omissions including Yukon Flats National Wildlife Refuge of east central Alaska, one of the largest and most productive migratory waterfowl regions of the world, as well as 3 million acres in Gates of the Arctic National Park and Preserve, Denali National Park and Preserve, and the Alaska Maritime National Wildlife Refuge.[40]

In 1991 the Wilderness Society called for the following improvements in ANILCA: repealing Title X, which permits drilling for oil and gas in the 1.4-million-acre coastal plain of the Arctic National Wildlife Refuge*; repealing Section 705, which mandates the annual cutting of 450 million board feet

* For a more detailed report on the Arctic National Wildlife Refuge, see chapter 5, pages 160-165.

from Tongass National Forest; establishing the National Petroleum Reserve in Alaska as a unit of the National Wildlife Refuge System; and restoring the width of Alaska's Wild and Scenic River corridors to 2 miles on each side of the river, as had been originally proposed.

Preserving Wildlife

The first federal effort to control the capture or killing of species was the prohibition of hunting in Yellowstone National Park in 1894. Six years later Congress passed its first legislation regulating wildlife, the *Lacey Act*.[41] Based on congressional authority to regulate interstate commerce, this act prohibited the transportation of wild animals or birds killed in violation of state law. It authorized the secretary of agriculture to preserve, distribute, introduce, and restore game birds, subject to existing state laws. This marked Congress's first official recognition that the loss of species was a matter of national concern. In asserting the existence of interstate commerce in wildlife, the act strengthened the hand of states to prohibit the export of game lawfully killed within their boundaries.

When President Theodore Roosevelt designated Penguin Island, Florida, as a national wildlife refuge in 1903, the federal government began a steady process of habitat acquisition that required legislative protection. Three years later the "taking"* of birds on federal lands reserved as breeding grounds was banned. A broader law, the *Migratory Bird Act of 1913*, declaring all migratory and insect-eating birds to be within federal custody, prohibited their hunting except when permitted by federal regulations. Although this law was held unconstitutional by federal district court, similar regulations on the taking of birds were included in the *Migratory Bird Treaty Act of 1918*, which was upheld by the Supreme Court in 1920.

Federal Funding for Wildlife Programs

Since the 1930s federal and state wildlife programs have been funded in part from specially designated taxes and license fees as well as from general tax revenues. The first major federal statute setting up a special fund exclusively for wildlife conservation was the *Migratory Bird Hunting Stamp Act* of 1934. The proceeds of the fund came from the sale of federal migratory bird hunting stamps, required from any adult person taking migratory waterfowl. Until 1958 the fund was used mainly for the operation of refuges, but at that time the law was amended so that revenues could be spent only to acquire refuges and "waterfowl production areas."[42]

Commonly known as the Pittman-Robertson or P-R program in honor of its sponsors, the *Federal Aid in Wildlife Restoration Act* has been described as the single most productive wildlife undertaking on record.[43] It created a special aid-to-wildlife fund whose revenues come from an 11 per-

* "Taking" includes collecting, trapping, hunting, killing, or harming individuals of an endangered species; it also includes destruction of habitat. When applied to an ecosystem, it includes logging, road building, and other ecologically harmful operations.

cent manufacturers' excise tax on the sale of sporting rifles, shotguns, ammunition, and archery equipment used in hunting, and a 10 percent tax on handguns. In 1985 those tax receipts amounted to more than $120 million. Up to 8 percent of the funds may be retained for administrative expenses by the Fish and Wildlife Service; the remainder goes to state wildlife agencies. For each $3 of P-R funds received, the states add at least $1. Since 1937 the program has pumped well over $2 billion into wildlife restoration programs throughout the United States. In 1950 Congress enacted the *Federal Aid in Fish Restoration Act* (more commonly known as the *Dingell-Johnson Act*), which is basically identical with the P-R act except that it provides federal aid to states for projects relating to fish. In the late 1970s pressure from environmentalists to broaden the P-R and Dingell-Johnson programs to include nongame* wildlife resulted in the *Fish and Wildlife Conservation Act of 1980*, which encompasses both nongame and other wildlife but excludes invertebrates because their potential number "could quickly exhaust the money and overwhelm... the proposed programs."[44]

Endangered Species

In 1966 the first *Endangered Species Preservation Act* envisioned a comprehensive program of saving endangered plant and animal species that would "protect, restore, and where necessary... establish wild populations [and] propagate selected species of native fish and wildlife... found to be threatened with extinction."[45] The act was significant because it officially recognized the right of nonhuman species "to share in the glory of this planet."[46] In addition to directing the federal land-managing agencies to preserve the habitats of vertebrate species found by the secretary of the interior to be threatened with extinction, it gave legislative recognition to the National Wildlife Refuge System, which had existed since 1903 as a kind of federal orphan.

Unfortunately, the act lacked clearly stated guidelines for the identification of endangered species and provided little or no authority for enforcement. It was replaced in 1969 by the *Endangered Species Protection Act*, which extended the Lacey Act's ban on interstate commerce to include reptiles, amphibians, mollusks, and crustaceans, and recognized the international dimensions of the extinction crisis, directing the interior secretary to coordinate an international effort to save endangered species and to prohibit the importing of most such species and their furs, skins, tusks, etc. This led to the creation of the Convention on International Trade in Endangered Species (CITES), designed to impose trade restrictions based on a species' vulnerability to extinction. In the U.S. CITES is implemented through the Endangered Species Act of 1973.

The Endangered Species Act[47]

The *Endangered Species Act of 1973* (ESA) was intended to remedy the weaknesses of the previous legislation and to make a clear statement of the federal gov-

* Animals not ordinarily valued for sport hunting or commercial purposes.

ernment's commitment: "The Congress finds and declares that various species of fish, wildlife, and plants in the United States have been rendered extinct as a consequence of economic growth and development untempered by adequate concern and conservation... These species are of esthetic, ecological, educational and scientific value to the Nation and its people... The purposes of this Act are to provide means whereby the ecosystems upon which endangered species and threatened species depend may be conserved..."

Among its key provisions are: prohibition of the taking of endangered species anywhere in the U.S.; extension of protection to include species likely to become threatened within the foreseeable future; elimination of limits on acquisition funds that could be used to purchase endangered species habitat; and a requirement that no federal agency may jeopardize the continued existence of endangered species. It was the latter provision that led to the famous snail-darter case, in which the courts held up completion of the Tellico Dam on the Little Tennessee River on the grounds that it would endanger the existence of this 3-inch-long rare fish.*

ESA requires the two agencies responsible for its enforcement — the U.S. Fish and Wildlife Service (FWS) and the National Marine Fisheries Service — to draw up recovery plans for listed species.† In the initial listing the secretary of the interior or the Commerce Department is required to make determinations "solely on the basis of the best scientific and commercial data available". In reaching this decision the secretary may consider only whether a species is endangered ("in danger of extinction throughout all or a significant portion of its range") or threatened ("likely to become an endangered species within the foreseeable future"), not whether it would be too costly or inconvenient to protect.[48]

The ESA contains provisions allowing economic and social costs to influence how listed species are treated, what form their recovery plans take, and whether or not to permit actions that would lead to their extinction; however, the provisions may be considered only after a species' listing status has been scientifically evaluated.

In 1982 the act was amended to allow private individuals (including developers) and government agencies to cooperate in the development of "habitat conservation plans" in areas where endangered species are jeopardized by proposed activities. These plans represent a trade-off between the government and developers who agree to take certain steps such as modifying or relocating their projects in exchange for assurance that they will not be prosecuted for "incidental takes."[49]

Although it has remained the law of the land, the ESA is subject to considerable attack every time it comes up for reauthorization. According to those who oppose its continuance or at least its enforcement, it can conflict with human interests when it implements its mandate to save species lower than humans on the food chain — snail darters, spotted owls, and Lange's metalmark butterflies, to mention but a few.

* For more on the snail-darter case, see next page.

† For more on the FWS recovery plans, see page 291.

As the Wilderness Society and other supporters of the ESA point out, it is precisely this "interference" with human goals that is the key issue: "It is the opinion of the Congress of the United States as expressed in this act that the continuing existence of a given species of life is more important than the transient desire of human beings to build things up, tear things down, move things around, and exterminate things in the process."[50]

The Snail Darter Case*

In the 1930s President Franklin D. Roosevelt charged the newly created Tennessee Valley Authority (TVA)† with the broad mission of upgrading the quality of life in one of America's most impoverished regions. With that objective the TVA in 1966 initiated the process of building the Tellico dam, creating a deep reservoir over 30 miles of the Little Tennessee River in eastern Tennessee. The snail darter is a three-inch-long minnow, which until 1981 was known to exist only in that river. The river's shallow, turbulent freshwater environment provided the only habitat that the small fish can live and reproduce in. It was a threat to this habitat that was at heart of the snail darter case.

While the snail darter was and is not an extraordinary fish — there are roughly 100 identified species in Tennessee alone — it was considered unique because of its limited range, its even more limited spawning area, and its susceptibility to habitat destruction and/or loss. In January 1975 conservationists petitioned the secretary of the interior to declare the snail darter endangered. The following November it was so classified and a crucial 17-mile stretch of the Little Tennessee was declared a critical habitat of the snail darter.

In January 1976 a group of environmentalists filed suit against the TVA, charging that they were violating the ESA because they had not consulted with the Department of the Interior as required by the act. TVA argued that abandoning the dam project supported by Congress for more than a decade would result in considerable financial loss. The district court agreed, but the appellate court reversed the decision on the grounds that once a species is placed on the endangered list, any violation of ESA calls for an automatic injunction. The effect of this ruling was to stop the dam. In response to a TVA petition to reinstate the district court's decision, the Supreme Court in June 1978, in a 6 to 3 opinion, ruled that the agency was in violation of ESA's Section 7, which requires federal agencies to take the steps needed to prevent any destruction of habitat of an endangered species.

Pressed by the Tennessee congressional delegation and developers of the dam, the Senate passed an amendment to ESA, which set up a Cabinet-level committee that could grant federal construction agencies exemptions from the act, but only on a "super majority" vote by at least five of its seven committee members. Because of its power of granting life-or-death

* Condensed from Alex J. Naar, "The Snail Darter," *Environmental Law: An Examination and Analysis of How It Has Been Used by Selected Environmental Organizations* (Cambridge, Mass.: Lesley College, 1990), pp. 34-46.

† The Tennessee Valley Authority is a major independent federally run organization that oversees a large network of hydro and hydroelectric dams and reservoirs in Tennessee.

sentences on animal and plant species, it was tagged with the appellation "God Committee."* [51] Soon thereafter this committee was convened by President Carter to consider such an exemption in the snail darter case, deciding in favor of the fish. However, the game was far from over. Proponents of the dam, led by Tennessee Senator Howard Baker, deviously buried a rider in an omnibus energy and water appropriations bill that authorized the continuation of the dam Thus, the Senate unknowingly passed a measure to continue construction of the dam.[52]

Saving and Improving ESA

One major priority regarding the Endangered Species Act is to work out new guidelines for the Fish and Wildlife Service (FWS) for deciding which species should be included. At present some four thousand species are waiting to be listed. The five hundred or so species now listed as endangered or threatened thus represent only about 15 percent of native species believed to be nearing extinction. Meanwhile, as many as three hundred species either on the list or awaiting inclusion have probably gone extinct. Another problem is that the annual $8 million budget for species recovery programs is woefully inadequate. A recent Interior Department report estimated that it would take $4.6 *billion* just to get all currently listed species on the road to recovery. More money must be appropriated from Congress for this work, the Wilderness Society and other conservation groups urge, to prevent the act from becoming meaningless. According to Jasper Carlton, director of the Biodiversity Legal Foundation, which is suing the FWS for its failure to protect "hundreds, if not thousands, of biologically threatened and endangered species," it will take between thirty-eight and forty-eight years, at current rates of listing, for FWS to list just those species that by their own estimation qualify for protection.[53]

Above all, strategies need to be developed to prevent the degradation of species habitat and to strengthen the ability of the FWS to enforce all of the provisions of the ESA. One strategy increasingly favored by supporters of the ESA is to use the act not to save species one at a time but to save whole ecosystems by finding endangered species within them. For example, when conservation groups sue to stop clear-cutting in the old-growth forests of the Pacific Northwest, they are not concerned primarily with the endangered northern spotted owl; they want to save the whole forest, which is the habitat of untold numbers of unlisted, but no less important, other species.

ESA and the Northern Spotted Owl[54]

As we observed in chapter 6, the fate of the northern spotted owl is closely linked with the destruction of the old-growth forests in the Pacific Northwest. The dramatic decline of the spotted owl's population indicates that it may soon fall below what biologists call "minimum viable popula-

* It consists of the secretaries of the interior, agriculture, and the army, the chairperson of the Council of Economic Advisers, and the administrators of the EPA and the National Oceanic and Atmospheric Administration.

tion." Responding to this danger, the Audubon Society and other environmental groups have filed lawsuits to enjoin the sale or logging of old-growth timber. In August 1987 the Sierra Club Legal Defense Fund, on behalf of twenty-nine environmental groups, petitioned the FWS to list the owl as *endangered* along the Oregon coast and on the Olympic Peninsula, and as *threatened* over the remainder of its range. Claiming that it did not qualify for listing, the FWS denied the petition and then struck a deal with the Forest Service, the BLM, and the National Park Service to "protect spotted owls informally."[55] Meanwhile, additional suits were filed in federal district court by environmental groups charging the FWS with violation of the ESA. The judge found that the FWS had indeed ignored the advice of its own experts and had acted in an "arbitrary and capricious" manner, "contrary to law."[56] In response, the FWS notified the court in April 1990 that it had begun steps to designate the owl a threatened species because of modification and loss of its habitat. On June 26, 1990, the agency listed the northern spotted owl as such under the ESA. However, as we have already noted, the bitter controversy is far from ended.*

Critical Habitat Determination

The Fish and Wildlife Service is given some discretion in selectively listing a species geographically. The bald eagle, for example is listed as endangered throughout much of the United States but as threatened in Washington, Oregon, and certain Great Lakes states, and is not listed at all in Alaska. When the FWS lists a species with limited exceptions, it is supposed to accompany the listing with a designation of "critical habitat." However, its listing of the spotted owl did not include such a designation, and in February 1991 a federal district court held that it had abused its discretion and ordered the service to publish its proposed critical-habitat plan by May of that year. The FWS met the deadline with a plan covering 11.6 million acres in Oregon, California, and Washington, satisfying neither conservationists nor the timber industry. According to the latter, the proposal would eliminate 100,000 jobs in the Northwest, a figure that is sharply contested by the government and many others. National Audubon's Mark Liverman in Portland called the plan "the first step toward the birds' survival." Other environmentalists, however, complained that the proposal did not go far enough, because it still left the door open for tree cutting.[57] It is only in designating critical habitat that the FWS may consider economic factors, balancing the costs and benefits of excluding certain areas from the designation. In the case of the spotted owl, for example, the Service weighs the harm caused by destruction of habitat against the economic benefits gained by logging the area. If the habitat loss will not lead to extinction of the species, it can be argued that the habitat does not have to be protected. On the other hand, when the species is threatened or endangered by habitat loss (as is the spotted owl), it is antipreservationist to exclude critical habitat. Therefore, many conservationists believe, the FWS should only exclude habitat when it lists a species for reasons other than habitat loss.

* See chapter 6, pages 199-203, and Appendix B, pages 360-361.

Gutting the ESA

The ESA controversy does not end with habitat preservation. As we have seen in both the snail darter and the spotted owl cases, dam builders and timber companies go to extraordinary lengths to influence Congress to act on their behalf. The Supreme Court decision to protect the snail darter "drove home to Congress the financial implications of saving endangered species, frightening Congress enough to amend the ESA to include the 'God Committee' bailout provision."[58] In passing the Hatfield-Adams Act, for example, Congress bowed to Northwestern legislators who negated the ESA in the belief that protecting the spotted owl would seriously harm the regional economy. Similarly, the White House policy on the spotted owl and other endangered species has minimized the cost of protecting species while offering them little protection. As concludes a comprehensive report on the ESA and the spotted owl in *Ecology Law Quarterly*, "The northern spotted owl now faces more than dwindling habitat. In the Bush administration, the owl faces a political predator determined to avoid paying the necessary cost of owl protection."[59]

Recovery Plans

In addition to designating critical habitat, the FWS is required by the ESA to develop recovery plans for the conservation and survival of listed species.[60] Under the informal arrangement mentioned above, an interagency committee chaired by Jack Ward Thomas, an internationally recognized wildlife biologist with the Forest Service, was set up to develop a "scientifically credible conservation strategy" for the owl. In May 1990 the committee issued a 428-page report concluding that the owl was imperiled over significant portions of its range because of continuing losses of habitat from logging and natural disturbances. To save the species the committee recommended creating habitat-conservation areas (HCAs) — blocks of habitat protecting a minimum of twenty pairs of owls and spaced no more than 12 miles apart. Within the HCAs logging and other silvicultural activities, excepting regeneration of tree stands, should cease.[61] The Thomas Committee's plan, which would set aside nearly 8 million acres for HCAs in the Northwest and prohibit logging on 30 to 40 percent of available public forests, triggered a new round of economic and political debate with the Fish and Wildlife Service favoring the plan and the BLM, its parent agency, the Department of the Interior, and the Bush administration supporting a scenario that would have less impact on logging and other commercial interests.

Legislating for Biodiversity

Many conservationists believe that the spotted-owl case underlines an inherent weakness in the Endangered Species Act and other existing environmental laws; that, even if enforced rigorously, they deal with problems on a piecemeal basis rather than in the broader ecological context that cuts across jurisdictional, state, and national boundaries. A leading proponent of the ecosystems approach is Dr. Reed F. Noss, ecologist and science editor of the

quarterly *Wild Earth*. Are species too tangible for their own good? is one of several "difficult" questions raised by Noss as he considers the results achieved by more than a decade and a half of the ESA.

We are emotionally attracted to wildlife-conservation causes, Noss says, by the appeal of the "proud" stance of the eagle, the "cute" face of the panda, or the "majesty" of the tiger. Although most Americans may be willing to make some economic sacrifice to protect certain endangered mammals, birds, or butterflies, they have a harder time with endangered invertebrates, plants, or snakes. We can relate to certain species as individuals, *"but ecosystems composed of nutrient cycles, energy flows, and disturbance regimes in addition to populations are unlikely to capture the public imagination."*[62]

> *"American conservation is, I fear, still concerned for the most part with show pieces. We have not yet learned to think in terms of small cogs and wheels."*
> — Aldo Leopold, *A Sand County Almanac*

Another important question posed by Noss is, How does a species-centered approach to conservation affect the crucial issue of biodiversity? The answer, he says, lies not in a species-by-species approach but in a strategy that transcends individual species. National parks, for example, are generally too small for viable populations of many species, and their legal boundaries do not conform to ecological boundaries. Another problem is politics, controlled by powerful economic interests, which often determines the listing of species. "High-profile 'glamour species' such as the bald eagle and the California condor receive millions of dollars for high-tech recovery efforts while hundreds of lesser-known cogs and wheels (especially plants and invertebrates) silently disappear."[63]

A Native Ecosystems Act

Many scientists and conservationists agree with Noss on the need for legislation covering endangered ecosystems. A Native Ecosystems Act (NEA), as he proposes, would be designed to protect fully and restore the *entire* spectrum of native plant and animal communities, ecosystems, and landscapes across the United States.[64] The NEA would focus on two different aspects of ecosystems: endangered and "representative." First, it would prohibit logging, road building, livestock grazing, and other ecologically destructive activities affecting ecosystems that have declined substantially (by 80 percent or more) over their range and would set recovery goals for restoring these ecosystems. Secondly, the law would avoid the necessity for such last-ditch measures in the future by setting aside "exemplary" or representative areas of various ecosystem types in large, intact units. Representing *all* native ecosystem types in a network of protected areas, regardless of their current rarity, across the full range of their natural variation would complement the recovery process of the first section and prevent further degradation of both threatened and nonthreatened ecosystems.

Virtually the only space for this kind of operation can be found in large public land holdings, which, as we have seen throughout this book, are being increasingly fragmented by clear-cutting, road building, ORVs, and other human intrusions. Working for an NEA is an ambitious project, but, Noss reminds us, we should take courage from the fact that it took eight years to pass the Wilderness Act. If we can pass a Native Ecosystems Act within five to ten years, he says, a significant portion of our natural heritage can be saved.

Conserving the Future

> *"Conservation was simpler when all we had to worry about were such straightforward problems as egret plumes on ladies' hats. Then came the Dust Bowl, silted streams, and belated recognition of the effects of the plow, the axe, and the cow on once healthy land."*
> — Reed F. Noss

By the 1960s and '70s the problem had become more complex as we had to contend with DDT, acid rain, toxic wastes, habitat fragmentation, and "edge effects." The numbers of endangered species escalated. Today we suffer additional perils from widespread air pollution, destruction of the ozone layer, and global climate change. Worse yet is the interaction of environmental threats in what Noss terms "a multiplicative, synergistic fashion with unpredictable effects that have no easy remedy."[65]

Biodiversity is the incredible multitude of different forms of life on this planet. According to entomologist Terry Erwin, just one group of animals — insects — may account for fifty million species. Tragically, what has taken evolution three billion years to produce may take the human species a few generations to wipe out. Erwin estimates that human activity in our generation alone may destroy twenty to thirty million species. The consequences will profoundly affect all other species, including humans.[66]

What You Can Do to Encourage an Ecosystems Approach to Preservation

- Expand your knowledge on this subject by reading *Wild Earth*, a quarterly magazine that does the best job of all publications at presenting the ecosystems approach. Contact them at 68 Riverside Drive, Canton, NY 13617.
- Support conservation groups that place emphasis on protecting habitat and entire ecosystems rather than individual species. One recent example of this approach is the "Last Great Places" initiative of the Nature Conservancy.* Also to be commended are the wildlife sanctuary programs of the National Audubon Society.
- Support the work of the Greater Ecosystem Alliance (P.O. Box 2813, Bellingham, WA 98227); the Biodiversity Legal Foundation, a nonprofit organization dedicated to the preservation of native wild plants and animals, communities of species, and naturally functioning ecosystems (P.O. Box 18327, Boulder, CO 80308-8327); and other ecologically minded organizations in the forefront of promoting an ecosystems approach to endangered species and

* For more about their work, see chapter 5, pages 75 and 151-152.

biodiversity. For details on these organizations, see the Directory, pages 331-355. *Note: Although mainstream environmental organizations pay lip service to this concept, they sometimes oppose proposed legislation such as the Wild Rockies National Lands Act that would implement it. Check out in particular the groups to which we have awarded two bullets!*

- Urge your congressional representatives to strengthen the Endangered Species Act and, where necessary, educate them to the importance of preserving and protecting ecosystems. Give them a copy of *This Land Is Your Land* as a start in the right direction!

Key Environmental Laws

The following environmental laws are included here because they impact on virtually all of the federal land and resources legislation discussed earlier and because they provide important opportunities for citizen participation in the legislative process.

The National Environmental Protection Act of 1969 (NEPA)

The Magna Carta of environmental legislation, NEPA requires federal agencies to take into account the environmental consequences of all their plans and activities. Specifically, whenever the government plans to initiate, finance, or permit a "major" action, it must prepare an "environmental impact statement" (EIS) assessing the project's environmental effects. As part of this process, the appropriate agency must hold hearings to take public comment and provide a public review period of a draft EIS. Until a final EIS is released, the project cannot proceed. Among other things, the EIS must review alternatives to the proposal that would have less impact on the environment, including forgoing the project altogether ("the no-action alternative"). As a case in point, more environmentally benign alternatives to building a new highway might include building a smaller one, slight expansion or improvement of existing roadways, or development of new public-transit facilities that would eliminate the need for any highway. All of these would have to be explored in an EIS before work got under way on the proposed highway.

Since 1970 lawsuits challenging the adequacy of an EIS, or the decision not to prepare one, have become a staple of environmental litigation and have often held the process up long enough to allow for legislative intervention or to cause the project to fall of its own weight. Because NEPA applies only to actions of the federal government, not of states or private parties, it is not a complete environmental-protection statute. Efforts to give it more teeth have been shot down by the Supreme Court. Thus, NEPA requires the feds to stop and think, but never simply to stop.*

The Environmental Quality Improvement Act (1970)

The Environmental Quality Improvement Act (1970), aimed at "combatting pollution and degradation of our environment,"[67] declared that while the federal government would encourage and support this goal, the primary respon-

* For more on NEPA, see *Design for a Livable Planet*, pages 231-232.

sibility for its implementation rested with state and local governments. The act also mandated the staffing of the Council on Environmental Quality, which was established in the executive office of the President.

The Equal Access to Justice Act (1980)

The Equal Access to Justice Act provides that the United States or any federal agency may be liable for a prevailing party's legal fees to the same extent that any other party would be liable under the common law or under the terms of any statute that specifically provides for such an award.[68] Included among these statutes are the Marine Protection Research and Sanctuaries Act, the Clean Air Act, and the Water Pollution Control Act.

Freedom of Information Act (1967)

The FOIA established a judicially enforceable public right to secure official information.[69] There are, however, nine exemptions to the government's obligation, which include: classified information relating to national security; internal personnel rules of federal agencies; personal files that would constitute invasion of privacy; trade and commercial secrets; inter- or intra-agency documents that would not be available in litigation with the agency; and geological and geophysical information concerning wells.

How to File an FOIA Request

Any citizen or group can make an FOIA request; processing fees will generally be waived for individuals and nonprofit groups. To file an FOIA request, telephone the appropriate government agency and ask to whom to address your letter, then begin by stating that you are submitting "a request under the Freedom of Information Act, 5 U.S.C. 522." Be as specific as you can about the documents you want. If you are turned down, the decision can be appealed within the agency and ultimately in a federal court.

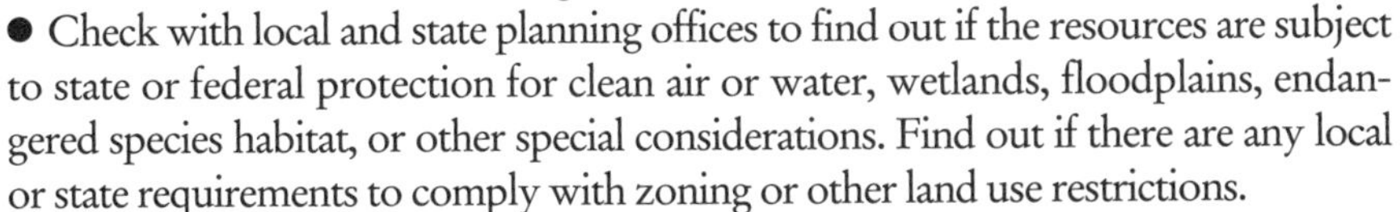

- Educate yourself on the key laws described in this chapter and follow up the leads provided on the "Resources" and in the Directory.
- Research newspapers, magazines, journals, and other historical records for information relating to the issue of concern.
- Check with local and state planning offices to find out if the resources are subject to state or federal protection for clean air or water, wetlands, floodplains, endangered species habitat, or other special considerations. Find out if there are any local or state requirements to comply with zoning or other land use restrictions.
- In cases of land use, check the history of the title to the property; look for easements, deed restrictions, or any other limitations.
- Before considering bringing a suit of your own, consult with a conservation or citizen's organization in your area or nationally that is already engaged with public resources or environmental litigation. (For details, see the Directory.)

• Before bringing action, bear in mind that it is likely to be a drawnout affair that will drain your time, energy, and — depending on whether you obtain a lawyer or expert witness — money.

• Explore the possibility of getting a lawyer to work for your case on a *pro bono publico* basis — i.e., for the public good. You can ask any attorney if she or he would be willing to do this or to recommend a colleague who would.

• Contact a community legal service or community-oriented law organization that might be interested in supporting your case.

• *Note: Before undertaking any legal action relating to natural resources, land, and the environment it is highly advisable to consult with an attorney who has experience in those areas.*

RESOURCES

1. U.S. Government

Distribution of Congressional Publications

Senate bills, reports, and documents are distributed through the *Senate Documents Room*, B-04, Hart Senate Office Building, Washington, DC 20510. House equivalents can be obtained from the *House Documents Room*, H-226, U.S. Capitol, Washington, DC 20515. Public laws are distributed by both document rooms.

Telephone inquiries on the status of legislative items may be made by calling the Senate room at (202) 224-7860 or the House room at (202) 225-3456.

Daily proceedings of Congress are published in the *Congressional Record.* To find out the status of a House bill, telephone (202) 225-1772; for a Senate bill, (202) 224-2971. To receive copies of bills, write (enclosing a self-addressed mailing label) to either the House or the Senate documents room at the addresses given above.

Directly concerned with all proposed legislation are the *House Appropriations Committee*, H-218, Capitol Building, Washington, DC 20515, (202) 225-2771; and the *Senate Appropriations Committee*, SD-128, Capitol Building, Washington, DC 20510, (202) 224-3471.

— *U.S. Government Printing Office*, Superintendent of Documents, Washington, DC 20402, (202) 275-3030; sells copies of all U.S. laws, treaties, and implementing regulations. Copies of these documents may also be obtained from the main or regional offices of the federal agency responsible for enforcement.

— *The EPA Public Information Center*, 401 M Street, SW, Washington, DC 20460, (202) 382-2080, will provide copies of laws of which the EPA is the regulatory agency, as well as information on all EPA-related activities.

2. State Laws

Documents on state environmental legislation can be obtained from the individual state governments, most of which have departments of environmental protection or natural resources. Most state environmental laws are collected in the *Environmental Reporter*.

3. Environmental and Conservation Organizations

Litigation is undertaken by a wide variety of local, regional, and national organizations, some formed expressly for the purpose of bringing an individual suit.

— *Center for Marine Conservation*, 1725 De Sales Street, NW, Washington, DC 20036; (202) 429-5609. Devoted to marine conservation and the fight against illegal wildlife trading.

— *Center for the Study of Responsive Law*, 2000 P Street, NW, Washington, DC 20036.

— *Citizens for Ocean Law*, 1601 Connecticut Avenue, NW, Washington, DC 20009.

— *Environmental Law Foundation*, California Building, 1736 Franklin Street, 8th Floor, Oakland, California 94612; (510) 208-4555. See also the Directory, page 346.

— *Environmental Law Institute*, 1616 P Street, NW, Washington, DC 20036. Publishes the *Environmental Law Reporter* and the *National Wetlands Newsletter*, books, and research materials. It also runs a continuing legal education program.
— *The Environmental Litigation Fund*, P.O. Box 10836, Eugene, OR 97440; (503) 683-1378. This is a project of Earth Island Institute.
— *Freedom of Information Clearinghouse*, P.O. Box 19367, Washington, DC 20036. Provides assistance on how to use federal and state freedom-of-information laws.
— *League of Women Voters Education Fund*, 1730 M Street, NW, Washington, DC 20036; (202) 429-1965. Concerned with citizens' right to know, election laws, voting rights, government processes, and other issues.
— *Sierra Club Legal Defense Fund*, 2044 Fillmore Street, San Francisco, CA 94115. Although SCLDF maintains a close working relationship with the Sierra Club, it is an independent organization devoted exclusively to legal matters.

4. Further Reading

— *Balancing on the Brink of Extinction: The Endangered Species Act and Lessons for the Future*, Kathryn A. Kohm, editor. Washington, D.C.: Island Press, 1991.
— *Biodiversity*, E. O. Wilson, editor. Washington, D.C.: National Academy Press, 1988.
— *Biological Diversity*, Richard Tobin. Durham, N.C.: Duke University Press, 1990.
— *A Conservation Strategy for the Northern Spotted Owl: Report of the Interagency Scientific Committee to Address the Conservation of the Northern Spotted Owl*, Jack Ward Thomas, chairperson. Portland, Ore.: U.S. Department of Agriculture Forest Service, 1990.
— *Crossroads: Environmental Priorities for the Future*, Peter Borrelli, editor. Contains "Environmentalists at Law," Frederic P. Sutherland and Vawter Parker; and "Legal Eagles," Tom Turner. Washington, D.C.: Island Press, 1988.
— *Directory of State Environmental Agencies*, Kathryn Hubler and Timothy Henderson, editors. Washington, D.C.: Environmental Law Institute, 1989.
— "Ecological Foundations for Big Wilderness," issue theme, *Wild Earth*, vol.1, no.1 (Spring 1991).
— *Ecology Law Quarterly*. Berkeley: University of California, Boalt Hall School of Law.
— *The Environmental Impact Statement Process*, Neil Orloff. Washington, D.C.: Information Resources Press, 1978.
— *Environmental Law in a Nutshell*, Roger Findley and Daniel Farber. St. Paul: West, 1988.
— *Environmental Law Reporter*. Monthly reporting service on environmental issues. Environmental Law Institute, Washington, D.C.
— *Environmental Law — Twenty Years Later*, J. William Futtrell. Washington, D.C.: Island Press, 1988.
— *The Evolution of Wildlife Law*, Michael J. Bean. New York: Praeger, 1983.
— *Federal Public Land and Resources Law*, George Cameron Coggins and Charles F. Wilkerson. Mineola, N.Y.: Foundation Press, 1987.
— *Federal Statutes on Environmental Protection*, Warren Freedman. New York: Quorum Books, 1987.
— *Guide to State Environmental Programs*, Deborah Hitchcock Jessup. Washington, D.C.: Bureau of National Affairs, 1990.
— *A Guide to the Clean Water Act Amendments*, OPA 129/8. Washington, D.C.: EPA Office of Public Awareness, 1988.
— *Handbook on Environmental Law*, William Rodgers. St. Paul: West, 1977 (with a 1984 supplement).
— *How You Can Influence Congress*, George Anderson and Everett Sentman. New York: E. P. Dutton, 1979.
— *In Defense of the Land Ethic*, J. Baird Callicot. Albany: State University of New York Press, 1989.
— *Law of Environmental Protection*, Sheldon Novick, Margaret Mellon, and Donald Stever, editors. New York: Clark Boardman, 1987.
— *Legislative Sourcebook on Toxics*, David Jones and Jeffrey Tryens, editors. Washington, D.C.: National Center for Policy Alternatives, 1986.

— "Life List, U.S.A.: The Endangered and Threatened Species of the United States," Liz Broussard, *Wilderness*, Summer 1991.
— *Preserving Communities and Corridors*, Gay Mackintosh, editor. Washington, D.C.: Defenders of Wildlife, 1989.
— *Resolving Environmental Disputes: A Decade of Experience*, Gail Bingham. Washington, D.C.: Conservation Foundation, 1986.
— *Should Trees Have Standing? Towards Legal Rights for Natural Objects*, Christopher D. Stone. Palo Alto, Calif.: Tioga Publishing, 1988.
— *U.S. Environmental Laws*. Washington, D.C.: Bureau of National Affairs, 1991. Indispensable 1,200-page documentation of the major environmental legislation.

NOTES

1. Source material for this section includes George Cameron Coggins and Charles F. Wilkerson, *Federal Public Land and Resources Law* (Mineola, N.Y.: Foundation Press, 1981); Paul W. Gates, *History of Public Land Law Development* (Washington, D.C.: Public Land Law Review Commission, 1968); Marion Clawson, *The Land System of the United States* (Lincoln: University of Nebraska Press, 1972); David J. Simon, editor, *Our Common Lands* (Washington, D.C.: Island Press, 1988); Dyan Zaslowsky, *These American Lands* (New York: Henry Holt, 1986); and Joseph L. Sax, *Mountains Without Handrails* (Ann Arbor: University of Michigan Press, 1980).
2. For a more detailed discussion of public trust and common law, see Simon, *Our Common Lands*, pp. 96-105.
3. Coggins and Wilkerson, *Federal Public Land*, pp. 289–92.
4. Winters v. United States, 207 U.S. 564 (1908).
5. Arizona v. California, 373 U.S. 546 (1963).
6. Ibid.
7. 13 U.S.C. § 1291; 16 U.S.C. § 1601, § 1801; 33 U.S.C. §§ 431-37; et seq.
8. 33 U.S.C. §§ 1251–1376.
9. 43 U.S.C. § 1251.
10. Sources include Kevin J. Coyle, *The American Rivers Guide to Wild and Scenic Rivers Designation* (Washington, D.C.: American Rivers, 1988); Tim Palmer, *Endangered Rivers and the Conservation Movement* (Berkeley: University of California Press, 1986); Olson W. Kent, *Natural Rivers and the Public Trust* (Washington, D.C.: American Rivers, 1988); Simon, *Our Common Lands*, pp. 331–86.
11. 16 U.S.C. § 1274(a).
12. 16 U.S.C. §§ 1361–1407.
13. 33 U.S.C. §§ 1401–34.
14. 16 U.S.C. §§ 1451–64.
15. 33 U.S.C. §§ 1221–36.
16. 33 U.S.C. §§ 1901–11.
17. 33 U.S.C. §§ 1002–15.
18. 33 U.S.C. §§ 1371–87.
19. 43 U.S.C. § 1811 et seq.
20. 33 U.S.C. § 401 et seq.
21. 33 U.S.C. §§ 1901–12.
22. Zaslowsky, *These American Lands*, p. 123.
23. Ibid.
24. Katherine W. Bueler, "The Great Terrain Robbery," *The Wings of Conservation, Audubon Activist*, May 1991; and Dan Dagget, "Old Mining Law Leaves Deadly Legacy," *Audubon Activist*, June 1991.
25. Sierra Club special bulletin, January 1991.
26. 30 U.S.C. § 1201 et seq.
27. Marc Reisner, *Cadillac Desert* (New York: Penguin Books, 1987), p.43.

28. Cited in Zaslowsky, *These American Lands*, p. 103. For commentary on the Taylor Act, see also Coggins and Wilkerson, *Federal Public Land*, pp. 539-47.
29. The main source for this section is D. Michael Harvey, "FLPMA," in Simon, *Our Common Lands*, pp. 127–42.
30. FLPMA § 103(e), 43 U.S.C. § 1702(c).
31. FLPMA § 102(a)(8), 43 U.S.C. § 1701(a)(8).
32. FLPMA § 603(c), 43 U.S.C. § 1782(c).
33. FLPMA § 202(c)(9), 43 U.S.C. § 1712(c)(9).
34. An excellent review of the attempts by national forests to prepare management plans is given in Keith Ervin, *Fragile Majesty* (Seattle: The Mountaineers, 1989), pp. 185–204.
35. Zaslowsky, *These American Lands*, p. 16.
36. 16 U.S.C. § 1.
37. 16 U.S.C. § 1(a)1 (1978), cited in *Our Common Lands*, p. 17; emphasis added.
38. 16 U.S.C. §§ 1131–36.
39. California v. Block, 690 F. 2d 753 (9th Circ. 1982).
40. Ibid., pp. 276–78.
41. The historical background of endangered species legislation is documented in Kathryn A. Kohm's introduction to *Balancing on the Brink of Extinction* (Washington, D.C.: Island Press, 1991), pp. 3–22.
42. Act of August 1, 1958, Public Law No. 85-585, § 3, 72, Stat. 486.
43. Lonnie L. Williamson, "Evolution of a Landmark Law," in *Restoring America's Wildlife* (Washington, D.C.: U.S. Department of Interior Fish and Wildlife Service, 1987), p. 4.
44. Senator John Chafee (R-Rhode Island) cited in Michael J. Bean, *The Evolution of National Wildlife Law* (New York: Praeger, 1983), p. 229.
45. Letter from Stewart Udall, Secretary of the Interior, to John McCormack, Speaker of the House, June 5, 1965, cited in Kohm, *Balancing on the Brink*, p. 12.
46. T. H. Watkins, "The Protocols of Endangerment," *Wilderness*, Summer 1991, p. 8. Virtually the whole issue of this quarterly magazine is devoted to a discussion of endangered species.
47. 16 U.S.C. §§ 1531-1543.
48. 16 U.S.C. § 1532(6) (1988).
49. Gordy Slack, "Natural Law," *Pacific Discovery*, Spring 1992, p. 24.
50. Watkins, "Protocols of Endangerment," p. 8.
51. "The God Committee," *Audubon*, May 1979, p. 10.
52. Congressman John D. Dingell, "The Endangered Species Act: Legislative Perspectives on a Living Law," Kohm, *Balancing on the Brink*, p. 27.
53. Slack, "Natural Law," p. 26.
54. One of the best sources of information on the ESA and the spotted owl is Mark Bonnett and Kurt Zimmerman, "Politics and Preservation: The Endangered Species Act and the Northern Spotted Owl," *Ecology Law Quarterly*, vol. 18, no. 1, pp. 105–71.
55. Ibid., p. 125.
56. *Northern Spotted Owl v. Hodel*, 716 F. Supp. 479, 483 (W.D. Wash. 1988).
57. "FWS Proposes Critical Habitat for Northern Spotted Owl," *Ecology USA*, May 6, 1991, p. 83.
58. Bonnett and Zimmerman, "Politics and Preservation," p. 164.
59. Ibid., p. 169.
60. 16 U.S.C. § 1533(f)(1)(A) (1988).
61. *A Conservation Strategy for the Northern Spotted Owl*, report of the Interagency Scientific Committee to Address the Conservation of the Northern Spotted Owl (Portland, Ore.: U.S. Department of Agriculture Forest Service, 1990), p. 1.
62. Kohm, *Balancing on the Brink*, p. 229.
63. Ibid., p. 228.
64. Reed F. Noss, "A Native Ecosystems Act (Concept Paper)," *Wild Earth*, Spring 1991, pp. 24-25.
65. Kohm, *Balancing on the Brink*, p. 239.
66. E. O. Wilson, editor, *BioDiversity* (Washington, D.C.: National Academy Press, 1988), p. 127.
67. 42 U.S.C. §§ 4371–74 amended by Public Law 94–52 (1975) and Public Law 94–298 (1976).
68. 28 U.S.C. § 2412(b).
69. 5 U.S.C. § 552.

10

RESTORING THE EARTH

"If the creature destroys its environment, it destroys itself."

—Gregory Bateson

Land, air, water, and wildlife resources are succumbing to the effects of uncontrolled industrialization, urbanization, and agriculture, and widespread human encroachment into wilderness areas. The natural resources are endangered by pollution, acid rain, and major climatic disturbances caused directly or indirectly by incinerating fossil fuels, cutting and burning rain forests, and releasing ozone-destroying chemicals into the atmosphere. Habitat and species are being lost on an unprecedented scale. Many experts believe the planet may be facing a wave of species extinction comparable with the one 65 million years ago in which the dinosaurs perished along with three quarters of the species on earth. According to two prominent biologists, Paul Ehrlich and Edward O. Wilson, a quarter or more of all species could be exterminated within fifty years.[1] This means that some 50,000 species a year, or about six each hour, are doomed to extinction, Wilson estimates.[2] The current rate of species and ecosystem loss exceeds by far that at which evolution is creating new diversity, and the rate continues to rise.

Why Biodiversity?

The biological diversity (biodiversity) crisis goes far beyond losing large numbers of animals and plants. It involves more than species; it encompasses the full range of biological systems from genes to the entire planet. Genetic diversity refers to the spectrum of genes found in a species or population. The greater the genetic diversity within a population, the greater its ability to adapt to future changes in its environment. Species diversity, sometimes called species richness, accounts for all the kinds of species in a given area. Ecosystem diversity is the variety of different interacting systems across a region or the earth itself. Why protect biodiversity? First, there are ethical reasons. As the dominant species on earth, we have the moral responsibility to protect all other forms of life. Second, there is self-interest. Organisms provide us with great benefits, including food, medicines, and industrial products. Wheat, corn, rice, beef, pork, and lamb, for example, were all domesticated from wild species. Plant-derived chemicals are major components of more than a quarter of all prescriptions written in the United States. One compound, quinine, is the treatment of choice against malaria. The bark of the Pacific yew, found in the Pacific Northwest, contains taxol, a chemical with great promise in cancer therapy.[3] Other plants contain commercially valuable drugs, waxes, and oils. There are thousands, perhaps millions of species not yet explored for medicinal or other valuable properties.

The third basis for protecting diversity is esthetic. Millions of tourists, campers, bird-watchers, gardeners, conservationists, photographers, and other nature lovers find beauty, recreation, and solace in the natural world. The fourth reason, discussed throughout this book, is the wide range of "services" performed by ecosystems, of which diverse species are the key working parts. As we have seen, ecosystems influence the climate by regulating atmospheric gases and recycling rainfall (critical to rain forests, for example). Biodiversity is vital to the creation and maintenance of soil, keeping ecosystems healthy by cycling nutrients and disposing of wastes.* The diversity of species also helps

* For a more detailed description of the role of soil ecosystems, see Introduction pages 6-7.

protect crops and animals from pests, as well as pollinating plants.

The case for preventing further disruption of ecosystems is graphically stated by Dr. Terry L. Erwin, curator of entomology at the National Museum of Natural History: "Within a few hundred years this planet will have little more than lineages of domestic weeds, flies, cockroaches, and starlings, evolving to fill a converted and mostly desertified environment left in the wake of nonenvironmentally adaptive human cultural evolution."[4]

Dissenting Views

Not all scientists accept Erwin's scenario. A minority, including Dr. Michael A. Mares, a zoologist at the University of Oklahoma who is an expert on neotropical habitats, and Dr. Jared Diamond, an ecologist at the University of California at Los Angeles, say that the true dimensions of the problem cannot be assessed on present evidence. First of all, they argue, no one knows the exact number of species in the world. Only 1.4 million species have been so far identified out of a possible 10 to 100 million. "When you deal with that kind of error, it's hard to say what's happening," states Mares.[5] Nevertheless, most scientists would agree that, however many kinds of species there may be in the world, the *number* that are severely endangered by human activities today is unprecedented. Even those like Julian Simon, an economist at the University of Maryland who maintains that the world's resources can support a vastly expanded human population, agree with the conclusion that we must nevertheless protect species as a "valuable endowment."[6] "We would be remiss," warns Harold J. Morowitz, professor of biology and natural philosophy at George Mason University in Fairfax, Virginia, "not to repeat the assertion that as human population goes up, biological species go down. We might be able to moderate the rate of decline, but we cannot fend off the inevitable."[7]

The Conservation Dilemma

Granted that conservation of species must take high priority on our environmental and political agendas, a number of perplexing questions remain, including the overriding one of who decides which species are to be conserved. Leaving aside the economic and resource arguments in favor of conservation, which could be said to be self-serving for humans and thus subject to compromise, the ultimate reason for conservation of species and ecosystems is moral: *"They should be conserved because they exist and because this existence is itself but the present expression of a continuing historical process of immense antiquity and majesty."* — David Ehrenfeld, *The Arrogance of Humanism*[8]

Should Trees Have Standing?

The idea of rights for nonhuman entities has a long history. In the Hindu song of god *Bhagavad-Gita*, Brahman, the total godhead, is declared to exist within all creatures and objects. Buddhism teaches compassion for all Nature. And in the New Testament we are told: "Consider the lilies of the field; they toil not, neither do they spin. And yet I say unto you, that even Solomon in all his glory was not arrayed like one of these."[9] The concept has

contemporary expression in the book *Should Trees Have Standing?*, in which the author, Christopher Stone, claims that forests, rivers, and other natural systems are entitled to legal rights apart from the interests of people associated with them. If corporations can have legal rights, responsibilities, and access to the courts, why not trees, rivers, or other inanimate objects?, he asks, raising a principle that was cited by Justice William Douglas in a celebrated minority opinion in the U.S. Supreme Court.[10] Much is made in the media of the saving of the whooping crane, the bald eagle, and other endangered species. And, to a certain extent, these can be called victories for conservation. But in some cases the species becomes a victim of its own success. News of the comeback of the whooping crane, for example, has brought such an invasion of tourists to their once-remote marshland habitat in the Aransas National Wildlife Refuge on the Gulf of Mexico that the need to protect the birds from human contact is now as urgent as the threat from erosion, pollution, or barge accidents spilling toxic substances in shipping lanes that pass through the refuge.[11]

What Price Salvation?

Even if saving a species is considered desirable, is the economic cost justified, or indeed feasible, in every case? In response to a call to reintroduce the Mexican gray wolf (which probably number fewer than forty, all in captivity) into New Mexico and Arizona, biologist James Brown, author of a landmark study of mammal extinctions, said: "Their former habitat is so fragmented that reestablishing a viable population may not be possible. I'm not sure there's a place for the wolf in the Southwest anymore."[12] More critical then than loss of species is loss or degradation of habitat, which is affecting many other species including the Florida panther (fewer than twenty left), the northern spotted owl, and even the California condor, whose dwindling population (down to four in 1987) had to be captured, bred in captivity, and supplemented by another condor species imported from the Andes. The peregrine-falcon recovery program, started in 1970 after virtually the entire population in the eastern U.S. had been wiped out by DDT and related pesticides, has succeeded only at a cost of several million dollars and with the involvement of numerous government agencies, conservation groups, universities, and others.[13] It is highly doubtful that such an effort could be mobilized for less charismatic species such as the Tecopa pupfish, which once enjoyed the warm outflow from two California hot springs, or the yellow-legged frog that has completely disappeared from Yosemite and other national parks, even though these creatures are an integral part of the ecosystems to which they belong.[14]

Another difficulty is that scientists do not fully understand how species and ecosystems interact. One case described by the conservation biologist David Ehrenfeld is that of *Calvaria major*, a tree found on the island of Mauritius in the Indian Ocean. Only recently was it discovered how the species could survive. Its seeds, which the old trees still drop in abundance, must pass through the gizzard of a dodo before they can germinate. An early victim of species extinction, the last dodo disappeared in 1681![15]

Endangered and Threatened

American Crocodile: Listed as endangered in 1975. Found in the Florida Keys and tidal marshes of the Everglades. Hunting for their skins is the major cause of their decline. Estimated five hundred adults remaining.

Bald Eagle: Listed as endangered in the lower forty-eight states since 1967. Population declined as result of habitat loss, shooting, and DDT and other pesticides that weakened shells of their eggs. Now making a comeback and may be reclassified as endangered only in certain areas areas.

Black-footed Ferret: Listed as endangered in 1967. Formerly ranged throughout the Great Plains from northern Texas to Montana. Breeding program began in 1987 with a captive population of about 325. First ferrets were returned to the wild in September 1991.

California Condor: Listed as endangered in 1967. Ranged from Baja California to British Columbia. Fifty-two were kept in captivity; two individuals were released in the wild in a sanctuary north of Los Angeles.

Grizzly Bear: Listed as threatened in 1975 in the lower forty-eight states. Once ranging throughout the West and the Great Plains, they are now, outside of Alaska, confined to primarily in and around Glacier and Yellowstone national parks, where their population is estimated to be fewer than four hundred. Decline due to trapping, shooting, and poisoning.

Northern Spotted Owl*: Listed as threatened in June 1990. Located in the old-growth forests of the Pacific Northwest. Decline due to clear-cutting and other habitat loss. May be saved by Fish and Wildlife Service designation of critical habitat areas limiting the areas used for logging.

Snail Darter: Listed as endangered in 1974, reclassified as threatened in 1984. Their only known population was in the Tennessee River, halting construction of the Tellico Dam. Although the decision was upheld by the Supreme Court, it was overruled by Congress. Since then populations have been found elsewhere.

West Indian Manatee: Listed as endangered in 1978. Concentrated in Florida during winter, but is found in summer in coastal Louisiana, Virginia, and the Carolinas. It also ranges in the coastal and inland waterways of Central and South America as far south as Recife, Brazil. Severely reduced by hunting, boat traffic, floodgates and canal locks, and pollution, the U.S. population is estimated at about twelve hundred.

Whooping Crane: Listed as endangered in 1967. Populations ranged over most of North America but are now severely reduced by hunting and agricultural development. Population reached a low point of 16 in 1941 and is now estimated to be 150.

* The case of the spotted owl is examined at greater length in chapter 6, pages 199-203, and chapter 9, pages 289-291.

Who Gets into the Ark?

If plant and animal species have standing, who should decide which to preserve? Are certain animals more important than others? Are mammals more important than invertebrates? Are animals more important than plants? These are questions now being pondered by conservation biologists, conservationists, and members of Congress. As we noted in our discussion of the Endangered Species Act,* the legislative branch even set up a "God Committee" to rule on the fate of our nonhuman kin. Perhaps the most equitable principle was one applied in an earlier world calamity, as the quote above details. No species was excluded on the basis of low priority, and, it will be recalled, not a single species was lost in the operation.

> *"Of clean beasts, and of beasts that are not clean, and of fowls, and of everything that creepeth upon the earth, there went in two and two unto Noah into the ark, the male and the female."*[16]
> — Genesis 7:8-9

Conservation Biology

Conservation biology, the science of scarcity and diversity, is a discipline that grew out of environmental consciousness and activities of the 1960s and '70s. One of its leading proponents, Michael Soulé, professor of conservation biology at the University of California at Santa Cruz, describes it as a crisis discipline in contrast to "normal" science, stating that "it is sometimes imperative to make an important tactical decision before one is confident in the sufficiency of the data."[17]

Seeking ways "to protect living nature from humanity," Soulé proposes a tactical approach that responds to the extent of human intrusion in a given system. First on his conservation list are systems based on bounded wild areas with relatively little human disturbance; they include most protected areas from wilderness parks to the core areas of biosphere reserves. Even in the best of circumstances these protected areas tend to degrade. Most of them are too small to maintain viable populations of large predators and omnivores without supplementation from the outside.[18] Second are conservation systems or activities in regions where native species still persist but which are outside the boundaries of protected areas. Relatively infertile, cold, steep, rocky, or arid, most of these lands in the United States are administered by the Bureau of Land Management and the U.S. Forest Service. Next are extractive reserves, where certain kinds of resource harvesting, including rubber tapping, and limited hunting and logging, are permitted. In practice, Soulé believes, there may be little difference between these and the preceding systems, except that the latter are more circumscribed. The remaining systems are those subject to varying amounts of

> *"There is one thing stronger than all the armies in the world: and that is an idea whose time has come."*
> — Victor Hugo

* See chapter 9, pages 286-291.

human management — zoo parks, highly managed agroecosystems and agroforestry projects, botanical gardens, zoos, aquariums, and ecological restoration projects intended to increase species richness or productivity in degraded habitats.[19]

Restoration Ecology

"An epochal development has begun: For the first time in human history, masses of people now realize not only that we must stop abusing the earth, but that we must restore it to ecological health." — John J. Berger, director, Restoring the Earth

In the introduction to the proceedings of a landmark conference on restoration ecology he organized in 1988, Berger points out that ecologists are not alone in developing this relatively young science. They are joined by consultants, students, environmentalists, corporate officials, members of sporting and youth organizations, and ordinary citizens. "These people have heard the earth's cry for help, and they have responded."[20]

Defining Restoration

Ecological restoration is an effort to return an ecosystem to a close approximation of what is was before it was disturbed. The process involves recreating both the structure and the function of the original system. In addition to reconstructing the original physical conditions, it may require chemical cleanup of the environment and biological management, including revegetation and the reintroduction of native species. The goal is to recreate a natural, self-regulating system that is integrated with the ecological environment in which it occurs. However, as Berger cautions, no restoration can ever be perfect; it is impossible to reproduce the complex sequence of "biogeochemical and climatological events over geological time that led to the creation and placement of even one particle of soil, much less to exactly reproduce an entire ecosystem. Therefore, all restorations are exercises in approximation and the reconstruction of naturalistic rather than natural assemblages of plants and animals with their physical environment."[21]

In the following pages we give examples of restoration projects being undertaken in different regions of the United States. Although it is too early to claim long-term successes, results indicate that restoration techniques, along with traditional conservation efforts of preservation and pollution control, are an important means of helping natural healing processes.

Eco-Pioneering

According to Berger, the roots of restoration ecology go back to the early twentieth-century school of naturalist landscape architecture, exemplified by Jens Jensen and others.[22] Restoration of soil, forest, and rangeland was extensively carried out from 1933 to 1943 by the Civilian Conservation Corps, which was established as part of the New Deal's program to provide employment. Some stream restoration was done in the early 1930s. The earli-

est planned ecosystem restoration began at about the same time at the arboretum of the University of Wisconsin in Madison when botanist Theodore M. Sperry and a crew of Civilian Conservation Corps workers restored a prairie planted on an old horse pasture as part of a plan by Aldo Leopold to have all the native biotic communities represented at the arboretum. Since then the project has provided a steady stream of restoration ecology research, most notably the 1940s work on the rejuvenating effects of prairie fire. A more recent University of Wisconsin arboretum study focuses on why certain prairie plants known as "conservative species" are difficult to restore. Considered the hallmarks of a healthy prairie, these species are the first to disappear when a prairie is disturbed and may take many years to come up in a restoration, if they appear at all. At the University of Nebraska studies are under way on the relationship between rare plants and their more common relatives. And at the University of Illinois in Chicago, ecologist Henry F. Howe is using prairie restoration techniques to test whether a "pristine" prairie is actually the result of human intervention.*

The Fermilab Experiment[23]

Under the direction of Dr. Robert F. Betz, a prairie ecologist at Northeastern Illinois University, a 1,000-acre prairie has been created inside the ring formed by the circular 4-mile tunnel of the Fermi National Accelerator Laboratory at Batavia, Illinois, near Chicago. It is a functioning ecosystem that now supports some 125 native plants and a wide range of previously missing fauna including falcons, meadowlarks, bobolinks, foxes, and coyotes that now make Fermilab their permanent home. The experiment shows how an ecosystem develops in a precise manner and sequence, starting with what Dr. Betz calls a few "matrix species" of plants. "You cannot put a prairie in backwards... you have to start with some basic plants that prepare the way for others."[24] Big bluestem grass, often reaching 10 feet tall, and other matrix plants prepare the ground by changing the soil, adding organic matter and antibiotic chemicals that in turn nourish successive waves of flowering plants.

> *"It was not our intention to rediscover a lost ecosystem. We were trying... to restore a tallgrass prairie landscape in an aggressive, non-compromising way."*
> — Steve Packard, Illinois director of science and stewardship, the Nature Conservancy

Once the way is prepared, the process of a prairie coming back to its native condition is gradual over years and decades, says Betz. "You don't get it all at once." In order to restore an ecosystem, you have to keep tending it until it runs on its own. This means beating back invading nonprairie plants and animals and continuing to plant native vegetation that will attract native creatures. Today, more than fifteen years after the first big

* As, for example, the burning of the prairie by Native Americans to open up pasture to attract buffalos (see page 147).

bluestem was sown at Fermilab, the dynamics of the prairie ecosystem are becoming apparent, with new native species appearing each year. The restorationists would like to see grazing bison taking their place alongside the foxes and coyotes that have already reappeared. But they must first determine whether the site is large enough to support a viable population. It will take decades before the prairie restoration reaches its prime, Betz says. Meanwhile, he and his coworkers keep reseeding the plot to accelerate the process of ecological development and contend with deer, pheasants, and other invaders. The long-range hope would be for the humans to depart and leave the cleansing process to naturally occurring fires, bison, and other native species.

*Restoring the Tallgrass Savanna**[25]

Starting with a handful of individuals Packard recruited at a Sierra Club membership meeting in 1977, the North Branch Prairie Project today involves scientists from eight institutions and hundreds of volunteer workers. Aside from its scale, what makes the project especially interesting is that it is being carried out *within* the city of Chicago and its northern suburbs. The initial goal was to restore the few remnants of prairie that had survived the encroachments of urbanization — seven sites along the Chicago River, ranging from a few acres to a part of an acre. Most were little more than small openings in thickets of young brush. The idea was to enlarge the remnants by replacing the brush around them with prairie species gathered from rapidly vanishing original prairie patches along railroads and other "odd spots."

The plan met with criticism over the need to cut large numbers of trees, but as Packard explained, there were 67,000 acres of brushland in the forest preserve's district, compared with only a few acres of extremely valuable prairie relics. Like the prairie, tallgrass savanna should be preserved because it is a natural, open "parkland" rich in rare species. At the same time it makes an excellent recreational landscape. This happy coincidence between the wishes of ecologists for biological conservation and of the general public for recreation is, Packard emphasizes, an important factor in the success of the ongoing restoration. "Right in the metropolis we would restore something of real cultural and ecological significance."[26]

The question then arose of what should be done with the oak trees after the surrounding brush had been cleared. Did the restoration planners envision forest or savanna? How close to the trees would the fires get when the restorationists burned the prairies? Despite a strong sentiment for protecting the "forest," Packard and his coworkers wanted to restore "something that no longer existed anywhere — the rich grassland running up to, under, and through the oaks. The tallgrass savanna — a prairie with trees."[27] Along the North Prairie, the existing oak trees were separated from the prairies by a bank of thicket ranging from 50 feet to a quarter of a mile. Packard's plan was to use natural forces — in this case fire — to reunite

* Savannas are intermediate in structure between open prairies and woodlands. They have some trees, but they are too widely separated to shade all the ground beneath them.

prairie and woodland. But most of the green wood of the brush patches would not burn or, where it did, quickly grew back. The question now, said Packard, was, "Did we have enough determination and patience to give natural processes two or three hundred years to work themselves out?"

Speeding Up the Process

> *"Thick stands of switch, Indian, and Canada wild rye grasses stood shoulder-to-shoulder with rattlesnake master and yellow coneflower."*

In April 1980 Packard and five colleagues cut away the brush — cherry, hawthorn, and elm trees — from under the oaks on one of the prairie remnants so that they could sow half a bushel of choicest prairie seed mix. By 1982 the results looked promising: "Thick stands of switch, Indian, and Canada wild rye grasses stood shoulder-to-shoulder with rattlesnake master and yellow coneflower." Buoyed by its apparent success, the team then cut the brush away from many other oak edges and planted much more prairie seed. Although at first a few species "did tolerably well," the plantings remained thin and burned poorly. Even under the oaks, where the fires burned well, none of the seeds grew. Frustrated in their efforts to expand the prairie into the areas near and beneath the oaks, the team decided to leapfrog the persistent brushy border and bring the fire in behind the brush, into the heart of the woodlands — that is, to attack the brush from both sides. In the springs and summers of 1984 and 1985 they burned deep in the oak groves, but nothing much appeared except, during the second summer, an exponential growth of native weeds. "The prairie species just wouldn't move in under the oaks," Packard observed. Part of the problem, he began to realize, was that "we were thinking too much about prairie and weren't picking up on what the other community — the savanna — was trying to tell us."[28]

During the quieter winter months of 1985 Packard researched the work of earlier ecologists, especially as they related to the species of oak-edge sites. One important clue came from an 1863 report by Henry Engelmann, indicating that in the absence of fire the savanna community required some other type of disturbance to maintain itself. Thus, concluded Packard, after 150 years of protection from fire, the best savanna remnants were to be found in "disturbed" areas. At this time Packard recalled Betz's prairie restoration work at Illinois University* and, after reading the 800-page *Plants of the Chicago Region* (by Floyd Swink and Gerald Wilhelm) from cover to cover, began to search for savanna species in old cemeteries, golf-course roughs, railroad rights-of-way, and at the edges of horse paths in forest preserves. Gradually he found the plants he was seeking as well as other unfamiliar but important species.

Another insight into the savanna ecology came in the fall of 1985 when Packard's team was preparing its first major batch of "savanna mix." Prairie

* See pages 308-309.

seed is small, dry, and hard. As they were sorting through the many different savanna plant species, they identified more than three dozen types of fruits and nuts, "dramatically making the point that we were dealing with a community that did not function by prairie principles, but depended, for one thing, not on wind to disperse a seed, but on animals — no doubt the turkeys, deer, squirrels, doves, and other species known to have been abundant on the old savanna."[29] Now the team mixed the piles of fruits and nuts with little sacks of other native plant seeds — rare grasses, asters, snapdragons, gentians, and sedges — found in the semishaded areas similar to where the restoration had so far been unsuccessful. This was then mixed half-and-half with the regular prairie seed and thrown by handfuls to the wind through the oaks in the 90-acre Northbrook tract. In spring 1986 came the first signs of success — countless tiny new green cotyledons thriving where only black dirt and thistles had been. By the summer of 1987, Packard reported, the space under and near the oaks "waved with hundreds of thousands of blooming rare and uncommon grasses and flowers. An oak grassland was unfolding, almost entirely from seed we'd held in our hands." Uncommon species of butterflies, like the great-spangled fritillary and the Edwards hairstreak, appeared as if from nowhere to feed on the plants. A pair of rare eastern bluebirds returned in 1989 to establish a family in what had become for them an ideal habitat. Commenting on the striking beauty of the resurrected ecosystem, Packard says, "It was like finding a Rembrandt covered with junk in someone's attic."[30]

"It was like finding a Rembrandt covered with junk in someone's attic."

Prairie, savanna, and woodland habitats have been restored by the Nature Conservancy and other groups throughout the Chicago area. Their reappearance is particularly gratifying because many of the plants and creatures flourishing there belong to species that are listed as endangered or at least locally rare. Although Packard's findings remain to be fully confirmed by the passage of time, they support the theory that ecosystems do not assemble and develop haphazardly, but in special patterns and sequences, according to locally specific conditions and affinities and that, when even a few key species are removed, the ecosystem will not work.

Living Nets in a New Prairie Sea[31]

As we have seen from the previous examples, the naturally diverse ecosystem of the prairie is based on *poly*culture, the development of many species of plants and animals. Within this natural system, the alliance of soil and perennial roots of plants creates what Wes Jackson, plant geneticist and founder of the Land Institute in Salina, Kansas, describes as "the holding power of the living net and the nutrient recharge managed by nosing roots of dalea, pasqueflower, and bluestem."[32] The agriculture that replaced the prairies is *mono*cultural, based usually on single-species annual crops such as corn, wheat, or soybeans, whose roots do not carry over from year to year and thus contribute to soil erosion and lower productivity. As a long-term

response to this problem, the Land Institute is studying the development of mixed perennial grain crops as the basis of "domestic" prairies that would have high-yielding fields of crops that are planted only once every twenty years or so. "After the fields had been established, we would need only to harvest the crop, relying on species diversity to take care of insects, pathogens, and fertility," says Jackson.[33]

> *"Return as much of your land* as you can afford *to diverse native prairie. Do not add improved varieties that are products of the tools of modern technology, lest you pollute the landscape. Do not try to improve on this patch of native prairie, for it will serve as your standard by which to judge your agricultural practices. There is no higher standard of your performance than the land and its natural community."*
>
> — Wes Jackson, *Altars of Unhewn Stone*

Although practical applications of Jackson's theory may take twenty years or more, there are encouraging signs that some of the prairie perennials might produce high yields. For example, Land Institute studies of Illinois bundleflower, a legume already laden with seeds, suggest that one strain might be able to yield more than 3,000 pounds of seeds per acre, confirming similar findings elsewhere on buffalo grass and other prairie plants.[34] Jackson and his coworkers at Salina do not think their approach represents the entire answer to our total agricultural problem. They believe, however, that breeding new crops from nature's abundance and simulating the botanical complexity of a native region should make it easier for us to solve many agricultural problems, especially in the "million square miles that was turned under to make our Corn Belt and Breadbasket."[35]

Bring Back the Buffalo

Restoring the Great Plains frontier has long been a dream of conservationists. It may be coming closer to reality as the result of changing meteorological and economic climates and the pioneering efforts of an East Coast husband-and-wife team. Creating a buffalo commons is advocated by Frank Popper, chairman of the urban studies department at Rutgers University and international expert on land-use planning, and his wife, Deborah Epstein Popper, a Rutgers geographer. The idea is to convert much of the prairie outback into public domain for its original inhabitants — the buffalo. Over the last century there have been three separate attempts to settle the Plains. Despite increasing federal support, they have not been conspicuously successful. Those of the 1890s and 1930s were largely uprooted, and many of those of the 1990s seem likely to do likewise, say the Poppers. Although there have been numerous remedies proposed, including genetically engineering new species of corn and wheat, creating llama, goat, and donkey industries, or establishing East African-style hunting reserves, none of the plans seems likely to help the Plains on a regional basis.

The Popper concept, first outlined in a 1987 article in the professional journal *Planning*, favors a buffalo commons replacing private development with public stewardship of the land, allowing Nature gradually to take back an area one and a half times the size of California. They say that many of the building blocks for creating such a commons are already in place, including existing national parks and grasslands, public grazing lands, wildlife refuges, Indian lands, and private and land conservancy holdings, as well as thousands of properties foreclosed by federal credit agencies and private banks, and farms operating under the U.S. Department of Agriculture program, which pays farmers not to cultivate marginal land.[37]

"We believe that over the next generation much of the Plains will, as the result of the largest, longest-running agricultural and environmental miscalculations in American history, become almost totally depopulated."
— Frank J. Popper and Deborah E. Popper[36]

To those who fear their plan spells the end of human habitation in the region, the Poppers reply that the small cities of the Plains will thrive as urban islands in a shortgrass sea. Other settlements and self-contained "service centers" such as Bismarck and Cheyenne will survive the general depopulation and, in fact, prosper. Agriculture and some mining activities will continue, but in "more economically appropriate areas. For example, the Plains will still remain the nation's granary of last resort." In the Poppers' vision, the government, the private sector, and individual citizens must start planning to keep the region from turning into "an American Empty Quarter," an area the nation has deserted.[38] Ultimately, the Poppers conclude, land use must be a deliberate national choice. "The Plains are about to change, maybe for the worse. But they still retain a magnificent, terrifying sense of possibility."[39]

Restoring Aquatic Ecosystems

In December 1991 the National Research Council of the National Academy of Sciences called for prompt action on a large-scale program to restore damaged and polluted wetlands, rivers, streams, and lakes throughout the United States to prevent permanent ecological damage. "We can repair damaged ecosystems to a close approximation of the condition they were in before they were disturbed, even with present knowledge," said John Cairns, Jr., director of the University Center for Environmental and Hazardous Materials Studies at Virginia Polytechnic Institute in Blacksburg. "Although this is a new field, the results from restoration are already dramatic."[40]

In urging the federal government to establish a long-term, comprehensive strategy for restoring aquatic ecosystems, the council recommended that damaged or destroyed wetlands be restored at a rate that offsets any further loss and contributes to an overall gain by the year 2010. It also proposed a

national restoration effort targeting 400,000 miles — about 12 percent — of U.S. rivers and streams. This restoration should begin with improved land-management practices that would allow natural restoration to occur. High priority should be given to protecting undisturbed large river and floodplain ecosystems, such as portions of the Atchafalaya and Upper Mississippi rivers, that could be used as "templates" for restoration efforts.

Restoring 2 million acres of the 4.3 million acres of degraded U.S. lakes is an achievable goal, the council said, recommending that by the year 2000 at least 1 million acres should be restored. Noting that the severity of the pollution is "grossly inadequate" for determining the extent of lake damage and assessing the progress of cleanup efforts, the council suggested that funding should come from federal and nonfederal sources and be administered through state lake programs.[41] The following case history provides such an example.

*Resurrecting Delavan**

The 2,200-acre Lake Delavan is just one of fifteen thousand in the Land of Lakes, but it provides an example of positive action that can be taken when there is the right combination of citizen concern, local, state, and federal government involvement, and thoughtfully integrated resource management. In 1980 Delavan was dying. Suffering from more than four decades of abuse — wastewater discharges, destroyed wetlands, nonpoint source pollution — this popular resort lake near Milwaukee and Chicago turned a putrid green. Once rich with healthy vegetation and the habitat of the much sought-after northern pike, walleyes, and bluegills, all that could now survive in its murky depths were "rough fish" — carp, which upset the lake bottom by rooting up valuable newly growing plants, and wide-mouth buffalo fish, which eat the zooplankton that otherwise would consume the algae. Some two thousand property owners around the lake were outraged because toxic algae scum made the lake unswimmable.

> *"The key to our success was realizing that all the components of an ecosystem fit together like a jigsaw puzzle."*
> — Richard Wedepohl, lake management coordinator, Wisconsin Department of Natural Resources[42]

Linking the decline in water quality to pollution from inadequately treated sewage, the community agreed to pay higher taxes for a new $40 million wastewater plant to keep human waste out of the lakes. This, local residents hoped, was all they needed to clean up the water and restore the natural balance of fish and plants. But 1983, it turned out, was the worst year ever for algal blooms. The Delavan community then persuaded the Wisconsin Department of Natural Resources (DNR) to investigate why the water was still in such bad shape. The study, done in cooperation with the

* We are especially indebted to Richard Wedepohl, and to Jana M. Suchy, water-resources media specialists, Wisconsin Department of Natural Resources, for their generous assistance on this report.

U.S. Geological Survey, showed that the causes of the problem went far beyond the shores of Lake Delavan, extending into the entire watershed. They included urban runoff discharging directly from storm sewers, unchecked soil erosion from fields, chemical runoff from a nearby fertilizer plant, and the loss of valuable natural filtration from the destruction of wetlands, which further accelerated the decline in water quality.

In 1984, with the town of Delavan taking the lead, the DNR and the University of Wisconsin worked out a $7 million comprehensive lake restoration plan, involving the cooperation of Delavan Lake Sanitary District, the Walworth County Land Conservation Committee, the Wisconsin Department of Agriculture, the U.S. Fish and Wildlife Service, and the U.S. Environmental Protection Agency. Sixty percent of the costs were borne by the local community, $1 million came from the EPA's Clean Lakes Program, $750,000 from the Fish and Wildlife Service, $400,000 from the Wisconsin DNR, and $350,000 from the Geological Survey.

Draining the Lake

The Delavan plan was nothing if not ambitious. It meant draining the lake, killing off the undesirable fish, and preventing the lake from returning to its polluted state. According to the EPA, it was one of the largest and perhaps most significant water-rehabilitation projects ever undertaken. In addition to replacing the fish population, the project called for: restoring 90 acres of natural wetland at the upstream end of the lake to filter out plant nutrients before they get into the lake; building a barrier peninsula in the same area to divert nutrient-carrying water around the deepest part, especially when ice covers the lake; dredging many shallow spots to remove several feet of polluted silt; improving street cleaning in Elkhorn just north of the to-be-restored wetland; temporarily lowering the level of the lake to allow stabilization of the bottom, and killing off the rough fish; and spreading aluminum sulfate (alum) on the deeper part of the lake to create an impervious layer between the water and the nutrient-rich part of the silt.

Killing the Fish

An important part of the Delavan project was getting rid of the turbidity-creating carp and algae-eating buffalo fish in the lake. The agent of choice was rotenone, a natural, biodegradable substance derived from the root of a South American plant that is in short supply worldwide. It is selective in killing only gilled creatures, by inhibiting their uptake of oxygen, without harming other organisms. Because treating Delavan's 2,000 acres would have been too costly, the lake was drawn down to 1,300 acres over months of dredging and pumping. Using ten tanker trucks to deliver 50,000 gallons of rotenone, advance work crews treated all tributaries, wetlands, and ponds in the 45-square-mile watershed one week before tackling the lake itself so that the rough fish could not evade the treatment. On November 6, 1989, the main operation began and lasted four days. "We got 'em all," reported area fish manager Doug Welch.[43] Follow-up surveys by gill net, scuba diving, and

sonar showed no live fish. By weight, more than 99 percent of those killed were carp and buffalo fish.

In 1990 the lake fishery was restocked with game fish, the dam was reconstructed to bypass high flows, the barrier peninsula was built, and inlet dredging was completed. In April 1991, lake sediments were treated with 660,000 gallons of alum in order to eliminate algae growth and contain the huge concentration of phosphorus on the lake bed bottom. A start was made on restoring the wetlands that had been formerly drained and farmed, and nonpoint-source controls were completed. To restore the wetlands, the local community purchased 150 acres of previously farmed lands. Phosphorus levels were reduced by 99 percent from preproject levels, and the fish that have been restored are showing very good signs of becoming reestablished. An indication of the progress being made, Wedepohl said, was that northern pike had grown from fry size to a length of 17 inches in the first year.[44] Although it will be some years before the final evaluation can be made, the present indications are very promising, according to reports from the DNR and the EPA.[45] One important sign of hope was that in midsummer 1991 Delavan Lake experienced water clarity of 26 feet — its best ever. The dramatic turnaround at Delavan is a case of successful environmental protection and a compelling argument for the merits of preventing pollution *before* it happens.

Living Machines

Located on Cape Cod, Massachusetts, OAI is a small but growing nonprofit research organization that incorporates the Center for the Restoration of Waters. The family of technologies used for this purpose, known as living machines, was developed by John Todd, founder and president of OAI. An example of how these can be applied, in this case to wastewater purification, is Solar Aquatics™. By harnessing biological processes, solar aquatics imitates the way nature purifies water, but more quickly and thoroughly. "We use large, translucent cylindrical tanks, placed in rows inside a greenhouse, positioned and piped so that gravity creates a stream through each line." Todd explains, "The first tanks contain bacteria, algae, and snails. Subsequent tanks downstream contain more complex life forms, including higher plants, other mollusks, and fish. The wastewater [to be purified] is pumped into the first tanks, where microscopic bacteria attack and consume organic matter or nutrients causing the population to grow. The algae thrive on the nutrients released by the bacteria and grow rapidly from the abundant food source. The snails then consume the algae, and on it goes: one organism thriving in another, consumed and being consumed by something else, and all the while transforming the wastewater. Further on, we add on rafted plants to the top of the water. Their oxygen-rich roots reach beneath the surface for the higher organisms to graze. By the last

> *"It must be possible to use sunlight and ecology to purify water the way Nature does."*
> — John Todd, Ocean Arks International (OAI)

tanks, fish such as tilapia and bass swim around in clean water. Final purification takes place in an engineered marsh."[46]

The center's flagship research station is located in Providence, Rhode Island. Opened in July 1989, it now treats municipal sewage to advanced wastewater standards. Current projects range from energy efficiency to marketing of the plants and animals produced from the cleansing process. The center serves as a school for ecological design and stewardship and is developing living machines in other schools to help students understand nature in a new way. At Marion on Buzzards Bay, Massachusetts, the center is exploring alternatives for treatment of wastes from boats. In a small greenhouse-enclosed system Marion sewage water is mixed in varying dilutions with the boat waste. The mixture is then fed into a series of tank and marsh ecosystems designed to break down and assimilate the pollutants. The system is expected to process 250 gallons of water a day, which will be tested and discharged into the town of Marion's sewage-treatment ponds. Other Ocean Arks projects include developing an ecological method for treating the factory wastes of Ben and Jerry's Ice Cream Company and the restoration of Flax Pond in Harwich, Massachusetts.

Restoring Flax Pond

*John Todd and Karen Schwalbe**

Fresh water ponds are a central part of the landscape of Cape Cod. They are used for town water supplies, recreational boating, and, oddly enough for maritime communities, as favorite fishing places. The ponds are a source of water for cranberry growers. Cranberry bogs use enormous amounts of water and pond levels can drop up to a foot in a few hours when neighboring bogs are flooded. Flax Pond is such a pond. Fifteen acres in size, it is ringed by trees, mostly pitch pine, with cranberry bogs at each end and is home to the endangered plant, the beautiful Plymouth Gentian. But the pond is sick. It is contaminated with human sewage. In 1988 it contained toxic organic compounds including the suspected carcinogens toluene and chlorobenzene. In the fall the eastern end turns an eerie, bright bloody red color due to the presence of iron in the water. Nitrogen and phosphorus were found in the sediments at levels vastly greater than those found to be toxic to fish and other aquatic life. The normal metabolic processes that sustain ponds were shut down. There was a 10-foot blanket of sludgelike sediments on the bottom. The pond water was having an adverse effect on the cranberry crops and local farmers were threatening to sue the town.

The major cause of the pollution is thought to be a large landfill situated 100 yards from the pond's northern perimeter. At one corner of the landfill there are six pits about 10 feet deep into which untreated cesspool

* Condensed and updated with permission from "The Restoration of Flax Pond," *Annals of Earth*, vol. IX, no. 1 (1991). *Annals* can be ordered from OAI, One Locust Street, Falmouth, MA 02540; (508) 540-6801.

and septic tank wastes have been dumped at a rate of over 10,000 gallons a day for many years. The wastes percolate 10 to 15 feet farther down into the water table. Thus loaded with heavy metals, toxic organics, and sewage, the ground water hits a clay layer, which funnels the liquid underground toward Flax Pond. The basic problem we had to solve first was the sediments that had to be "unlocked" and at a slow enough rate so that much of the pond life would not be killed in the process. The conventional solution would have been to dredge the sediments out and landfill them, which would have cost millions of dollars simply to put the problem somewhere else.

Our plan was to create pockets of upwellings within the eastern end of the pond, followed by a sequence of processes we call ecological augmentation. This involves adding missing mineral elements necessary because Cape Cod waters are replenished by rain and snow that may lack critical elements required by restoration organisms. Next we planned to introduce over half a dozen bacterial types, which were either absent or present in insufficient numbers in the pond. Later, in years two or three of the project, we intend to reintroduce missing higher life forms, organisms like freshwater mussels, which are abundant in many healthy Cape ponds. One of our basic objectives was to base restoration strategies upon renewable energy. After twenty years of designing with the wind and sun at New Alchemy Institute and elsewhere we are beginning to discover the value of pulsing, cyclic, and sometimes chaotic forms of energy. As a means of creating upwellings and gently lifting the sediments up into a water column, we installed floating windmills of a Savonius rotor design. Simple, reliable, and inexpensive, the Savonius wind-driven rotor turns a propellor-like blade that is submerged in the water above the sediments. As the wind speed increases, the propellor turns faster and sucks more sediments into the water column. During periods of calm, the upwellings cease and the sediments settle. We believed this periodic pulsing would help start the restoration process, bringing the sediments up into a light and oxygen-rich environment.

On September 20, 1990, we placed three floating Savonius windmills on the pond located 40 feet apart and held in position by metal pipes driven into the sediments. From then through October they created the hoped-for upwellings. During this period, ammonia levels in the eastern end of the pond where we were working rose four- to five-fold. But nitrification, the process where ammonia is oxidized and detoxified, did not occur. The windmills alone couldn't complete the biochemical cycles we wanted to establish. By late October temperatures in the pond were dropping fast. Because the biology we wanted to enhance shuts down when the temperatures drop into the lower 40 degrees Fahrenheit, we decided to race against the weather and undertake the next phase of our experiment. Under the most southerly windmill we lowered a bag of dolomite limestone onto the bottom. It was perforated so that the contents initially would be localized there. Next, on a once a week basis, under the southerly and central windmills we injected eight strains of natural soil bacteria into the sediments. The most northerly windmill, treated with neither bacteria nor dolomite limestone, was the control.

What happened over the next two weeks was extraordinary. Nitrification began and the pond woke up. In the eastern zone ammonia plummeted to nontoxic levels. Concurrently, nitrate-nitrogen, a safe chemical for fish and aquatic life, skyrocketed in the water column, most notably under the windmill with mineral additions and bacteria. The conditions under the bacteria-only windmill hovered midway between these nitrogen levels and the control windmill. There was a modest drop in ammonia and concurrent rise in nitrates throughout the whole pond. After the middle of November when pond temperatures cooled and hit the thermal barrier, microbial processes and the pond slept again. Our analysis of the sediment data revealed an even more remarkable story. The total nitrogen in the sediments had dropped by as much as 67 percent. At the two sampling stations in the zone of maximum intrusion of pollutants from the septage lagoons, nitrogen levels dropped by 14 and 40 percent respectively. This meant that we were purifying the pond at a rate greater than the rate of recharge of new pollutants. Equally remarkable and harder to explain is that nitrogen reduction was greatest at the station adjacent to but not in the eastern zone of the pond. The organic forms of nitrogen in the sediment were somehow affected by a process that began under the windmills. The pond's progress continues — a strong indication of a reviving body of water. We are just beginning to see the potential of the partnership between human stewardship and living machines in restoring polluted ecosystems.

For more information: Contact Ocean Arks International, 1 Locust Street, Falmouth, MA 02540; (508) 540-6801.

Marsh Builder

"Does this look like a wetland?" is a question often asked by Dr. Ed Garbisch, a chemist and founder of Environmental Concern Inc. (ECI) in St. Michaels, Maryland, whose avowed goal is to make a business of fooling Mother Nature. To all appearances it *is* a wetland with springy peat soil, stands of six-foot-tall greenish-yellow cordgrass reaching out into the water, and just inside the high-tide line ranks of finer, darker salt-marsh hay. If you remain still, you'll see typical coastal marsh birds — nesting mallards, blue herons, and white egrets seeking the baby fish that use the wetland as their nursery. Since 1972 ECI has reconstructed some 300 acres of wetlands from Maine to South Carolina and as far inland as Ohio, including 25 acres of marshes strung out for erosion control in 20-foot-wide shoreline buffer strips along tributaries of the Chesapeake Bay. In the course of more than 350 projects, Garbisch has had a success rate of 95 per-

cent, losing only a handful of wetlands destroyed mainly by disease or storms in the 1970s.[47]

More than half of ECI's work consists of damage mitigation, in which, for example, 5 acres of marsh destroyed by a utility in one location might be replaced by 6 new acres of marsh elsewhere, with the extra acre a bonus to compensate for any difference in the quality of the new marsh. In the small community of wetland restorers Garbisch is sometimes called an optimist because he believes that many kinds of wetland can be restored in their basic functions and, in certain situations, be made to function even better than they did originally. Thanks to ECI's track record, the company is at the forefront of what has become a wetlands creation industry. In his book *Restoring the Earth*, John Berger defines the essence of Garbisch's restoration method as "planning all aspects of a marsh project at once, considering which plants to use, their correct elevations, and the on-site stresses."[48] The successful outcome, says Garbisch, combines gardening and hydrology. Thus ECI maintains an extensive nursery of wetland plants from which they draw the raw materials for their restoration projects. As for hydrology, "if [it] is on the button, you can't fail. But if the water is too deep or too shallow or is not there at the right time of year, the project is doomed."[49]

> *"I'm not a conservationist or environmentalist who wants to get back the system that's been destroyed over the years."*
> — Ed Garbisch

Restoring Our Urban Environments*

A key to restoring the earth is enabling a broad base of people to be involved in such efforts. The models for restoration and stewardship of the earth and our communities should be the responsibility of all people. The programs and projects created to accomplish this need to reflect a participatory process, not leaving the task in the hands of professionals and "experts" alone and not expecting broad public participation by wishing it alone. Sharing the information and skills with the public are necessary to ensure that they can take part meaningfully.

If humans are to be part of environmental solutions, we must recognize that an increasing proportion of the earth's population is found in urban areas. Moreover, these areas produce their own unique and often unjust ecological problems. And, finally, because we exist in one interconnected ecosystem, urban environmental problems have direct impact on suburban and rural areas. This means not only paying attention to ecological concerns directly but to understand and make the

* Contributed by David Levine, founder and director of the Learning Alliance.

links to other social justice issues such as health, economics, and community empowerment, thus creating an ecological philosophy and practice based on environmental justice.

(The Learning Alliance, a nonprofit organization based in New York City, has developed numerous educational and community-oriented programs including Restoring New York's Environment, a training program related to urban restoration and natural resources management. For further information, contact The Learning Alliance, 494 Broadway, New York, NY 10012; (212) 226-7171.)

A Tree Grows in Brooklyn (and Other Cities)

As Steve Packard, John Berger, and other ecologists remind us, a great deal of grassland, forest, and wetland acreage survives in urban America. Although most of these original ecosystems are degraded and neglected among a jungle of rubble, trash, and automobile wrecks, their potential for revitalization is literally breathtaking. When restored, they can provide not only fresh air but significant habitat for many native plants and animals, ample recreational and educational opportunities for children and adults, and an important source of vegetables, fruits, flowers, and grains for the estimated 90 percent of us who will live in urban areas by the turn of the century.

Going with Gaia

For the past two thousand years most of what we call civilization has lived with the belief that earthly matter is essentially 'dead,' and that life is fundamentally different from the stuff of the earth, which can be arranged for human convenience. Today we reap the bitter fruits of this perspective: global warming, acid rain, ozone depletion, and urban worlds where most of its human inhabitants live in sterile isolation from the natural systems that make life possible.

The Gaia hypothesis, named after the Greek goddess of Earth, offers a way out of this dilemma in suggesting that organisms, in conjunction with the atmosphere, oceans, and rocky substratum of the earth, comprise a self-regulating, homeostatic feedback system that regulates global temperature and the concentration of gases and salts much in the same way as the bodies of individual organisms are regulated. If we creatures are self-regulating complexes of matter and energy flow, so is the earth; then the earth is, like us, alive.

In the forefront of applying this hypothesis to the problems of urban communities is the Gaia Institute at the Cathedral of St. John the Divine in New York City. Its ongoing research programs include: recycling organic wastes, wastewater, and sewage treatment to revegetate urban environments; the urban rooftop greenhouse project adapting greenhouse and composting technologies to make city roof space capable of providing organic produce for all city dwellers; ecological restoration of meadow, wetland, and salt marsh sites extending from the central Bronx and Bronx River to Long Island Sound and the Hudson River. The restoration work is done in conjunction with local schools and universities, Boy Scouts, and community and

environmental groups. Among other Gaia Institute projects are the detoxification and bioremediation of a severely polluted landfill in Pelham Bay on Long Island Sound, the long-term study of the evolution of regulatory systems in biological systems, and the integration of architecture and ecology in the design of human communities. (For further information on the institute, see the Directory, page 347.)

From Rubble to Restoration

This is the title of a report by Karl Linn, a leading landscape architect, showing how unused land in many cities can be reclaimed for urban agriculture and recreation. Since the 1950s urban renewal projects have been based on heavy capital investment (which is no longer available) and on limiting the use of vacant land to expensive small-scale parks or community gardens. Removing garbage and weeds from vacant land with bulldozers and other heavy equipment has become a monumental task for many municipal governments and usually destroys many acres of wild grasses, which stabilize the soil, as well as burying the topsoil under layers of unproductive fill. Even more disturbing is the indiscriminate use of toxic weed killers, often by untrained crews in the presence of children. Removal of weeds is environmentally harmful: It exposes bare earth, which results in rainwater runoff, soil erosion, dust flurries, and other effects.

An ecologically sound approach described by Linn is to improve the quality of the soil on vacant land by cultivating a green cover of grasses and wildflowers. This protects the soil from baking dry in the sun and becoming too hard to till. If left to grow, the cover crop will become fully established within three years. The sudden presence of "vibrant, colorful meadows in the midst of squalor," he writes, ". . . will transform the appearance of the neighborhoods rapidly, uplift the spirit of its inhabitants and attract new people and their investment into the area."[50] If the land is designated for agricultural production, the plowing under of cover crop creates fertile soil in which you can grow a wide range of crops such as sunflower and safflower, triticale (a rugged wheat-rye hybrid), kenaf (a bamboolike plant for making paper), and perennials that reseed and fertilize themselves like a prairie. Among the successful urban meadow- and wetland-restoration projects documented by Linn are those of the Newark West Side Development Corporation, Andropogon Associates in Philadelphia, and the Green Guerrillas in New York City. An innovative pilot compost operation was carried out by Bronx Frontier at Hunt's Point, close to New York City's fruit and vegetable market from which contractors delivered waste at half the price it would have cost them at the regular dump. Initially the operation used a large machine to turn the "windrows" (long piles of compost) and aerate them three to five times a week for a month to six weeks. However, the large quantity of nondegradable plastic bags mixed in with the waste upset the composting process so badly that Bronx Frontier was forced to use leaves and manure from the Bronx Zoo as their sources of organic matter.* Building

* For more on Bronx Frontier, see Jon Naar, *The New Wind Power* (New York: Penguin, 1982), pages 156-161.

on the Bronx Frontier experience, American Soils Inc., a compost enterprise in Freehold, New Jersey, recycles some 25,000 tons of compost a year with a gross revenue of $625,000 and a net of about $200,000. Using a simple front-end loader to turn its compost piles every two to four weeks for six to seven months, it operates on an 18-acre tract of land on which it recycles all biodegradable waste, including food, paper, leaves, grass, and brush. The finished compost is successfully marketed to organic farmers in Philadelphia and New Jersey, as well as being used in ecosystem restoration at Superfund cleanup sites in New Jersey.

There are many other similar composting enterprises nationwide, including Organic Recycling, Inc., in Valley Cottage, New York, and Woods End Research Laboratory, Inc., in Mt. Vernon, Maine, which among other projects is working with plant pathologists at Disney World to make organic fertilizer from grass clippings and sewage waste.*

Reintroducing Native Species

> *"We know more about the tops of mountains than we know about the forests of New York City."*
> — Dr. Mark J. McDonnell, ecologist, Institute of Ecosystem Studies, Millbrook, New York[51]

Contrary to preconceived notions, naturally occurring urban forests can adapt to many types of natural and human-induced stress. Studies of the 6,000 acres of forests in New York City indicate that the city woodlands may be as ecologically efficient as other forests and may even retain more of their original character. A possible explanation is that in the urban environment nitrogen from fossil fuel–burning automobiles and power plants fertilizes the soil and causes microbes to break down plant litter at a faster rate than they otherwise would. Also, because city forests are, in effect, islands in a sea of concrete, they may be less susceptible than suburban woodlands to invasion from alien plant and tree species.[52] In their study of a 90-mile corridor of forests extending northeast from New York City, McDonnell's team documented a gradual decrease in pollution and ecological damage as they moved away from the city. In the rural areas, less than 1 percent of the trees died annually, compared with between 2.5 and 5 percent in the city. In the 1935–1975 period, when hemlock and oak were dominant, a combination of human-induced and natural stresses killed about 65 percent of very large specimens, which were replaced by a dominant birch-cherry-maple forest. "I think it's terrific, because the forest is regenerating itself with native trees," McDonnell said. However, in the long run the excess nitrogen might result in the trees not putting out so many of the fine roots they otherwise use to obtain water and nutrients. Thus a prolonged drought could endanger the forest's existence, he warned. "Down the road, it might do them in."[53]

In order to reintroduce native species and give them an edge against the rapidly encroaching exotic trees like ailanthus (of *A Tree Grows in*

* You can order *From Rubble to Restoration* from Urban Habitat Program, Earth Island Institute, 300 Broadway, Suite 28, San Francisco, CA 94133.

Brooklyn fame) or oriental bittersweet, foresters in a Manhattan park are planting tulip trees among a grove of Norway maples that they later plan to cut down. Although a few two hundred-year-old tulip trees remain, they and the ancient oaks have trouble reproducing. When their seeds germinate and take root, they are shaded out by the faster-growing maples. As described by Tony Emmerich, the forestry program's director, the reintroduction process begins with planting oaks, tulip trees, hickories, and other desired natives. When they are established, the crews thin out the nonnatives, many of which were originally brought in by wealthy landowners in the eighteenth century.[54] The crews have collected 40,000 seeds of native trees from all over New York, which they are planting along with azalea, blueberry, chokeberry, and other native shrubs in the parks. Mostly rescued from construction sites, these shrubs increase plant diversity and help attract different species of animals and birds. White pine and pitch pine are planted to increase the nesting sites for owls and hawks. Eventually 100,000 new native trees will be planted. Around the perimeter, 20 miles of fence was put up to stop the dumping of cars, many of which are driven in, stripped, and torched by thieves. It is a major threat to the ecosystem, said Emmerich. "Fires destroy the saplings and the trees can't regenerate. They also wound the trunks of the mature trees and open them up to fungus and rot."[55] After barricades were placed in Pelham Park in 1988, there hasn't been a significant fire there. However, it may take some time to change the attitude of many people that the city forests are a dumping ground. "Preservation," remarked one of the program's environmental advisers, "goes against the very principles on which the country was founded."[56]

Restoring the Earth

The directory *Ecological Restoration in the San Francisco Bay Area*, published by the Berkeley-based Restoring the Earth (RTE), documents 155 projects in that region — 16 forest, grassland, and coastal projects, 3 mined land, 11 wildlife, 12 native plant, 6 watershed, 24 creek, 22 riparian and lake, and 56 wetland projects. Fittingly, its first report concerns the work of Magic Inc., a nonprofit group in Palo Alto dedicated to addressing environmental degradation and its uses, by teaching how to live in an environmentally gentle way that will "perpetuate diverse and abundant life on earth for the indefinite future."[57] In 1981 Magic formed the Evergreen Park Neighborhood Association as a first step to what became a citywide program to upgrade the street trees in Palo Alto (whose name means "tall tree" in Spanish). This was done under the guidance of a city task force on which one of Magic's staff served. At the same time Magic began to work on the regeneration of native California oak trees on Stanford University land in the Palo Alto foothills. In this project the group enlisted the help of several hundred volunteers and recruited student interns who received academic credit for their work in the reforestation program. A third Magic project is the coordinating of Peninsula ReLeaf, a grass-roots tree-planting effort that is part of a national campaign by the American Forestry Association to add 100 million trees to American cities by 1993.

Among the other restoration projects described in the RTE directory are the California Department of Fish and Game's Inland Fisheries Habitat Restoration Program, East Bay Citizens for Creek Restoration, the Nature Conservancy's Volunteer Habitat Restoration Program, the San Francisco Estuary Project, Trout Unlimited's Operation Rescue, and the Urban Creeks Council. In addition to publishing the directory, RTE, which organized a national ecological restoration conference in 1988 at the University of California at Berkeley, is engaged in a wide range of activities that include Project Restore, designed to restore damaged ecosystems in the San Francisco Bay area. In this project RTE has worked with the East Bay Regional Park District to eradicate artichoke thistles (*Cynara cardunculus*). It has recruited and organized community volunteer teams for native grassland revegetation, riparian restoration, eucalyptus removal, and native plantings on other public lands. *For more information*, contact Restoring the Earth, 1713C Martin Luther King Jr. Way, Berkeley, CA 94709; (415) 843-2645.

Who Says We Can't Change the World?

The restoration projects we have described in this chapter are a small sampling of thousands of other efforts being undertaken throughout the United States and Canada. Their importance cannot be underestimated. They are helping to repair damage inflicted on ecosystems by human and natural causes, showing the way to even greater restoration programs that must be initiated at state, federal, and *global* levels. But restoration alone will not stem the tide of ecological destruction. What is required is an ecorevolution. By revolution we mean change on an unprecedented scale, far greater in scope (and cost) than the Marshall Plan, credited with saving Western Europe after the devastation of World War II.

How will we find the resources to pay for such a program? Clearly, the end of the cold war offers a golden opportunity to drastically reallocate resources. At close to a trillion (one thousand billion) dollars a year — $190 for every person on the planet — military spending is an enormous source of funding that can be turned over to the defense of natural ecosystems and human ecosystems, for infant and maternal health care, education, family planning, and other equally pressing social priorities on which a sustainable global future depend. Think of how wonderful it would be to mobilize the armies of the world to clean up toxic waste dumps and polluted rivers, just for starters! If we can launch Operation Desert Storm virtually overnight, why can't we mount Operation Desert Bloom?

There are precedents to follow and examples to adapt: the Civilian Conservation Corps in the United States of the 1930s, and in 1991 the case of Germany paying for a new public transportation system in what was East Germany by boosting the gasoline tax fifty cents a gallon (from $1.35 to $1.85). Such a tax also discourages private gas use, helping reduce air pollution, acid rain, and global warming. This was done at a time when the

U.S. Congress refused to add a paltry nickel to our federal gas tax of 14.45 cents a gallon.*

Ecorevolution calls for building a sustainable future based on restructuring the global economy and changing the way we use energy, the way we grow food, the numbers of children we produce. Above all, it means changing the way we think — about ourselves and about the cosmos. Nothing less than radical changes are needed to restore and preserve the earth's ecological systems on which we all depend. The changes will entail radically different life-styles for large numbers of consumers and a reworking of our basic values. We should avoid the temptation to shift the blame to someone else's backyard — the Brazilian rain forest, for example, or India with its seemingly ever increasing population — but we should remind ourselves that with 5 percent of the world's population the United States produces 25 percent of the world's carbon dioxide, which is a major contributor to global warming. This we do mainly through our profligate use of fossil fuels. Yet many of us view the prospect of limiting the amount of gasoline we use or the meat we eat as a denial of our constitutional rights.

However, if we want to become part of the solution rather than remaining part of the problem, we must see the connections between the way we live and the world we live in. To the extent we do this, we can make a difference. As we have shown throughout this book, there are many ways to make your influence and your vote count in terms of both specific problems and broader ecological issues. Thus we hope that the examples, case histories, and solutions given in *This Land Is Your Land* will inspire and encourage you to take the necessary action as individuals and with others working to make our land more sustainable for all the species that inhabit it. As you become increasingly part of the solution, we invite you to get in touch with us directly so that we can help make your contribution available to an even wider audience.

RESOURCES

1. U.S. Government

The main federal agencies whose work bears on ecological restoration are: the Bureau of Land Management, especially through its habitat program; the Fish and Wildlife Service, through its divisions of endangered species and habitat conservation and of national fish hatcheries; and the National Park Service. All of these come under the jurisdiction of the Department of the Interior. Within the Department of Agriculture's orbit are the U.S. Forest Service and the Soil and Conservation Service. (For details about these agencies, see the Directory, pages 352-353).

2. State Agencies

Most states now have departments of environmental conservation and natural resources usually listed in your telephone directory. For a more comprehensive listing, check with the *Conservation Directory*, published annually by the National Wildlife Federation (see page 340 for details).

* Combined U.S. federal and state gasoline taxes average about 39 cents nationwide.

3. Research and Educational Organizations

Among the leading organizations are the Cenozooic Society; the Center for Plant Conservation; the Ecological Society of America; the International Center for the Solution of Environmental Problems; the Land Institute; Restoring the Earth; the Society for Ecological Restoration; the Soil and Water Conservation Society; the Water Science and Technology Board (National Research Council, 2101 Constitution Avenue, NW, Washington, DC 20418, (202) 334-3422); the Wildlife Habitat Enhancement Council; and the Xerxes Society. (For details on these and other groups, see the Directory, pages 331-355.)

4. Environmental and Conservation Organizations

Among the leading general membership organizations working on different aspects of ecological restoration are American Rivers; the Earth Island Institute; the Izaak Walton League; the National Audubon Society; the Nature Conservancy; the National Arbor Day Foundation; the Sierra Club; Trout Unlimited; and the Wilderness Society. (For more information on these and other groups, see the Directory.)

5. Further Reading

— *Altars of Unhewn Stone: Science and the Earth*, Wes Jackson. San Francisco: North Point Press, 1987.
— *Biodiversity*, E. O. Wilson, editor. Washington, D.C.: National Academy Press, 1988.
— *Conservation Biology: An Evolutionary-Ecological Perspective*, Michael E. Soulé and Bruce A. Wilcox, editors. Sunderland, Mass.: Sinauer, 1980.
— *Conservation Biology: The Science of Scarcity and Diversity*, Michael E. Soulé, editor. Sunderland, Mass.: Sinauer, 1986.
— *Design with Nature*, Ian L. McHarg. New York: Doubleday, 1971.
— *Ecological Restoration in the San Francisco Bay Area: A Descriptive Directory and Sourcebook*, John Berger, editor. Berkeley, Calif.: Restoring the Earth, 1990.
— *Environmental Restoration*, John J. Berger, editor. Washington, D.C.: Island Press, 1990.
— *The Evolution of National Wildlife Law*, Michael J. Bean. New York: Praeger, 1983.
— *The Fragmented Forest*, Larry D. Harris. Chicago: University of Chicago Press, 1984.
— *Gaia: The Human Journey from Chaos to Cosmos*, Elisabeth Sahtouris. New York: Pocket Books, 1989.
— *A Green City Program*, Peter Berg, Beryl Magilavy, and Seth Zuckerman. San Francisco: Planet Drum, 1989.
— "Helping Nature Heal: Environmental Restoration," Special issue. *Whole Earth Review*, Spring 1990.
— *Powerful Peacemaking*, George Lakey. Santa Cruz, Calif.: New Society, 1987.
— *The Redesigned Forest*, Chris Maser. San Pedro, California: R. and E. Miles, 1988.
— *Reforesting the Earth*, Sandra Postel and Lori Heise. Washington, D.C.: Worldwatch Institute, 1988.
—*The Resilience of Ecosystems: An Ecological View of Environmental Restoration*, René Dubos. Boulder, Col.: Colorado Associated Press, 1978.
— *Restoring the Earth*, John J. Berger. New York: Knopf, 1985.
— *Where Have All the Birds Gone?* John Terborgh. Princeton, N.J.: Princeton University Press, 1989.

NOTES

1. Paul R. Ehrlich and Edward O. Wilson, "Biodiversity Studies: Science and Policy," *Science*, August 16, 1991, pp. 758-62.
2. William K. Stevens, "Species Loss: Crisis or False Alarm?" *New York Times*, August 20, 1991.
3. *Journal of the National Cancer Institute*, 1122 (1989); see also "Taxol Update," *Nature Conservancy*, January/February 1991, p. 13.
4. Terry L. Erwin, "An Evolutionary Basis for Conservation Strategies," *Science*, August 16, 1991, p. 751.

5. Stevens, "Species Loss."
6. Ibid.
7. Harold J. Morowitz, "Balancing Species Preservation and Economic Considerations," *Science*, August 16, 1991, p. 754.
8. David Ehrenfeld, *The Arrogance of Humanism* (New York: Oxford University Press, 1981), pp. 207-8.
9. Matt. 6:28-29.
10. Supreme Court of the United States, no. 70-34, April 19, 1972.
11. Roberto Suro, "Whooping Cranes Coming Back, But Nearby Barges Pose a Threat," *New York Times*, February 13, 1990.
12. Cited in James R. Udall, "Launching the Natural Ark," *Sierra*, September/October 1991, p. 80.
13. Holly Doremus, "Patching the Ark: Improving Legal Protection of Biological Diversity," *Ecology Law Quarterly*, vol. 18., no. 2 (1991), pp. 316-17.
14. See Michael Milstein, "Unlikely Harbingers," *National Parks*, July/August 1990, pp. 19-24.
15. Ehrenfeld, *The Arrogance of Humanism*, p. 191.
16. Gen. 7:8-9.
17. Michael Soulé, editor, *Conservation Biology* (Sunderland, Mass.: Sinauer, 1986), p. 6.
18. Michael F. Soulé, "Conservation: Tactics for a Constant Crisis," *Science*, August 16, 1991, p. 747.
19. Ibid.
20. John J. Berger, editor, *Environmental Restoration* (Washington, D.C.: Island Press, 1990), p. xvii.
21. Personal communication.
22. Berger, *Environmental Restoration*, p. xix.
23. Material in this section is based mainly on Christine Mlot, "Restoring the Prairie," *BioScience*, vol. 40, no. 11 (December 1990), pp. 804-9, and John J. Berger, "The Prairie Makers," *Sierra*, November/December 1985, pp. 64-70.
24. Stevens, "Species Loss."
25. Based on information provided by Steve Packard, director of science and stewardship for the Illinois Nature Conservancy.
26. Steve Packard, "Restoration and the Rediscovery of the Tallgrass Savanna," *Whole Earth Review*, Spring 1990, p. 5.
27. Ibid.
28. Ibid., p. 8.
29. Ibid., p. 11.
30. William K. Stevens, "Green-Thumbed Ecologists Resurrect Vanished Habitats," *New York Times*, March 19, 1991.
31. This title is taken from an essay of the same name appearing in Wes Jackson, *Altars of Unhewn Stone: Science and the Earth* (San Francisco: North Point Press, 1987), on which this report is based.
32. Jackson, *Altars of Unhewn Stone*, p. 79.
33. Ibid., pp. 81-82.
34. Jon R. Luoma, "Prophet of the Prairie," *Audubon*, December 1989, pp. 59-60.
35. Jackson, *Altars of Unhewn Stone*, p. 82.
36. "The Restoration of the Great Plains Frontier," *Earth Island Journal*, Spring 1991, p. 34.
37. Ibid., p. 36; See also Anne Matthews, "The Poppers and the Plains," *New York Times Magazine*, June 24, 1990.
38. Frank J. Popper and Deborah E. Popper, "The Restoration of the Great Plains Frontier," *Earth Island Journal*, Spring 1991, p. 36.
39. Matthews, "The Poppers and the Plains."
40. Opening statement of press conference to release the report *Restoration of Aquatic Ecosystems: Science, Technology, and Public Policy*, prepublication copy. National Research Council, Washington, D.C., November 1991.
41. Ibid.
42. From "Diamond in the Rough: The Lake Delavan Story," a videotape produced by the Wisconsin Department of Natural Resources, P.O. Box 7921, Madison, WI 53707-7921, 1991.

43. Jana M. Suchy, "Delavan Lake: Where the Buffalo Roamed," *DNR Digest*, Wisconsin Department of Natural Resources, January 1990, p. 7.
44. Personal communication.
45. Personal communications from Richard Wedepohl, DNR, and Frank Lapensee, EPA, June 1991.
46. "About Ocean Arks," *Annals of Earth*, vol. IX, no. 2 (1991), p. 5.
47. Bill McAllen, "Environmental Concern, Inc. — Ecological Farmers," *Maryland Magazine*, Autumn 1990; Pat Emory, "Company Teams with Developers to Ease Harm to Fragile Wetlands," *Baltimore Sun*, January 31, 1990.
48. John J. Berger, *Restoring the Earth* (New York: Knopf, 1985), p. 64.
49. William K. Stevens, "Restoring Lost Wetland: It's Possible But Not Easy," *New York Times*, October 29, 1991.
50. Karl Linn, *From Rubble to Restoration*, (San Francisco: Earth Island Institute, Urban Habitat Program, 1991), p. 3.
51. William K. Stevens, "Amid Insult and Injury, Urban Forests Hang On," *New York Times*, November 12, 1991.
52. Ibid.
53. Ibid.
54. Anne Raver, "A Tree Grows in Brooklyn, but Maybe Not for Long," *New York Times*, January 11, 1992.
55. Ibid.
56. Ibid.
57. Cited in Carl Wittenberg, "Magic's Urban and Oak Woodland Restoration," in *Ecological Restoration in the San Francisco Bay Area* (Berkeley: Restoring the Earth, 1990), p. 3.

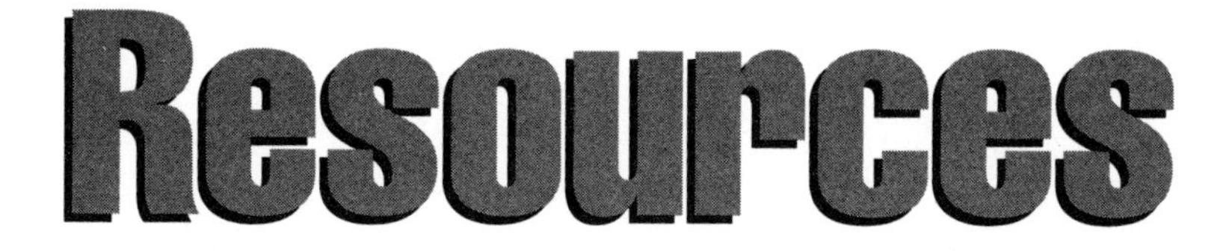
Resources

A
DIRECTORY OF ENVIRONMENTAL RESOURCES

This directory will guide you through the maze of environmental organizations, describing who they are, their aims, and their track records so that you can judge for yourself which ones to support. Part 1 covers general membership (and primarily national) groups, including the "Big Ten," Part 2 covers professional and research organizations, Part 3 lists key federal agencies, Part 4 lists congressional committees and subcommittees with environmental responsibilities, and Part 5 — computer networks. Where available, budgets and revenues for the most recent year are supplied. DND indicates "did not disclose." Because of space considerations, only larger regional organizations and coalitions are listed here; references to other such groups can be found throughout the book and in the Index.

A bullet before a listing denotes that the group merits special consideration; two bullets are awarded for outstanding conservation performance. All of the groups will provide further information about their operations and send you copies of their annual reports and other publications on request.

General Membership Organizations

AMERICAN FORESTRY ASSOCIATION, 1561 P Street, NW, Washington, DC 20005; (202) 667-3300. President, Richard M. Hollier. **Who they are:** Founded in 1875, a lobbying group with a broad constituency. **Aim:** Stimulate interest in forestry through action-oriented programs. **Membership:** 115,000. **Fees:** $24 and up. **What they do:** Organize conferences and "Global ReLeaf," an international campaign to help plant trees to improve the environment and combat global warming. **Publications:** *American Forests; Urban Forest FORUM; Global ReLeaf Report.*

● **AMERICAN HIKING SOCIETY**, P.O. Box 20160, NW, Washington, DC 20041; (703) 385-3252. Executive director, Susan Henley. **Who they are:** Premier hiking organization with 100+ affiliates. **Aims:** Protect hikers' interests and preserve America's footpaths. **Membership:** 300,000. **Fees:** $25; $15 for students and seniors. **What they do:** Maintain public information service; encourage volunteerism in trail building and maintenance; lobby to raise funds for trails; organizing National Trails Day and the American Discovery Trail project to stretch 5,500 miles from coast to coast. **Publications:** *American Hiker* (quarterly magazine and separate newsletter); *Pathways Across America; AHS Directory of Technical Assistance* (for trails development); *Helping Out in the Outdoors*, a 120-page directory of more than 2,000 volunteer jobs on public lands. **Comment:** What better way to get in touch with Nature than hiking?

AMERICAN LITTORAL SOCIETY, Sandy Hook, Highlands, NJ 07732; (201) 291-0055. Executive director, D. W. Bennett. **Who they are:** Founded in 1961, a national organization of pro-

fessional and amateur naturalists, divers, fishers, boaters, beachcombers, birders, and others, with regional offices in Sarasota and Miami, Fla., Lambertville, N.J., Broad Channel, N.Y., Woods Hole, Mass., and Olympia, Wash. **Aims:** Study and conserve coastal habitat, barrier beaches, wetlands, estuaries, and near-shore waters. **Membership:** 10,000. **Fees:** $25; $15 for students and seniors. **What they do:** Sponsor baykeeper program in Delaware River; field trips, dive and study expeditions, and a fish tag-and-release program. Special activities for scuba divers and underwater photographers. **Publications:** *Underwater Naturalist; Coastal Reporter.*

●● **AMERICAN RIVERS,** 801 Pennsylvania Avenue, SE, Suite 400, Washington, DC 20003; (202) 547-6900 or (800) 783-2199. President, Kevin J. Coyle. **Who they are:** Founded in 1973 (as American Rivers Conservation Council), the premier river conservation organization. **Aims:** Preserve and restore America's river systems and foster a river stewardship ethic, including protecting nationally significant rivers, reforming national hydropower policy, protecting endangered aquatic and riparian species, western water allocation, clean water protection, and urban river protection. **Membership:** 17,000. **Fees:** $20 and up. **What they do:** AR has preserved some 20,000 river miles on 600 rivers, totaling over 3.5 million acres of crucial riparian and aquatic habitat; helped prevent construction of $25 billion worth of environmentally destructive dams and unnecessary public-water resource development projects. Programs include lobbying, volunteer river activist projects, and litigation. **Publications:** *Ten Most Endangered Rivers List* annually (see chapter 1, page 19) and the quarterly *American Rivers.* **Comment:** Even if you don't live near a river, you depend on one in more ways than you think. Help support their programs.

AMERICAN TRAILS, 1400 Sixteenth Street, NW, Washington, DC 20036; (202) 483-5611. Chair, Charles Flink. **Who they are:** Founded in 1971, AT is composed of trail clubs, agencies, individuals, and landowners interested in trails. **Membership:** DND **Fees:** DND **Aims:** Promote and support planning, development, and maintenance of trail systems on public and private land. **What they do:** Give out information on trail issues at federal, state, and local levels and on trails generally. **Publication:** *Trail Tracks.*

● **AMERICAN WILDLANDS,** 655 South Revere Parkway, Suite 160, Englewood, CO 80111; (303) 649-9020. President, Sally A. Ranney. **Who they are:** Founded in 1977, a group of professionals and activists supported by memberships, grants, and contributions. With a Northern Rockies office in Bozeman, Mont., and field reps in Alaska, Idaho, and Nevada, they focus primarily on the interior western states and Alaska. **Aims:** Responsible management and protection of forests, wildlife, wilderness, wetlands, watersheds, and fisheries. **Membership:** 3,500. **Fees:** $25 and up. **What they do:** Coordinate public education, research, litigation, coalition building, and lobbying with hands-on experience. Programs include timber management, river defense, wildland research, organizing conferences and "eco-adventures" — wild-country educational tours in some of the world's finest natural areas. **Publications:** *Wild America; On the Wild Side* (quarterly); and *It's Time to Go Wild.* **Comment:** Along with Citizens' Clearinghouse for Hazardous Wastes and World Wildlife Fund, one of the few environmental groups headed by a woman.

APPALACHIAN MOUNTAIN CLUB, 5 Joy Street, Boston, MA 02108; (617) 523-0636. Executive director, Andrew J. Falender. **Who they are:** Founded in 1876, a service organization for members and nonmembers, primarily in the northeastern states. **Aims:** Trail and shelter maintenance, outdoor education, and other trail-related services. **Membership:** 44,000. **Fees:** $25. **What they do:** Operate public huts, self-service and full-service camps on the Appalachian Trail (AT); publish guidebooks, maps, and educational materials. **Publications:** *Appalachia Journal; Appalachia Bulletin*; AMC guidebooks. **Comment:** See also next listing.

APPALACHIAN TRAIL CONFERENCE, P.O. Box 807, Harpers Ferry, WV 25425; (304) 535-6331. Executive director, David N. Startzell. **Who they are:** Founded in 1925, overall coordinating group for the AT. **Aims:** Preservation and management of the AT from Maine to Georgia. **Membership:** 23,000. **Fees:** DND. **What they do:** Coordinate the work of members and other groups interested in the AT; publish trail guidebooks and other trail-user information. **Publications:** *Appalachian Trailway News; Trail Lands; The Register.*

CENTER FOR MARINE CONSERVATION, 1725 DeSales Street, NW, Suite 500, Washington, DC 20036; (202) 429-5609. President, Roger McManus. **Who they are:** Founded in 1972, nonprofit organization with regional offices in Austin, Tex. (Gulf Coast), Hampton, Va. (Chesapeake Bay), San Francisco, and St. Petersburg, Fla. **Aims:** Protect marine wildlife and habitats, especially endangered species, and conserve coastal and ocean resources. **Membership:** 125,000. **Fees:** $15 and up. Annual revenue: $3.5 million. **What they do:** Run programs on marine sanctuary advocacy, fisheries conservation, species recovery, pollution prevention, and habitat conservation, focusing primarily on research, education, and public information. **Publications:** *Marine Conservation News; Coastal Connection; Sanctuary Currents.*

CHESAPEAKE BAY FOUNDATION, 162 Prince George Street, Annapolis, MD 21401; (301) 368-8816. President, William C. Baker. **Who they are:** Founded in 1966 by a group of citizens who wanted to save the bay, a broad-based group of concerned baywatchers, with offices in Richmond, Va., and Harrisburg, Pa. **Aim:** Restore Chesapeake Bay, particularly re water quality, agricultural and urban runoff, land-use development, toxic waste, wetlands, and wildlife. **Membership:** 75,000. **Fees:** $20. **What they do:** Run a wide range of on-the-water educational awareness programs for students, teachers, and general public; legal advocacy; scientific research; and environmental planning. **Publications:** *CBF News; Baywatchers Bulletin;* many other policy papers and reports.

●● **CHILDREN OF THE GREEN EARTH,** Box 31087, Seattle, WA 98103; (206) 523-6279. Directors: Dorothy Craig and Michael Soulé. **Who they are:** Founded by Dr. Richard St. Barbe Baker (the "Man of the Trees") and friends at the United Nations, a worldwide association of people working with children through local communities. **Aim:** Regreen the Earth by planting and caring for trees and forests. **Membership:** 1,000 **Fees:** $15 and up. **What they do:** Help organize children around the world to do exactly that! **Publications:** *Tree Song* and *Children of the Green Earth* (newsletter). **Comment:** Give them all you can.

●● **CITIZENS CLEARINGHOUSE FOR HAZARDOUS WASTES,** P.O. Box 6806, Falls Church, VA 22040; (703) 237-CCHW. Executive director, Lois Marie Gibbs. **Who they are:** Founded in 1981 by Gibbs, a former Love Canal resident, CCHW is a no-compromise, bare-bones organization working with 7,000 grass-roots groups, especially in rural areas around the country, with offices in Atlanta, Riverside, Calif., Wendel, Pa., Spencerville, Ohio, and Floyd, Va. **Aims:** Clean up hazardous waste landfills, stop siting new ones, and work for environmental justice. **Membership:** 20,000. **Fees:** $15 and up. **What they do:** Help build strong community-based groups by providing organizing and technical skills, and working with local volunteers. CCHW has helped block 500 toxic waste sites and organized the McToxics campaign, mailing tons of styrofoam back to McDonald's and convincing them to pledge $100 million for recycling. **Publications:** *Everyone's Backyard* (bimonthly); numerous hands-on manuals and action guides. **Comment:** True grass-roots group worthy of your support.

● **CITIZENS FOR A BETTER ENVIRONMENT,** 501 Second Street, Suite 305, San Francisco, CA; (415) 788-0690. Program director, Michael Belliveau. **Who they are:** A California-based environmental team of activists, scientists, and lawyers with a growing citizen-supported base, with offices in Venice and Berkeley. **Aim:** Safeguard the environment and public health from toxic pollution. **Membership:** 30,000. **Fees:** $25. **What they do:** Work on specific issues such as cleaning up San Francisco Bay through technical research, policy advocacy, public education, and litigation. In 1990 CBE and the *Sierra Club Legal Defense Fund* (see below) successfully sued 5 major air-quality agencies in the Bay Area for failing to enforce the Clean Air Act. **Comment:** Tightly run, effective group.

CLEAN OCEAN COALITION, P.O. Box 505, Sandy Hook, NJ 07732; (908) 872-0111. Executive director, Cindy Zipf. **Who they are:** Founded in 1984, a coalition of 150 conservation, fishing, diving, boating, civic groups, and 275 businesses. **Aim:** Clean up degraded waters off the New York and New Jersey coasts. **Membership:** COC represents thousands of citizen members of the coalition. **Fees:** $25 for a participating membership; $15 to be put on the mailing list.

What they do: Research, lobbying, and citizen action to press legislators to enact and enforce laws protecting marine resources. Programs include: Anglers Call to Action; Women's Call to Action; Educators for a Clean Ocean. **Publication:** *Action.*

CLEAN WATER ACTION, 1320 Eighteenth Street, NW, Washington, DC 20036; (202) 457-1286. Executive director, David Zwick. **Who they are:** Founded in 1971, broad-based national citizens' coalition of diverse constituents. **Membership:** 600,000. **Fees:** $24. **Aims:** Clean, affordable water, control of toxic chemicals, and protection of natural resources. **What they do:** A leader in drafting tougher new water and toxics policies, including the Clean Water and the Clean Air acts. Conduct year-round intern programs. **Publication:** *Clean Water Action News* quarterly.

COASTAL CONSERVATION ASSOCIATION, INC., 4801 Woodway, Suite 220 West, Houston, TX 77056; (713) 626-4222. Executive director, Ray Poage. **Who they are:** Founded in 1977, national nonprofit corporation serving general public. **Aims:** Preserve and protect marine, animal, and plant life on and offshore along U.S. coastal areas. **Membership:** 35,000. **Fees:** $20. **What they do:** Organize educational and informational programs. **Publication:** *The Tide.*

COASTAL SOCIETY, P.O. Box 2081, Gloucester, MA 01930-2081; (508) 281-9209. Executive director, Thomas E. Bigford. **Who they are:** Founded in 1975, this is an international nonprofit organization. **Membership:** 400. **Fees:** $25 and up. **Aim:** Promote understanding and wise use of coastal resources. **What they do:** Organize conferences, workshops, and publications. **Publication:** *Coastal Society Bulletin* (quarterly) and proceedings of their meetings.

CONSERVATION INTERNATIONAL, 1015 Eighteenth Street, NW, Suite 1000, Washington, DC 20036; (202) 429-5660. President, Russell Mittermeier. **Who they are:** Founded in 1987, a private nonprofit group that broke away from the Nature Conservancy (see below). **Aim:** Preservation of tropical ecosystems. **Membership:** 55,000. **Fees:** $15. **Budget:** $4.6 million. **What they do:** Pioneered the concept of debt-for-nature swaps — reducing a developing country's national debt in return for sustainable forest management. Unlike World Wildlife Fund and other organizations, CI designs projects that economically benefit indigenous peoples. **Publications:** *Tropicus* newsletter; *The Debt for Nature Exchange.* **Comment:** CI organized the "Don't Bungle the Jungle" pop campaign in 1989. Some critics claim they spend too much time building up membership at the expense of doing conservation work.

● **COUSTEAU SOCIETY,** 930 West Twenty-first Street, Norfolk, VA 23517; (804) 627-1144. Executive vice president, Jean-Michel Cousteau. **Who they are:** Founded in 1973 by Jacques-Yves Cousteau, a nonprofit environmental educational society. **Aims:** Alerting and informing public to choose more environmentally friendly ways of living. **Membership:** 300,000. **Fees:** $20 and up. **What they do:** Produce films for television, books, lectures, and field study programs. Their ships, *Calypso* and *Alcyone*, sail the seas and rivers to study global ecosystems. **Publications:** *Calypso Log*; *Dolphin Log.* **Comment:** Along with Rachel Carson, Aldo Leopold, and David Brower, Jacques Cousteau is one of the conservation greats. But this tight, family-run ship may have identity problems when he moves into history.

● **DEFENDERS OF WILDLIFE,** 1244 Nineteenth Street, NW, Washington, DC 20036; (202) 659-9510. President, Dr. M. Rupert Cutler. **Who they are:** Founded in 1947, a national conservation organization run by "old-fashioned Rocky Mountain types who like animals with teeth,"[1] many of them disenchanted veterans of National Wildlife Federation and National Audubon. **Aim:** Species and habitat protection. **Membership:** 85,000. **Fees:** $20; students and seniors, $15. Budget: $4.6 million. **What they do:** Vigorously advocate citizen, legal, and governmental action in pursuing these aims. In promoting reintroduction of the gray wolf in the West, they set up a $100,000 fund to pay ranchers for livestock killed by the wolves. Other Defenders' projects include working to prevent entanglement of marine mammals in plastic debris, restoring the Everglades, and helping establish wildlife corridors. **Publication:** *Defenders.* **Comment:** Defenders stick close to their original purpose and integrity.

DUCKS UNLIMITED, INC., One Waterfowl Way, Memphis TN 38190; (901) 775-3825. Executive vice president, Matthew B. Connolly, Jr. **Who they are:** Founded in 1937, a nonprofit corporation made up largely of white, middle-class conservative male hunters. **Aims:** Conserve and enhance wetland ecosystems throughout North America so that they will have more ducks to shoot. **Membership:** 550,000. **Fees:** $20. Annual Revenue: $67 million. **What they do:** Hire contractors to build dikes and reflood drained wetlands; persuade farmers not to plow up duck habitats in prairie potholes; advocate hunters' access to these areas. **Comment:** A worthwhile conservation group that is becoming increasingly ecology-concerned.

● **EARTH FIRST!**, Box 5871, Tucson, AZ 85703; (602) 622-1371. **Who they are:** Founded in 1980, a high-profile, radical environmental "disorganization" with no formal hierarchical structure. **Aim:** No compromise in defense of Mother Earth. **Membership:** Estimated at 15,000. **Fees:** $20. **What they do:** Organize pro-environment demonstrations and campaigns such as Redwood Summer, a series of antilogging protests in the Northwest and other "monkey-wrenching" activities that often steal the media thunder from mainstream environmental groups. **Publication:** *EF! Journal.* **Comment:** *EF! Journal* is a valuable source of information on the destruction of wilderness and ecosystems.

●● **EARTH ISLAND INSTITUTE**, 300 Broadway, Suite 28, San Francisco, CA 94133; (415) 788-3666. Executive directors, John A. Knox and David Phillips. **Who they are:** Founded in 1982 by enviro great David Brower as a spinoff from Friends of the Earth (which he had founded as a spinoff from Sierra Club) because they had become too mainstream, EII is the center of a network of activist projects that include Save the Dolphins, the environmental litigation fund, Friends of the Ancient Forest, the urban-habitat program, and the climate-protection institute. **Aims:** Conserve, preserve, and restore global environment. **Membership:** 35,000. **Fees:** $25. **Budget:** $1 million. **What they do:** EII won a lawsuit requiring federal observers aboard Pacific tuna ships and pressured U.S. food companies not to buy tuna caught by setting nets on dolphins or by drift nets. Their sea turtle restoration project got the Mexican government to announce a new law banning the killing of this species. **Publications:** *Earth Island Journal* quarterly; *Race, Poverty, and the Environment* quarterly. **Comment:** They get things done. *EI Journal* is worth the price of admission alone.

ENVIRONMENTAL ACTION INC., 6930 Carroll Avenue, Suite 600, Takoma Park, MD 20912; (301) 891-1100. Director, Ruth Caplan. **Who they are:** Founded in 1970 and merged with Environmental Task Force in 1988, a membership-based lobbying organization. **Aims:** Political and social change. **Membership:** 20,000. **Fees:** $20. **What they do:** EAF is an effective lobbying voice for citizen action on toxic waste, energy-related matters, and other issues; its political action committee works to elect environment-friendly candidates to Congress by training volunteers as campaign organizers and helpers; has a foundation for supportive research and educational work. **Publication:** *Environmental Action* bimonthly. **Comment:** One of the few remaining radical voices on Capitol Hill.

●● **ENVIRONMENTAL DEFENSE FUND**, 257 Park Avenue South, New York, NY 10010; (212) 505-2100. Executive director, Frederic D. Krupp. **Who they are:** Founded in 1967 to save the osprey from the effects of DDT, EDF is one of the Big Ten in clout if not size. With 6 offices nationwide, it presents a top-flight multidisciplinary team of scientists, lawyers, and economists. **Aim:** Develop economically viable solutions to environmental problems. **Membership:** 170,000. **Fees:** $20 and up. Budget: $13 million. **What they do:** EDF's research and advocacy work encompasses global warming, acid rain, rain forest destruction, ocean pollution, and recycling. **Publication:** *EDF Letter.* **Comment:** EDF has shifted from traditional "sue-the-bastards" litigation to a free-market environmentalism based on influencing corporations like McDonald's and General Motors and top Washington politicos.

FRIENDS OF THE EARTH, 218 D Street, SE, Washington, DC 20003; (202) 544-2600. President, Jane Perkins. **Who they are:** Founded in 1969 by David Brower, FOE split apart over moving from San Francisco to D.C. and other policy differences. With none of the original staff

left, FOE in 1990 affiliated with the Environmental Policy Institute and the Oceanic Society. **Aims:** Protect the planet from environmental disaster and preserve biological, cultural, and ethnic diversity. **Membership:** 50,000. **Fees:** $25. Budget: $3.4 million. **What they do:** Focus on atmospheric ozone, groundwater contamination, and oil spills. **Publications:** *FOE Newsmagazine, Atmosphere Ozone Newsletter, Groundwater Newsletter.*

FRIENDS OF THE RIVER, Building C, Fort Mason Center, San Francisco, CA 94123; (415) 771-0400. Executive director, David Bolling. **Who they are:** Founded in 1973 during the fight to save the Stanislaus River, a public education and lobbying organization. **Aim:** Protect rivers and streams, especially in the San Francisco Estuary. **Membership:** 10,000. **Fees:** $25. **What they do:** Influence state and federal Wild and Scenic River programs through public education lobbying and legal action. **Publication:** *Headwaters* bimonthly.

GRAND CANYON TRUST, The Homestead, Route 4, Box 718, Flagstaff, AZ 86001; (602) 774-7488; Washington office, 1400 Sixteenth Street, NW, Suite 300, Washington, DC 20036; (202) 797-5429. Executive vice president, James B. Ruch. **Who they are:** Low-profile but influential advocacy group whose members are well informed and well placed on Capitol Hill. **Aim:** Conserve natural resources of the Colorado Plateau. **Membership:** 3,500. **Fees:** $25. **What they do:** Focus on air pollution in Grand Canyon, Glen Canyon dam operations, and Utah wilderness designation. **Publications:** *Colorado Plateau Advocate* quarterly; special reports.

● **GREATER ECOSYSTEM ALLIANCE**, P.O. Box 2813, Bellingham, WA 98227; (206) 671-9950. Executive director, Mitch Friedman. **Who they are:** Founded in 1989, a small, nonprofit group with an important ecological mission. **Aim:** Protect biological diversity through the conservation of greater ecosystems. **Membership:** 650. **Fees:** $25; students, senior citizens, and low-income, $15. **What they do:** Focus on conserving the greater North and Central Cascades, Selkirk, and Olympic ecosystems. **Publication:** *Northwest Conservation* quarterly.

●● **GREATER YELLOWSTONE COALITION**, P.O. Box 1874, Bozeman, MT 59771; (406) 486-1593. Executive director, Edward M. Lewis. **Who they are:** Founded in 1983, a coalition of nearly 100 regional and national organizations. **Aims:** Preservation and protection of the GY ecosystem. **Membership:** 5,000+. **Fees:** Basic, $25; patron, $500. **Revenue:** $600,000. **What they do:** Raise public and congressional consciousness against oil and gas development, mining, and excessive logging in this important region. **Publications:** *Greater Yellowstone Report* quarterly; *An Environmental Profile of the Greater Yellowstone Ecosystem.* **Comment:** The 132-page *Profile* is a landmark in environmental communication. Working with the Greater Yellowstone conservation community, GYC is a leader in saving the wilderness in the largest, essentially intact ecosystem in the temperate zones of the earth.

● **GREENPEACE USA**, 1436 U Street, NW, Washington, DC 20009; (202) 462-1177. Executive director, Steve D'Esposito. **Who they are:** Largest and only direct-action group in the Big Ten, with offices in Chicago, San Francisco, New York, Fort Lauderdale, Toronto, Montreal, Vancouver, and 23 countries overseas. **Aims:** Protect and conserve the environment and the life it supports. **Membership:** 2.5 million. **Fees:** $20. **Budget:** $50 million. **What they do:** Channel one-sixth of their money to Greenpeace International, which runs their commando operations against polluting factories, nuclear subs, etc. Although best known for their "Save the Whales" campaign, Greenpeace defends dolphins, sea lions, and other marine animals. Their campaign against nuclear testing led to the French government's blowing up the Greenpeace ship *Rainbow Warrior.* Greenpeace successfully persuaded Time-Life and other publishers to eliminate chlorine bleaching from the paper-making process. **Publications:** *Greenpeace Magazine*; special reports on toxic waste, deforestation, endangered species, and other subjects. **Comment:** Although Greenpeace's headline-grabbing tactics upset some mainstream organizations, it was the only one of the Big Ten to oppose the Gulf War, primarily on ecological grounds.

● **HUMANE SOCIETY OF THE U.S.**, 2100 L Street, NW, Washington, DC 20037; (202) 452-1100. Executive vice president, Paul G. Irwin, Sr. **Who they are:** Founded in 1954, a nonprofit organization experienced in animal cruelty investigation, laboratory animal welfare, and wildlife protection. **Aim:** Protect wild and domestic animals. **Membership:** 1.2 million. **Fees:** $10. **What they do:** Investigate cruelty to animals; wildlife and habitat protection; laboratory animal welfare; provide resources to local organizations, government, media, and general public. **Publications:** *HSUS News, Kind News, Children and Animals.* **Comment:** HSUS is a member of the James Bay Coalition, opposed to the James Bay project (see chapter 5, pages 166-169).

● **IZAAK WALTON LEAGUE,** 1401 Wilson Boulevard, Level B, Arlington, VA 22209; (703) 528-1818. Executive director, Jack Lorenz. **Who they are:** Founded in 1922, a nonprofit conservation group composed originally of 54 sport fishermen wanting cleaner streams to fish in, but now having 400 chapters with a broader environmental mandate. **Aim:** Defend the nation's soil, air, woods, waters, and wildlife. **Membership:** 55,000. **Fees:** $20. **What they do:** Protect public lands including Minnesota's Boundary Waters Canoe Area, the Everglades, the National Elk Refuge in Wyoming, and the Upper Mississippi wildlife and fish refuge. Has led the effort to create the Land and Water Conservation Fund for acquisition of public park lands. **Publications:** *Outdoor America; Outdoor Ethics Newsletter; Splash.* **Comment:** Prime example of enlightened self-interest.

LAND TRUST ALLIANCE, 900 Seventeenth Street, NW, Suite 410, Washington, DC 20006; (202) 785-1410. President/executive director, Jean W. Hocker. **Who they are:** Founded in 1982, a national service organization of regional and local land trust groups. **Aim:** Build public awareness of the role of land trusts in saving land resources. **Membership:** 1,500. **Fees:** $30 and up. **Annual Revenue:** $800,000. **What they do:** Provide services and programs to help land trusts and other land conservation professionals. **Publications:** *Starting a Land Trust; National Directory of Conservation Land Trusts; Conservation Easement Handbook.*

● **LEAGUE OF CONSERVATION VOTERS,** 1707 L Street, NW, Suite 550, Washington, DC 20036; (202) 785-8683. Executive director, Jim Maddy. **Who they are:** Founded in 1970, LCV is a nonpartisan political arm of the environmental movement with a board of directors of leaders from major environmental groups. **Aim:** Elect pro-environment candidates to Congress. **Membership:** 55,000. **Fees:** $25. Budget: $1.5 million. **What they do:** Support "green" candidates, hold press conferences, rate members of Congress on their environmental voting. **Publications:** *Environmental Scorecard; Election Report; Presidential Profiles.* **Comment:** An excellent source for keeping tabs on our "representatives."

● **LIGHTHAWK,** P.O. Box 8163, Santa Fe, NM 87504; (505) 982-9656. Executive director, Michael M. Stewartt. **Who they are:** A nonprofit membership organization that uses aircraft in its environmental sleuthing on behalf of 130 partner organizations in the U.S., Canada, and Central America. Advisory board includes John Denver, former Senator Timothy Wirth (D-Colorado), and other enviro heavies. **Aims:** Expose and correct environmental mismanagement. **Membership:** 3,500. **Fees:** $35. **Annual Revenues:** $800,000. **What they do:** Their environmental air force helps activists in many regions including the Northwest (forests) and Wisconsin (mining projects); they fly legislators, community leaders, and others to view deforestation and other ecological damage; and cosponsor annual Forest Reform Network powwow. **Publication:** *Lighthawk: The Wings of Conservation* (newsletter). **Comment:** Innovative and enterprising.

NATIONAL ARBOR DAY FOUNDATION, 100 Arbor Avenue, Nebraska City, NE 68410; (402) 474-5655. Executive director, John Rosenow. **Who they are:** Founded in 1971, a high-profile, public educational, membership organization. **Aims:** Tree planting and conservation. **Membership:** 1 million. **Fees:** $10. Annual revenues: $15 million. **What they do:** Organize National Arbor Day, Trees for America, Tree City USA, and the National Arbor Day Center; plan to establish an institute to improve tree planting and care practices and to promote conservation and environmental stewardship. **Publications:** *Arbor Day; Tree City USA Bulletin.*

●● **NATIONAL AUDUBON SOCIETY,** 700 Broadway, New York, NY 10003; (212) 979-3000. President, Peter A. A. Berle. **Who they are:** Founded in 1905 by conservationists outraged

by the slaughter of birds for hat plumes, one of the Big Ten, with Washington, D.C., Northeast, Southeast, Great Lakes, West Central, Southwest, Rocky Mountains, West, Alaska, and Hawaii offices. **Aim:** Protect air, land, water, and ecosystems critical to the planet's health. **Membership:** 600,000. **Fees:** $20. **Annual revenues:** $37 million. **What they do:** Scientific research; manage wildlife sanctuaries; public education (especially through TV specials); congressional lobbying; citizen action (with 60,000 activist members); litigation (often in conjunction with other environmental groups). **Publications:** *Audubon; Audubon Activist*; special reports on James Bay and other key issues. **Comment:** Audubon is getting more involved in social action. Stay tuned by reading their excellent publications and watching their TV specials.

NATIONAL PARKS AND CONSERVATION ASSOCIATION, 1015 Thirty-first Street, NW, Washington, DC 20007; (202) 944-8530. Executive committee chairperson, Norman G. Cohen. **Who they are:** Founded in 1919, a private nonprofit membership association independent of the Park Service. **Aims:** Defend, promote, and improve the National Park System. **Membership:** 95,000. **Fees:** $25. **What they do:** Provide forum for discussion of park problems and solutions through symposiums, workshops, and their magazine; provide a regular summary of key park legislation and calendar of events at national parks; sponsor March for Parks — the Celebration of the Outdoors; and, in cooperation with the NPS, develop curricula for schools and general public on conservation topics. **Publication:** *National Parks* (bimonthly).

● **NATIONAL TOXICS CAMPAIGN,** 1168 Commonwealth Avenue, Boston, MA 02134; (617) 232-0327. Executive director, John O'Connor. **Who they are:** Nationwide network of 1,300 community groups that combine protest with research and legislative work. **Aim:** Get the EPA, military, and industry to clean up Superfund and other toxic waste threats. **Membership:** 100,000. **Fees:** $15. **Budget:** $1.5 million. **What they do:** Get corporations to sign "good neighbor" agreements not to pollute, and if they renege, NTC collects evidence for prosecution — e.g., they got Conoco to pay $23 million to 400 families whose homes were contaminated by its refinery in Ponca City, Oklahoma. Have persuaded supermarket chains to phase out harmful pesticides used on produce. **Comment:** Less connected with grass-roots but more scientifically oriented than *Citizens Clearinghouse for Hazardous Wastes* (see above). Both would be even more effective if they worked together.

NATIONAL WILDLIFE FEDERATION, 1400 Sixteenth Street, NW, Washington, DC 20036; (202) 797-6800. President, Jay D. Hair. **Who they are:** Founded in 1936, second largest of the Big Ten, with offices in Vermont, Pennsylvania, North Carolina, Mississippi, Arkansas, Indiana, Minnesota, Texas, South Dakota, Colorado, Alaska, California, and Wyoming. **Aims:** Promote wise use of natural resources and protect the global environment. **Membership:** 5.5 million. **Fees:** $15 and up. **Budget:** $90 million. **What they do:** Distribute a huge array of periodicals and educational materials; sponsor outdoor education programs on conservation; litigate on behalf of natural resources and wildlife conservation. **Publications:** *International Wildlife; National Wildlife; Your Big Backyard* (preschool); *Ranger Rick*; and the invaluable *NWF Conservation Directory* (published annually). **Comment:** During the Reagan era NWF became more politically active, but it is hardly a grass-roots organization. Despite some corporate bashing, its top echelons are generally cozy with the old boys club of big business. NWF upset many of their members by bulldozing their modest 3-story building and surrounding trees to construct a $30 million "Taj Mahal" for their headquarters.[2]

● **NATIVE FOREST COUNCIL,** P.O. Box 2171, Eugene, OR 97402; (503) 688-2600. Director of Development, Deborah A. Ortuno. **Who they are:** Founded in 1988 by a group of professional people shocked at the extensive logging of our national forests, NFC is a nonprofit action-oriented organization. **Aim:** To promote solutions to protect all of the remaining forests. **Membership:** 3,000. **Fees:** $25; seniors, $20; low income, $15. **What they do:** NFC generates regular contact with more than fifty national and regional environmental organizations, including National Audubon, Greenpeace, the Sierra Club, Earth Island Institute, and the Wilderness Society, in developing strategies for saving the forest. Their work includes successfully developing a nationwide advertising campaign and advocating and drafting the Native Forest Protection Act. **Publication:** *Forest*

Voice newspaper, 1 million copies of which have been distributed nationally since 1988. **Comment:** A well-organized and highly effective action group.

●● **NATURAL RESOURCES DEFENSE COUNCIL,** 40 West Twentieth Street, New York, NY 10011; (212) 727-2700. Executive director, John H. Adams. **Who they are:** Founded in 1970, NRDC is, like EDF, on the lower end numerically of the Big Ten but highly regarded for the quality of its lawyers and scientists, with offices in Washington, D.C., Los Angeles, San Francisco, and Honolulu. **Aims:** Protect America's endangered natural resources and improve the quality of the human environment. **Membership:** 170,000. **Fees:** $10. **Budget:** $16 million. **What they do:** Lobby and litigate on air and water pollution, wilderness and wildlife protection, coastal-zone management, urban environment, Alaska, nuclear safety, energy conservation, and other key issues; monitor federal agencies and work with activists on specific issues such as plutonium processing and pesticides, scoring a big media coup in their anti-Alar campaign.* The NRDC report on the Arctic Wildlife Refuge helped turn public opinion against oil drilling before Exxon *Valdez*. **Publications:** *Amicus Journal* (quarterly); *NRDC Newsline* (newsletter). **Comment:** Similar in approach to EDF, NRDC has remained truer to its original purpose and is more aware of the problems of environmental racism. Because both do excellent work, we wonder if they wouldn't achieve more by working together rather than along separate tracks.

●● **NATURE CONSERVANCY,** 1815 North Lynn Street, Arlington, VA 22209; (703) 841-5300. President, John C. Sawhill. **Who they are:** Founded in 1951, NC, fourth in size and first in budget of the Big Ten, is an interesting mix of MBAs, lawyers, naturalists, and conservationists.[3] **Aim:** Protect biodiversity by protecting natural lands and wildlife. **Membership:** 600,000. **Fees:** $15. **Budget:** $156 million. **What they do:** Own and manage more than 1,600 nature preserves from the Peconic Watershed on New York's Long Island to the 500-square-mile Gray Ranch in New Mexico purchased by the conservancy for $18 million.† In May 1991, the Conservancy put together "The Last Great Places," an impressive plan for large-scale conservation of ecosystems in the U.S. and Latin America based on the bioreserve concept to safeguard larger landscapes where the full array of biological diversity is represented. **Publication:** *Nature Conservancy Magazine*. **Comment:** NC's efforts to protect biodiversity may be a model for the future of large-scale land conservation worldwide.

OCEANIC SOCIETY, see **FRIENDS OF THE EARTH**

● **PUBLIC FORESTRY FOUNDATION,** P.O. Box 371, Eugene, OR 97440-0371; (503) 687-1993. Executive director, Roy Keene. **Who they are:** A not-for-profit, citizens organization with certified foresters on staff. **Aim:** Help influence the management of public forests. **Membership:** 2,500. **Fees:** $15 and up. **Budget:** $300,000. **What they do:** Work with forest managers to encourage forestry that is healthy and sustainable. PFF monitors timber sales and conducts workshops, tours, and seminars for citizens, land managers, and conservationists. **Publications:** *The Citizen Forester* (quarterly) and periodic reports. **Comment:** Fills an important niche in the conservation movement.

RAILS-TO-TRAILS CONSERVANCY, 1400 Sixteenth Street, NW, Suite 300, Washington, DC 20036; (202) 797-5400. President, Louise Sagalyn. **Who they are:** Founded in 1985, a nonprofit membership group working with local conservation and recreation agencies nationwide. **Aim:** Convert 150,000 miles of abandoned railroad corridors to public trails for hiking, biking, horseback riding, cross-country skiing, nature and wildlife appreciation, etc. **Membership:** 65,000. **Fees:** $15. **Annual revenues:** $1.5 million. **What they do:** Work with local authorities, volunteers, conservation groups, and railroad agencies to convert more than 300 corridors in 31 states into trails and parks. **Publication:** *Trailblazer* quarterly.

* For a detailed report on NRDC's Alar campaign, see Jon Naar, *Design for a Livable Planet* (New York: Harper & Row, 1990), pages 259–261.
† For more on Gray Ranch, see chapter 5, pages 354-356.

• **RAINFOREST ACTION NETWORK,** 301 Broadway, Suite A, San Francisco, CA 94133; (415) 398-4404. President, Randall Hayes. **Who they are:** Founded in 1985 on a $12,000 Earth Island Institute grant, a national and international pressure group. **Aim:** Save world rain forests. **Membership:** 30,000. **Fees:** $25; $15, limited income. **Budget:** $1 million. **What they do:** Run major campaigns (letter writing, boycotts, demonstrations, media advertising) against corporations, banks, and public agencies contributing to rain forest destruction; produce educational materials for teachers and community organizers. **Publications:** *Action Alert; World Rainforest Report*. **Comment:** RAN has raised U.S. and world awareness of rain forest destruction.

RAINFOREST ALLIANCE, 270 Lafayette Street, New York, NY 10012; (212) 941-1900. Executive director, Daniel Katz. **Who they are:** Founded in 1986, a nonprofit, consciousness-raising membership group working in the U.S. and overseas. **Aim:** Save the world's tropical rain forests. **Membership:** 20,000. **Fees:** $20. **Budget:** $1 million. **What they do:** Work as information clearinghouse; influence public opinion through celebrities and private sector initiatives that include making agreements with companies to market eco-friendly rain-forest products. **Comment:** Rainforest Alliance favors negotiation over direct action, e.g., they oppose RAN's call for banning tropical wood imports, preferring to rely on regulation instead.

•• **SAVE AMERICA'S FORESTS,** 4 Library Court, SE, Washington, DC 20003; (202) 544-9219. Codirectors: Carl Ross, Chris Van Daalen, and Mark Winstein. **Who they are:** National coalition of 110 conservation groups and a half million individual members. **Aim:** Protect forests, biodiversity, and endangered species. **Fees:** $25 and up; students and seniors, $10. **What they do:** Lobby; run forest action conferences and seminars and other activities in Washington, D.C., and nationwide. **Publications:** *Save America's Forests*, a periodical newsletter, and special action bulletins. **Comment:** Impressive coalition, using 100 percent of its membership contributions for legislative action and grass-roots lobbying.

SAVE-THE-REDWOODS LEAGUE, 114 Sansome Street, Room 605, San Francisco, CA 94104; (415) 362-2352. Executive director, John B. Dewitt. **Who they are:** Founded in 1911 to rescue representative areas of primeval forests. **Aim:** Continue saving redwood forests. **Membership:** 50,000. **Fees:** $10. **Annual revenues:** $2 million. **What they do:** Purchase redwood groves to support reforestation and conservation; work with state and federal agencies to establish redwood and other parks. **Publications:** *Bulletin; California Redwood Parks and Preserves;* and other redwood-related pamphlets.

•• **SEA SHEPHERD CONSERVATION SOCIETY,** 1314 Second Street, Santa Monica, CA 90401; (213) 394-3148; P.O Box 48446, Vancouver, British Columbia, Canada V7X 1A2. Executive administrator, Lisa Lange. **Who they are:** Founded in 1977 by Greenpeace cofounder Paul Watson, a direct-action, all-volunteer group. **Aim:** Originally formed to stop illegal whaling, now protects all marine mammals and their habitat. **Membership:** 15,000. **Fees:** $25. **Annual revenues:** $500,000. **What they do:** Operate two ships to "actively interfere with the killing of marine mammals on the high seas where it occurs."[4] Veterans of scores of confrontations with pirate vessels from many countries, SSCS has sunk seven illegal whalers and scuttled two others, rescuing thousands of whales, seals, and dolphins.* **Publication:** *Sea Shepherd Log*. **Comment:** "The Sea Shepherd's crew of Canadians, Australians, Britons, and Americans are front-line soldiers in the war of the whales, environmentalists willing to risk their bodies as well as their time and energies . . . to save the great sea mammals." We couldn't put it better than the *New York Times*![5]

• **SIERRA CLUB,** 730 Polk Street, San Francisco, CA 94109; (415) 776-2211. Executive director, Michael L. Fisher. **Who they are:** Founded in 1882 by John Muir, a member of the Big Ten with 57 chapters and 340 groups. **Aim:** Conserve the natural environment by influencing Congress, the administration, and the legal and electoral processes. **Membership:** 650,000. **Fees:** $35; students and seniors, $15. **Budget:** $35 million. **What they do:** Maintain a heavy presence on Capitol Hill in part through a large political-action committee (PAC) fund supporting pro-

* For details see *Design for a Livable Planet*, page 279.

environment candidates and legislators; lobby at state and local levels on a wide range of environmental issues; programs include protecting old-growth forests, the Arctic National Wildlife Refuge, and public lands; global warming/energy conservation; and toxic-waste regulation. **Publications:** *Sierra* (bimonthly magazine); *National News Report*; and local chapter newsletters; the Sierra Club also publishes an impressive list of environmental books. **Comment:** Although the club is one of the enviro heavies, it seems to be losing contact with its own chapters and membership. See Margaret Hays Young, "It's 1992: Do You Know Where the Sierra Club Is?," *Wild Earth*, Spring 1992, pp. 65-66.

● **SIERRA CLUB LEGAL DEFENSE FUND,** 2044 Fillmore Street, San Francisco, CA 94115; (415) 567-6100. **Who they are:** Incorporated in 1970, SCLDF (which works with the Sierra Club but is not part of it) is a tax-exempt public-interest law firm supported by donations and bequests, working for its clients on a no-fee basis. **Aim:** Provide legal services to "groups that strive to protect parks and other public lands; others that concentrate on preserving wildlife and habitat; organizations that fight pollution of air, land, and water; even occasional commercial enterprises that use but do not abuse the natural environment — fishers, tour companies and the like."[6] **Membership:** 120,000. **Fees:** $15 and up. **Budget:** $7 million. **What they do:** Represent clients from the Adirondack Council of New York to Wyoming Outdoor Coordinating Council in administrative proceedings before state or federal agencies, litigating 100+ cases a year. SCLDF cases include winning a $5.5 million judgment against Union Oil Company over pollution by its refinery in San Francisco Bay, and getting a federal court in Texas to stop clear-cutting within 1,500 yards of the endangered red-cockaded woodpecker. **Publication:** *In Brief* quarterly. **Comment:** One of the environmental movement's most effective weapons.

SOUTHEAST ALASKA CONSERVATION COUNCIL, P.O. Box 021692, Juneau, AK 99802; (907) 586-6942. Executive director, Bart Koehler. **Who they are:** Founded in 1969, grassroots coalition of 13 conservation groups including native people, fishers, foresters, and community activists. **Aim:** Preserve integrity of southeast Alaska's natural environment. **Membership:** 2,500. **Fees:** $15 and up. **What they do:** Watchdog the Forest Service's management of the Tongass National Forest, monitoring timber sales, forest management practices, and land-use allocations; spearhead legislative efforts in Alaska and Washington, D.C., to preserve the Tongass. **Publication:** *Ravencall* newsletter.

STUDENT CONSERVATION ASSOCIATION, P.O. Box 550, Charlestown, NH 03603; (603) 826-4301. Executive vice president, T. Destry Jarvis. **Who they are:** Founded in 1957, a nonprofit educational and public-service organization. **Aim:** Provide educational opportunities for conservation work on public lands. **Membership:** 12,000. **Fees:** $15 and up. Annual revenues: $3.8 million. **What they do:** Provide 1,500 openings a year for expenses-paid field training and experience for high school and college students and adult volunteers in national parks, wildlife refuges, national forests, and other public and private lands. **Publications:** *Job Scan; Earth Work*. **Comment:** Check them out.

● **SURFRIDER FOUNDATION,** P.O. Box 2704, #86, Huntington Beach, CA 92647; (714) 960-8390 or (800) 743-SURF. Executive director, Jake Grubb. **Who they are:** Founded in 1984, a nonprofit, coastal activist membership organization with 20 chapters in California, Hawaii, New York, New Jersey, Virginia, North Carolina, Florida, and Australia. **Aims:** Protect and enhance world's waves and beaches through conservation, research, and education. **Membership:** 20,000. **Fees:** $25; students and seniors, $15. Annual budget: $500,000. **What they do:** Lobby local governments to open more shorelines to the public; organize the Blue Water Task Force gathering coastal water samples to determine patterns of pollution; take action against polluters such as the Louisiana-Pacific Corp. and Simpson Paper Co. that in September 1991 were forced to pay $5.8 million in fines and $50 million to reduce toxic discharges from their oceanfront pulp mills near Eureka, California. **Publication:** *Making Waves* quarterly. **Comment:** An early-warning monitoring and action-minded group that could be the wave of the future!

● **TREEPEOPLE,** 12601 Mulholland Drive, Beverly Hills, CA 90210; (818) 753-4600. President, Andy Lipkis. **Who they are:** Founded in 1973, a nonprofit environmental group promoting per-

sonal involvement, community action, and global awareness. **Aims:** Plant trees, save the world. **Membership:** 25,000. **Fees:** $25. **What they do:** Organize one of the most successful free, public tree-planting programs in the U.S. — they planted a million trees in L.A. for the 1984 Olympics. **Publications:** *Seeding News; TreePeople* newsletter; *The Simple Act of Planting a Tree*. **Comment:** If you haven't planted a tree recently, send these good people $5 or whatever you can afford.

TROUT UNLIMITED, 800 Follin Lane, SE, Vienna, VA 22180; (703) 281-1100. Executive director, Charles F. Gauvin. **Who they are:** Founded in 1959, the U.S. headquarters of an international organization with affiliates in Canada, Japan, Spain, France, and New Zealand. **Aims:** Protect clean water and enhance trout, salmon, and steelhead fishery resources. **Membership** (U.S.): 65,000. **Fees:** $25; students, $10; seniors, $20. **What they do:** Stream restoration. **Publications:** *Trout Magazine; News Flow*. **Comment:** Another example of enlightened self-interest.

● **U.S. PUBLIC INTEREST RESEARCH GROUP,** 215 Pennsylvania Avenue, SE, Washington, DC 20003; (202) 546-9707. Director, Gene Karpinski. **Who they are:** Founded in 1971, the national lobbying office for a nationwide network of PIRG. **Aim:** Promote public interest on a wide range of issues. **Membership:** More than 1 million nationwide. **Fees:** $20; students and seniors, $15. **What they do:** Coordinate national campaigns for state PIRGs; monitor corporate and government actions affecting environmental protection, energy, and legislative reform; priorities include strengthening the Clean Water Act and the Resource and Recovery Act, preventing pesticide contamination of food and water, reforming congressional election laws to reduce barriers to voter participation. **Publications:** *Citizen Agenda*; 50 investigative reports. **Comment:** PIRGs do excellent work at the grass-roots level, recruiting a lot of students and seniors. Check out your own state PIRG — there are 25 of them at last count.

WILD EARTH ASSOCIATION, see **CENOZOIC SOCIETY**

● **WILDERNESS SOCIETY,** 900 Seventeenth Street, NW, Washington, DC 20006; (202) 833-2300. President, George T. Frampton, Jr. **Who they are:** Founded in 1935, member of the Big Ten, with offices in Alaska, the Rockies, the Northwest, the Northeast, and the Florida region (including Puerto Rico and the Virgin Islands). **Aim:** "In wildness is the preservation of the world" — Thoreau. **Membership:** 375,000. **Fees:** New, $15; renewal, $30; students and seniors, $15. **Annual revenues:** $11 million. **What they do:** Heavy Capitol Hill lobbying, especially re Arctic Wildlife Refuge, national parks and forests, and ecosystem management; maintain well-researched and orchestrated public information and education programs, which include sponsoring events such as the twenty-fifth anniversary of the signing of the Wilderness Act. **Publications:** *Wilderness* quarterly; *Wilderness America*. **Comment:** A well-balanced organization that often succeeds in keeping up with the rapidly changing environmental scene.

● **WILDLIFE CONSERVATION INTERNATIONAL,** New York Zoological Society, 185th Street and South Boulevard, Building A, New York, NY 10460; (212) 220-5155. Director, John G. Robinson. **Who they are:** Founded in 1895, a branch of the New York Zoological Society, running the oldest conservation programs in America. **Aim:** Preserve biological diversity and endangered ecosystems. **Membership:** 50,000. **Fees:** $25; students and seniors, $15. **Annual revenues:** $4.5 million. **What they do:** Their field scientists take part in some 125 long- and short-term projects worldwide exploring conflicts between humans and wildlife and seeking sustainable solutions; issue grants for graduate students and professionals in wildlife sciences. **Publications:** *Wildlife Conservation Magazine; WCI News.* **Comment:** Although WCI doesn't do a good job of publicizing itself, its work is important to the conservation cause.

WORLD WILDLIFE FUND — U.S., 1250 Twenty-fourth Street, NW, Washington, DC 20037; (202) 293-4800. President, Kathryn Fuller. **Who they are:** Founded in 1961, a member of the Big Ten and affiliated with the *Conservation Foundation* (see above). **Aim:** Wildlife preservation, especially in tropical forests of Latin America, Asia, and Africa. **Membership:** 1 million. **Fees:** $15. **What they do:** Help protect national parks and nature reserves; monitor international wildlife trade; assist local groups in conservation projects; influence governments and private institutions to adopt more environmentally friendly policies. **Publication:** *Focus.* **Comment:** Critics claim WWF is sometimes patronizing, as when it suggests that developing countries should give up hunting in endangered species' habitat in favor of eco-tourism. However, if you love pandas and other cuddly creatures, WWF may be your cup of tea.

● **ZERO POPULATION GROWTH,** 1400 Sixteenth Street, NW, Suite 320, Washington, DC 20036; (202) 332-2200. Executive director, Susan Weber. **Who they are:** Founded in 1968 by Paul and Anne Ehrlich, ZPG is a national nonprofit membership organization. **Aim:** Sustainable balance between the earth's population, the environment, and its resources. **Membership:** 40,000. **Fees:** $20; students and seniors, $10. **Annual revenues:** $1.5 million. **What they do:** Develop in-school and adult educational programs on population issues; coordinate local and national citizen-action efforts on the relationship between population growth and global warming, energy, sustainability, and other population concerns; support voluntary family planning and individual's right to determine family size. **Publications:** *ZPG Reporter; ZPG Activist*; action alerts, fact sheets, and other materials. **Comment:** Given that human population growth is an overriding problem affecting this planet's sustainability, support for ZPG is imperative.

Professional, Research, and Educational Organizations

Note: In most cases these organizations do not have general membership categories. Although some of the societies have professional fees, others charge fees for services provided.

● **ABUNDANT LIFE FOUNDATION,** P.O. Box 772, 1029 Lawrence, Port Townsend, WA 98368; (206) 385-5660. Executive director, Forest Shomer. **Who they are:** Founded in 1975, a nonprofit educational and scientific institution. **Aims:** Acquire, propagate, and save the seeds of Northwest native plants, vegetables, and herbs. **Membership:** 15,000. **Fees:** $8. Donations of homegrown seeds may be acceptable in lieu of cash, but check first to see if needed. **What they do:** Run World Seed Fund, donating seeds internationally. **Publications:** *Abundant Life*, a catalog full of valuable information on seeds and books.

ACID RAIN FOUNDATION, 1410 Varsity Drive, Raleigh, NC 27606; (919) 828-9443. Executive director, Dr. Harriet S. Stubbs. **Who they are:** Founded in 1981, a membership organization whose directors are mainly scientists and educators. **Aim:** Better public understanding of global atmospheric issues, including acid rain, air pollution, climate change, and forest ecosystems. **Membership:** 750. **Fees:** $10 and up. **What they do:** Supply information to a wide audience; support research.

ALASKA CONSERVATION FOUNDATION, 430 West Seventh Avenue, Suite 215, Anchorage, AK 99501; (907) 276-1917. Vice president, Jim Stratton. **Who they are:** Founded in 1980, a grass-roots foundation supported by gifts and grants. **Aim:** Provide grant assistance, technical support, and advice to Alaska environmental community. **What they do:** Issue grants to save the Tongass forest, help native Alaskans join "Save the Arctic National Wildlife Refuge" campaign, and other vital conservation goals. **Publications:** *Alaska Conservation Directory; Alaska Issue Update.*

ALLIANCE FOR THE WILD ROCKIES, 415 N. Higgins Avenue, Missoula, MT 59802; (406) 721-5420. **Who they are:** Coalition of green groups and business owners. **Aim:** Protect 15 million acres of undesignated land in Montana, Idaho, Washington, and Wyoming. **Comment:** Low-budget, homegrown, and energetic, they may be prototypical of the new conservation movement.

AMERICAN CETACEAN SOCIETY, P.O. Box 2639, San Pedro, CA 90731-0943; (310) 548-6279. Executive director, Paul Gold. **Who they are:** Founded in 1967, a membership educational and research organization. **Aim:** Protect marine mammals, especially, whales, dolphins, porpoises, and their habitat. **Membership:** 3,000. **Fees:** $25. **What they do:** Support research, provide information, especially through their quarterly journal *Whalewatcher.*

AMERICAN CONSERVATION ASSOCIATION, 30 Rockefeller Plaza, Room 5402, New York, NY 10112; (212) 649-5600. Executive vice president, George R. Lamb. **Who they are:** A nonmembership educational and scientific organization, founded in 1958. **Aims:** Conservation and preservation of natural resources for public use. **What they do:** Give grants for conservation purposes to organizations already working in the field.

AMERICAN FOREST COUNCIL, 1250 Connecticut Avenue, NW, Washington, DC 20036; (202) 463-2455. President, Laurence D. Wiseman. **Who they are:** Founded in 1941, the voice of the U.S. forest-products industry. **Aim:** Promote that industry's interests. **What they do:** Promote commercial development of forest lands by government, industry, and private landowners. **Comment:** Don't confuse them with the *American Forestry Association* (see above).

AMERICAN SOCIETY OF LIMNOLOGY AND OCEANOGRAPHY, P.O. Box 1987, Lawrence, KS 66044. Publishes the scientific journal *Limnology and Oceanography.* Holds technical conferences and services its members.

AMERICANS FOR THE ENVIRONMENT, 1400 Sixteenth Street, NW, Washington, DC 20036; (202) 797-6665. Executive director, Roy Morgan. **Who they are:** Incorporated in 1972, an environmental training foundation with board representation from the Sierra Club, Nature Conservancy, National Wildlife Federation, NRDC, National Audubon, USPIRG, and others. **Aim:** Provide concerned citizens with the knowledge and skills needed to participate effectively in the electoral process. **What they do:** Organize training programs teaching specialized environmental skills.

ARCTIC INSTITUTE OF NORTH AMERICA, University Library Tower, 2500 University Drive, NW, Calgary, Alberta, Canada T2N 1N4; (403) 220-7515. Executive director, Michael Robinson. **Who they are:** Founded in 1945, a nonprofit organization of 23 associates whose work they sponsor. **Aim:** Research on arctic, subarctic, and low-temperature regions. **Membership:** 2,500. **Fees:** $25. **What they do:** Provide general reference services, organize research, run seminars. **Publication:** *Arctic Journal.*

● **ASSOCIATION OF FOREST SERVICE EMPLOYEES FOR ENVIRONMENTAL ETHICS,** P.O. Box 11615, Eugene, OR 97440; (503) 484-2692. **Who they are:** Founded in 1989 by Jeff DeBonis, a former timber planner for the Willamette National Forest, AFSEEE is a nonprofit association of Forest Service employees and concerned citizens. **Aim:** Promote a new vision of the U.S. Forest Service. **Membership:** 2,000. **Fees:** $20; limited income, $10. **What they do:** Educate and encourage government employees to exercise their free-speech rights; advocate forest planning based on biological facts from the ground up and restoration of ecosystems; work for protection of old-growth forests and roadless areas on public lands. **Publication:** *Inner Voice* quarterly. **Comment:** A growing and important voice for conservation.

ASSOCIATION OF MIDWEST FISH AND WILDLIFE AGENCIES, c/o Michigan Department of Natural Resources, P.O. Box 30028, Lansing, MI 48909; (517) 373-1220. President, David Hales. **Who they are:** Associated commissioners and directors of 18 agencies in the Midwest and including the provinces of Manitoba, Ontario, and Saskatchewan. **Aims:** Protect, preserve, restore, and manage fish and wildlife. **What they do:** Gather and disseminate information relating to their aims.

ATLANTIC CENTER FOR THE ENVIRONMENT, 39 South Main Street, Ipswich, MA 01938; (508) 356-0038. Director, Lawrence B. Morris. **Who they are:** A division of the Quebec-Labrador Foundation, a regional conservation organization conducting many of its programs through interns recruited from universities across North America. **Aim:** Promote public involvement in resource management. **What they do:** Provide technical assistance to public and private agencies; exchange ideas between their region (Atlantic Canada, eastern Quebec, and northern New England) and other regions. **Publications:** *Nexus, Atlantic Naturalist.* **Comment:** A conservation organization whose work cuts across political boundaries.

ATLANTIC SALMON FEDERATION, P.O. Box 429, St. Andrews, New Brunswick, Canada, E0G 2X0; (506) 529-4438. President, David R. Clark, Q.C.; U.S. executive director, John C. Philips, P.O. Box 684, Ipswich, MA 01938; (617) 356-0717. **Who they are:** Founded in 1982, an international organization dependent on contributions from individuals, foundations, and corporations in the U.S. and Canada. **Aims:** Preservation and wise management of the Atlantic salmon and its habitat. **Membership:** 500,000, including affiliates. **What they do:** Direct a regional network in a research and educational program.

AUDUBON NATURALIST SOCIETY OF THE CENTRAL ATLANTIC STATES, 8940 Jones Mill Road, Chevy Chase, MD 20815; (301) 652-9188. Executive director, Ken Nicholls. **Who they are:** Founded in 1897 and based in a 40-acre wildlife sanctuary at Woodend, Md., one of the original independent Audubon societies. **Aim:** Environmental education. **Membership:** 11,000. **Fees:** $25. **What they do:** Maintain an impressive educational program on conservation issues and natural science studies. **Publications:** *Atlantic Naturalist; Naturalist News;* and *Naturalist Review.*

●● **CENOZOIC SOCIETY,** 68 Riverside Drive, Apt. 1, Canton, NY 13617; (315) 379-9940. **Who they are:** Founded in 1991 by Mary Davis, John Davis, Tom Butler, and Dave Foreman, CS (initially known as Wild Earth Association) serves the biocentric grass-roots elements within the conservation movement, advocating restoration and protection of all natural elements of biodiversity. **What they do:** Serve as a networking tool for grass-roots wilderness activists. **Publication:** *Wild Earth,* $20 quarterly. **Comment:** *Cenozoic* is the age of mammals, just to remind us how fragile it is. *Wild Earth* is the best publication of its kind, providing important information and ideas that you won't easily find anywhere else. Put it on the top of your reading list!

● **CENTER FOR ENVIRONMENT, COMMERCE, AND ENERGY** (CE)2, 733 Sixth Street, SE, Suite 1, Washington, DC 20003; (202) 543-3939. President, Norris McDonald. **Who they are:** Founded in 1985 by McDonald, this is a nonprofit public-interest group whose board of directors includes Dr. Ruth Whitworth-Logan, Dr. Lenneal Henderson, Dr. Robert Bullard, Peggy Shepard, and Dick Gregory. **Aim:** Dedicated to environmental protection, human ecology, and the efficient use of natural resources. **Membership:** 5,000. **Fees:** $25. **What they do:** Provide opportunities for African Americans and other minorities in the environmental movement. **Publication:** *African American Environmentalist.* **Comment:** Given that the conservation and environment movement is dominated by white, middle-class males, it is salutary to find one organization working to change this seriously unbalanced representation.

● **CENTER FOR ENVIRONMENTAL INFORMATION,** 46 Prince Street, Rochester, NY 14607; (716) 271-3550. President, Elizabeth Thorndike. **Who they are:** Founded in 1974, CEI is a private, nonprofit organization. **Aim:** Provide comprehensive information on environmental issues and policies. **Membership:** 750. **Fees:** $25; students and seniors, $10. **What they do:** CEI is an environmental information clearinghouse sponsoring conferences and maintaining an extensive library, which is open to the public and provides access to worldwide databases. **Publication:** *CEI Sphere* (newsletter); *Global Change Digest* and *Acid Precipitation Digest* (bimonthly bulletins).

● **CENTER FOR PLANT CONSERVATION,** P.O. Box 299, St. Louis, MO 63166; (314) 577-5100. Executive director, Donald A. Falk. **Who they are:** Founded in 1984, a national network of 20 botanical gardens and arboreta. **Aim:** Create a national program of plant conservation, research, and education within existing institutions to preserve genetic diversity through habitat protection. Annual revenues: $850,000. **What they do:** Run the National Collection of Endangered Plants — collections of endangered species in regional gardens and seed banks as a resource for conservation and research work; work also with community groups such as the Garden Club of America on how to save plant species from extinction. **Publication:** *Plant Conservation Newsletter.* **Comment:** Saving plants is crucial to the survival of *all* species.

CENTER FOR THE GREAT LAKES, 35 East Wacker Drive, Chicago, IL 60601; (312) 645-0901; 320 1/2 Bloor Street, West, Suite 301, Toronto, Ontario, Canada, M5S 1W5; (416) 921-7662. President, William Brah. **Who they are:** Founded in 1983, a private binational policy-research

organization funded by foundation, corporate, government, and individual donations. **Aims:** Protect Great Lakes and St. Lawrence River and foster sustainable development of the regional economy. **What they do:** Provide information on waterfront development, shoreline management, water resources, hazardous waste, and pollution prevention; organize workshops, seminars, and conferences; developing a Great Lakes protection fund to finance toxic-pollution reduction in Canada to complement the $100 million fund created by U.S. state governors in 1989. **Publications:** *The Great Lakes Reporter*; fact sheets.

CONSERVATION FOUNDATION, 1250 Twenty-fourth Street, NW, Washington, DC 20037; (202) 293-4800. President, Kathryn S. Fuller. **Who they are:** Founded in 1948, nonprofit research and public education foundation affiliated with the World Wildlife Fund—U.S. (see above). **Aim:** Wise use of the earth's resources. **What they do:** Research and technical assistance in land use, water resources, wetlands, toxic substances, environmental dispute resolution, and other environmental issues in the U.S. and in developing countries. **Publications:** *CF Letter; Successful Communities.*

CONSERVATION LAW FOUNDATION, 3 Joy Street, Boston, MA 02108; (617) 742-2540. **Who they are:** Founded in 1966, a nonprofit environmental law organization. Executive director, Douglas I. Foy, Esq. **Aim:** Improve resource management, environmental protection, and public health in New England. **Membership:** 5,000. **Fees:** $30; students, seniors, and low-income, $15. **What they do:** Public and private land and marine resource preservation; energy and water conservation; transportation and environmental planning. **Publications:** *Power to Spare; Troubled Waters.*

● **ECOLOGICAL SOCIETY OF AMERICA,** 9560 Rockville Pike, Bethesda, MD 20814; (301) 530-7005. Director of public affairs, Marjorie M. Holland. **Who they are:** Founded in 1915, North America's professional society of ecologists. **Aims:** Study organisms in relation to their environment, exchange ideas among those interested in ecology, and instill ecological principles in decision making of society at large. **Membership:** 6,500. **What they do:** Promote an ecological perspective professionally and on issues of public policy for legislators and the general public; maintain an ecological information network, a data base of 1,500 scientists who can provide expert scientific information on environmental topics to Congress and other groups. **Publications:** *Ecology* bimonthly; *Ecological Monographs* quarterly; *Bulletin* quarterly; *Careers in Ecology* (a brochure). **Comment:** A valuable resource in an increasingly important field.

● **ENVIRONMENTAL LAW ALLIANCE (E-LAW),** 1877 Garden Avenue, Eugene, OR 97403; (503) 687-8454. **Who they are:** Founded in 1991, E-LAW is an international alliance of public interest attorneys, with office in Australia, Ecuador, Indonesia, Japan, Malaysia., Peru, the Philippines, and Sri Lanka. **Aim:** Protect the environment through law across borders. **What they do:** As a mutual support network, E-LAW provides pro bono litigation support, scientific and technical research, and other services to attorneys and citizen groups around the world. Because E-LAW doesn't charge for their services, they rely on individuals, law firms, and scientists to contribute either their services or tax-deductible contributions. **Comment:** Contribute to this important public interest alliance.

●● **ENVIRONMENTAL LAW FOUNDATION,** California Building, 1736 Franklin Street, 8th Floor, Oakland, California 94612; (510) 208-4555. President, James R. Wheaton. **Who they are:** Founded in 1991 by a group of leading environmental activists, ELF is a nonprofit public benefit corporation funded by public foundations and recovery of court awards from law violators. **Aims:** Explore new strategies for addressing environmental problems, specifically targeting investigations and public information to those with the least choice about and most vulnerable to toxic risks—children, inner-city dwellers, and workers. **What they do:** Execute investigations, public outreach and coalition building efforts, and law enforcement actions for individuals, community groups, and the public interest. **Comment:** Using the law to positive effect.

● **ENVIRONMENTAL LAW INSTITUTE,** 1616 P Street, NW, Suite 200, Washington, DC 20036; (202) 328-5002. President, J. William Futrell. **Who they are:** Founded in 1969, national center for research and education on environmental law and policy. **Aim:** Develop policy and

guidelines for implementation of existing and future legislation. **Membership:** 2,500. **Fees:** Government and nonprofit organizations, $50; private citizens, $75. **What they do:** Research environmental protection laws and natural resources use; extensive educational and informational programs, especially on wetlands (see also *National Wetlands Technical Council*). **Publications:** *Environmental Law Reporter; National Wetlands Newsletter; Law of Environmental Protection.*

FIRE RESEARCH INSTITUTE, P.O. Box 241, Roslyn, WA 98941-0241; (504) 649-2940. Director, Jason M. Greenlee. **Who they are:** Founded in 1983 as a nonprofit research center. **Aim:** Focus on role of wildland fires. **What they do:** Distribute a wide variety of related materials. **Publications:** International journal, directory, and bibliography of wildland fires.

●● **GAIA INSTITUTE**, the Cathedral of St. John the Divine, 1047 Amsterdam Avenue at 112th Street, New York, NY 10025; (212) 295-1930. Director, Paul S. Mankiewicz, Ph.D. **Who they are:** A nonprofit, research and community-oriented "collection of interested citizens, philosophers, systems theorists, artists, anthropologists, lawyers, molecular biologists and other scientists, musicians, architects, and educators." **Aim:** Explore the meaning of the integration of life and matter as they relate to ecology, economics, and the design of cities, towns, and villages. **Membership:** 200 active, with a mailing list of 1,000. **Fees:** $25. **What they do:** See chapter 10, pages 321-322. **Publications:** *The Gaia Newsletter*; and individual reports. **Comment:** A paragon of integrating scientific research with practical ecological engineering and design involving community participation.

GLOBAL TOMORROW COALITION, 1325 G Street, NW, Suite 1010, Washington, D.C. 20005; (202) 628-4016. President, Donald R. Lesch. **Who they are:** A coalition of nearly 100 non-governmental organizations, educational institutions, corporations, and concerned citizens. **Aim:** Build broader public understanding of long-term trends in population, resources, environment, and development. **Membership:** 10,000. **Fees:** $25; students and seniors, $15. **What they do:** Make available research and other materials produced by its member groups to communities nationwide; GTC's most ambitious program, Globescope, brings together national and international leaders from business, citizens groups, entertainment, scientific, environmental, and political organizations to generate a nonpartisan dialogue on global problems. **Publications:** *The Global Ecology Handbook*; guides on sustainable development; and the Global Issues Education Set.

GRASSLAND HERITAGE FOUNDATION, 5450 Buena Vista, Shawnee Mission, KS 66205; (913) 677-3326. President, Philip S. Brown. **Who they are:** Founded in 1976, a naturalist-oriented nonprofit organization. **Aims:** Preserve and acquire tracts of native American grassland. **Membership:** 1,300. **What they do:** Produce teaching materials for schools and the general public. **Publications:** *Prairie Center News; Gifts of Land for Conservation.*

● **HIGH DESERT MUSEUM**, 59800 South Highway 97, Bend, OR 97702; (503) 382-4754. President, Donald M. Kerr. **Who they are:** Founded in 1972, they focus on the Intermountain West — portions of eight western states and British Columbia. **Aim:** Increase knowledge of the natural and cultural history and resources of the high-desert region. **Membership:** 4,000. **What they do:** Open to the public since 1982, the museum offers "living" participation exhibits dedicated to high-desert themes and the wiser management of the region's resources. **Publication:** *High Desert Museum.*

●● **INFORM**, 381 Park Avenue South, New York, NY 10016; (212) 689-4040. President, Joanna Underwood. **Who they are:** Founded in 1974, an independent research and public education organization. **Aim:** Identify how industries and companies can lower their impact on natural resources. **Membership:** 2,500. **What they do:** Field research on chemical hazards, municipal solid waste, air quality, land and water conservation, and other subjects; identify issues of environmental concern, their causes, and possible solutions through an advisory board and a national network of legislators, regulators, public-interest leaders, and national and grassroots environmental groups. **Publications:** 50+ reports. **Comment:** INFORM's work fills an important gap in environmental knowledge, communication, and education.

●● **INTERNATIONAL CENTER FOR THE SOLUTION OF ENVIRONMENTAL PROBLEMS,** 535 Lovett Boulevard., Houston, TX 77066; (713) 527-8711. Technical director, Joseph L. Goldman, Ph.D. **Who they are:** Founded in 1975, a solution-oriented research network with a worldwide affiliation of more than 20 scientists, engineers, economists, ecologists, environmentalists, and other professionals, funded by fees charged for services and by tax-deductible contributions. **Aim:** Devise solutions to environmental problems that help align human activities with nature. **What they do:** Much of ICSEP's work deals with soil erosion, land subsidence, and other weather-related agricultural problems, including the relationship of weather and agriculture to the increasing urbanization of the planet. **Publications:** Reports on flood control, hazardous-waste disposal, movement of herbicides in the soil, climatic response to city design, and other environmental issues. **Comment:** Tightly run environmental think tank that often comes up with innovative solutions.

● **INTERNATIONAL ECOLOGICAL SOCIETY,** 1471 Barclay Street, St. Paul, MN 55106; (612) 774-4971. President, R. J. F. Kramer. **Who they are:** Founded in 1975, a volunteer-staffed international organization with representation in the U.S., Canada, and Central America, affiliated with the League to Save Lake Tahoe, Minnesota Committee to Protect the Mourning Dove, Voyageurs National Park Coalition, the Alaska Coalition, and other groups. **Aims:** Environmental protection and better understanding of all life forms. **Membership:** 6,000. **What they do:** Efforts include "Save the Whales" campaign, banning the leghold trap, campaigns against primate research, and preserving eastern timber wolves, Canadian harp seals, wild fur bearers, and other species. **Publications:** *Eco-Humane Letter; Sunrise*; and action alert brochures.

INTERNATIONAL OCEANOGRAPHIC FOUNDATION, 4600 Rickenbacker Causeway, Virginia Key, Miami, FL 33149; (305) 361-4888. President, Edward T. Foote II. **Who they are:** Founded in 1953, a nonprofit scientific and educational foundation. **Aim:** Extend knowledge of oceans in all their aspects. **Membership:** 55,000. **Fees:** $24. **What they do:** Run programs on oceanographic ecology. **Publication:** *Sea Frontiers* bimonthly.

●● **LAND INSTITUTE,** 2440 East Waterwell Road, Salina, KS 64701; (913) 823-5376. Codirectors, Wes Jackson and Dana Jackson. **Who they are:** Founded by the Jacksons in 1976, a private, nonprofit, ecologically-minded agricultural research center with an outstanding worldwide reputation. **Aims:** Practice and promote sustainable agriculture and good stewardship of the earth. **Membership:** 2,000. **Fees:** $15 and up. **What they do:** Research and develop cash crops that mimic natural prairie grasses, a polyculture of perennials that do not require annual plowing.* **Publications:** *The Land Report* (three times a year); and special reports. **Comment:** In addition to his work at the Land, Wes Jackson writes seminal books, including *New Roots for Agriculture* and *Altars of Unhewn Stone*.

MONITOR, 1506 Nineteenth Street, NW, Washington, DC 20036; (202) 234-6576. Executive vice president, Craig Van Note. **Who they are:** Founded in 1972, a consortium of 35 member conservation, environmental, and animal welfare organizations. **Aim:** Increase the impact of its constituent members. **What they do:** Serve as a clearinghouse on endangered species and marine mammals and their habitats.

● **NATIONAL ASSOCIATION OF BIOLOGY TEACHERS,** 11250 Roger Bacon Drive, #19, Reston, VA 22090; (703) 471-1134. Executive director, Patricia J. McWethy. **Who they are:** Professional teachers' association. **Aim:** Dedicated to the concerns of biology teachers. **Membership:** 7,000. **Fees:** $38. **What they do:** Inform teachers of innovative developments in environmental curricula; advocate alternative use of animals in classrooms; organize workshops and conferences. **Publications:** *American Biology Teacher; News and Views*, and other materials.

●● **NATIONAL AUDUBON SOCIETY EXPEDITION INSTITUTE** (AEI), P.O. Box 170, Readfield, ME 04355; (207) 685-3111. Director, Diana Becker. **Who they are:** Founded in 1978,

* For more on the work of the Land Institute, see chapter 10, pages 311-312.

an experiential school for high school, undergraduate, and graduate-level studies. **Aim:** Provide alternative to conventional schools, emphasizing environmental field studies. **Membership:** 80 to 100 students a year. **Fees:** $6,100 a semester; $10,400 a year. **What they do:** Groups of about 20 students travel in schoolbuses to self-selected locations in the U.S. where they study, e.g., marine biology in the Everglades, geology in the Grand Canyon, politics on Capitol Hill, folk music in Appalachia, interacting with natural and human resources. **Comment:** Children are the future, so we must give them the kind of exposure to the environment that AEI offers. It is a priceless investment. Support them.

NATIONAL PARK FOUNDATION, 1101 Seventeenth Street, NW, Suite 1008, Washington, DC 20036; (202) 785-4500. President, Alan A. Rubin. **Who they are:** Chartered in 1967 by Congress, a private nonprofit organization whose chairperson is secretary of the interior and whose secretary is director of the National Park Service. **Aim:** Provide private-sector support for the National Park System. **What they do:** Support educational outreach programs, encouraging participation by volunteers and assisting NPS employees.

NATIONAL WETLANDS TECHNICAL COUNCIL, 1616 P Street, NW, Suite 200, Washington, DC 20036; (202) 328-5150. Chairperson, Dr. Joseph Larson. **Who they are:** Independent council of leading wetlands scientists. **Aim:** Provide scientific assistance in wetlands conservation. **What they do:** Advise federal agencies and other institutions on wetlands.

NORTH AMERICAN LAKE MANAGEMENT SOCIETY, 1000 Connecticut Avenue, NW, Suite 300, Washington, DC 20036; (202) 466-8550. President, Richard S. McVoy. **Who they are:** Founded in 1980, an educational and research network of limnologists, lake managers, lake associations, and concerned citizens. **Aims:** Protect, restore, and manage lakes, reservoirs, and their watersheds. **What they do:** Run workshops, meetings, and an annual conference; disseminate citizens' booklets. **Publications:** *Lake Line Magazine; Lake and Reservoir Management Journal.*

● NUCLEAR INFORMATION AND RESOURCE SERVICE, 1424 Sixteenth Street, NW, Suite 601, Washington, DC 20036; (202) 328-0002. **Who they are:** A nonprofit clearinghouse on nuclear-power issues, supported by grants and membership. **Aim:** Provide information and materials to challenge nuclear facilities and policies, especially deregulation of radioactive waste. **Membership:** 1,500. **Fees:** $20. **What they do:** Provide updates on nuclear issues; publish manuals for university and other activist use; work to prevent a new generation of nuclear reactors. **Publications:** *NIRS Reports* and other materials. **Comment:** Valuable source of information.

●● OCEAN ARKS INTERNATIONAL, 1 Locust Street, Falmouth, MA 02540; (508) 540-6801. President, John Todd. **Who they are:** Founded in 1980 by John Todd and Nancy Jack Todd, a nonprofit ecology-driven organization forming part of the Center for the Protection and Restoration of Water and working in association with Ecological Engineering Associates. **Aim:** Ecologically solve water pollution and toxic waste-disposal problems. **Fees:** $30; students and seniors, $15. **What they do:** OIA has developed living machines, an innovative family of technologies for treating waste water naturally with successful systems operating in Harwich and Marion, Mass., Providence, R.I., and Moncie, Ind. OIA's training program is done at the center and at community boards of health and planning commissions. **Publications:** *Annals of Earth* (three times a year); *Bioshelters, Ocean Arks, City Farming—Ecology as the Basis of Design.** **Comment:** If anyone personifies thinking globally, acting locally, it is the Todds (who cofounded New Alchemy Institute with Bill McLarney in 1969). Stay in touch with their leading-edge work by subscribing to *Annals* and supporting their important work.

PACIFIC WHALE FOUNDATION, 101 North Kihei Road, Kihei, HI 96753; (800) WHALE-11. President, Gregory D. Kaufman. **Who they are:** Founded in 1980, a nonprofit research, public education, and conservation foundation. **Aims:** Study the ocean and its marine mammal inhabitants; preserve the marine environment. **Membership:** 5,000. **Fees:** $20; students

* For more on Ocean Arks International, see chapter 10, pages 316-319.

and seniors, $15. **Annual Revenues:** $700,000. **What they do:** Support marine mammal and coral-reef field studies in Hawaii, Alaska, Canada, and other countries; programs include Ocean Outreach, Adopt-a-Whale, and other interactive conservation projects. **Publications:** *Soundings; Fin and Fluke.*

RENE DUBOS CENTER FOR HUMAN ENVIRONMENTS, 100 East Eighty-fifth Street, New York, NY 10028; (212) 249-7745. Executive director, Ruth A. Eblen. **Who they are:** Founded in 1977 by the eminent scientist René Dubos as an independent nonprofit research and educational center. **Aim:** Carry on the humanistic tradition of environmental problem solving. **Membership:** 500. **Fees:** None. **What they do:** Develop environmental resources for the general public and for business and industry; conduct Dubos forums and related activities; maintain the Dubos library and archives. **Publications:** *Think Globally, Act Locally Newsletter*; reports, proceedings, and audio-video materials.

RIVER WATCH NETWORK, 153 State Street, Montpelier, VT 05602; (802) 223-3840. Executive director, John M. Byrne. **Who they are:** Founded in 1987, a national nonprofit organization funded by foundation grants, corporate donations, government support, contracts, and program revenue. **Aim:** River cleanup through a network of locally supported grass-roots programs. RWN is not a membership organization and does not have annual dues. It does, however, welcome contributions from trusts, businesses, municipalities, individuals, and other sources. **What they do:** Help local groups organize and conduct water quality protection programs for rivers, including the Connecticut, Lackawanna, Mississippi, Colorado, and Rio Grande; enlist students to work with local, state, and federal river protection efforts; serve as a water-quality information clearinghouse. **Publications:** Reports on river monitoring and water-quality programs. **Comment:** New group doing important work.

• **SOCIETY FOR ECOLOGICAL RESTORATION,** University of Wisconsin — Madison Arboretum, 1207 Seminole Highway, Madison, WI 53711; (608) 263-7889. President, John Rieger. **Who they are:** Founded in 1987, an interdisciplinary association of naturalists, foresters, biologists, political economists, administrators, and others concerned with ecological restoration. **Aims:** Research, education, and development of grass-roots support for restoration and management of natural areas in city, country, and wilderness. **Membership:** 1,750. **Fees:** $25; students and seniors, $15. **What they do:** Research and experiment with rebuilding entire ecosystems, including creating prototypical restoration projects and monitoring programs; hold annual conference bringing together leading experts in the field; publish technical reports and popular materials. **Publications:** *Restoration and Management Notes* (twice yearly journal); *SERM Newsletter* quarterly. **Comment:** Ecological restoration is increasingly recognized as having a crucial role to play in protecting ecosystems. This is the wave of the future.

SOIL AND WATER CONSERVATION SOCIETY (formerly Soil Conservation Society of America), 7515 NE Ankeny Road, Ankeny, IA 50021; (515) 289-2331 or (800) THE-SOIL. Executive vice president, Verlon K. Vrana. **Who they are:** Founded in 1945, a multidisciplinary society of farmers, scientists, administrators, and grass-roots (!) activists. **Aim:** Advocate conservation of soil, water, and related natural resources. **Membership:** 13,000. **Fees:** $30 and up. **What they do:** Identify, evaluate, and formulate workable solutions to land and water management issues. **Publication:** *Journal of Soil and Water Conservation.*

• **TRUST FOR PUBLIC LAND,** 116 New Montgomery Street, San Francisco, CA 94105; (415) 495-4014. Executive vice president, Ralph W. Benson. **Who they are:** Founded in 1972, a national nonprofit, nonmembership, nonadvocacy land conservation association with offices in Boston, New York, Santa Fe, Seattle, Tallahassee, Washington, D.C., Portland, Ore., and Morristown, N.J. **Aim:** Acquire land of scenic, historical, ecological, recreational, or cultural value for public use in wilderness, rural, and urban areas. **What they do:** Work with local and state land conservancies and citizens' groups in acquiring suitable lands; TPL has protected nearly half a million acres of land for public use in 38 states and Canada, much of it in cities underserved by parks and open space. **Publication:** *Land and People.*

● **UNITED NATIONS ENVIRONMENT PROGRAM,** Regional Office for North America, Room DC-2-0803, United Nations, New York, NY 10017; (212) 963-8138. Director, Dr. Noel J. Brown. **Who they are:** Established by the UN General Assembly in 1972 and with headquarters in Nairobi, Kenya, UNEP oversees the environmental work of all other UN agencies to ensure that an ecological perspective is incorporated in their development projects. **Aim:** Advocacy of global environmental concerns, especially relating to atmospheric pollution, fresh water, oceans, and coastal regions, land degradation, biological diversity, and toxic chemicals. **What they do:** Global monitoring and evaluation in these areas; train environmental experts; distribute information through briefings and publications; organize environmental conferences such as the Montreal Protocol on ozone depletion and the Earth Summit in Rio de Janeiro, 1992. **Publications:** *State of the Environment Report; UNEP News.* **Comment:** Along with UNICEF, UNEP is one of the most effective of the UN's agencies.

WATER POLLUTION CONTROL FEDERATION, 601 Whythe Street, Alexandria, VA 22314; (703) 684-2400. Executive director, Dr. Quincalee Brown. **Who they are:** Founded in 1928, an international nonprofit federation of those engaged in water pollution control. **Aims:** Preserve and enhance water-quality resources. **Membership:** 35,000. **What they do:** Provide information on water-quality resources to specialists and general public. **Publications:** *Research Journal; Water Environment and Technology.*

● **WILDERNESS COVENANT,** P.O. Box 5217, Tucson, AZ 85703; (602) 743-9524. President, Clarke Abbey. **Who they are:** Created in 1990 as a nonprofit foundation to seek and provide funding for individuals and groups committed to grass-roots environmentalism. **Aim:** Preserve natural environment through information, educational programs, and grass-roots efforts. **What they do:** Fund projects including *Wild Earth, Wildlife Damage Review*, the Sierra Madre Network, and Citizen Search. **Comment:** At the leading edge.

WILDLIFE HABITAT ENHANCEMENT COUNCIL, 1010 Wayne Avenue, Suite 1240, Silver Spring, MD 20910; (301) 588-8994. Executive director, Joyce M. Kelly. **Who they are:** Founded in 1988, a nonprofit, nonlobbying organization whose membership consists of corporations, conservation groups, and wildlife consultants. **Aim:** Create wildlife habitat on privately owned lands. **Fees:** Individual, $100; public-interest groups, $500; corporations, $1,500 for annual sales of less than $500 million, $3,000 for sales above $500 million. **What they do:** Link corporate managers with conservation experts to develop innovative wildlife projects; have created an international registry of corporate wildlife enhancement projects, evaluating corporate efforts on behalf of wildlife. **Publications:** Brochures and annual reports. **Comment:** New organization whose work should be worth watching.

● **WINDSTAR FOUNDATION,** 2317 Snowmass Creek Road, Snowmass, CO 81654; (303) 927-4777. Executive vice president, Steve Blomeke. **Who they are:** A nonprofit membership organization founded in 1976 by John Denver and Tom Crum. **Aim:** Inspire individuals to achieve an environmentally sustainable future. **Membership:** 5,000. **Fee:** $35. **What they do:** Programs include Aspen Global Change Institute, Windstar Biodome Project, and symposia. **Publication:** *Worldwatch/ Windstar Vision.*

WORLD RESOURCES INSTITUTE, 1709 New York Avenue, NW, Washington, DC 20006; (202) 638-6300. President, James Gustave Speth. **Who they are:** Founded in 1982 with funding from the MacArthur and other foundations as a policy research center. **Aim:** Help governments, international organizations, and the private sector to address vital issues of environmental integrity, natural resource management, and economic growth. **What they do:** Research and publishing. **Publications:** *World Resources Report* annual; *Research Report* series.

●● **WORLDWATCH INSTITUTE,** 1776 Massachusetts Avenue, NW, Washington, DC 20036; (202) 452-1999. President, Lester R. Brown. **Who they are:** Founded in 1976, a nonprofit research organization that identifies and analyzes global environmental problems. **Aim:**

Alert world leaders and general public to key issues such as energy, population growth, migration, food, and the changing role of women. **What they do:** Provide some of the most important information and research available on key issues. **Publications:** *State of the World* annual; *WorldWatch Magazine* bimonthly; *Worldwatch Papers* bimonthly.

• **XERXES SOCIETY**, 10 SW Ash Street, Portland, OR 97204; (503) 222-2788. **Who they are:** Founded in 1971, a nonprofit educational and scientific conservation organization named after the Xerxes, the first butterfly known to have become extinct as the result of human intervention. **Aim:** Protect invertebrates as major component of biological diversity. **Membership:** 3,500. **Fees:** $25 and up. **What they do:** Explain the role of invertebrates in preserving ecosystems; identify and protect critically threatened conservation sites; work with other conservation groups to bring invertebrates into global conservation planning. **Publications:** *Atala; Wings: Essays on Invertebrate Conservation* (membership magazine). **Comment:** Once more, small is beautiful!

Key Federal Agencies

— *Agriculture, U.S. Department of*, Fourteenth Street and Independence Avenue, SW, Washington, DC 20250; (202) 447-2791. Responsible for the protection of soil, water, forests, and other natural resources.

— *Agricultural Research Service*, U.S. Department of Agriculture, Public Affairs Office, 10300 Baltimore Avenue, Room 307, Building 005, Beltsville, MD 20705; (301) 344-2264. Research for the improvement of soil, water, and air.

— *Army Corps of Engineers*, U.S. Department of Defense, Office of the Chief of Engineers, Pulaski Building, 20 Massachusetts Avenue, NW, Washington, DC 20314; (202) 272-0001. Has jurisdiction on dams, reservoirs, flood control, and any work or structure in "navigable waters" including wetlands.

— *Bureau of Indian Affairs*, U.S. Department of the Interior, Interior South Building, 1951 Constitution Avenue, NW, Washington, DC 20245; (202) 208-5116. Charged with the protection and enhancement of Indian lands and the conservation and development of natural resources.

— *Bureau of Land Management*, U.S. Department of the Interior, Office of Public Affairs, 1849 C Street, NW, Room 5600, Washington, DC 20240; (202) 343-5717. Has management responsibility for 272 million acres of public lands and mineral management responsibility for an additional 370 million acres where these rights were reserved to the federal government.

— *Bureau of Reclamation*, U.S. Department of the Interior, Office of Environmental Affairs, Eighteenth and C Streets, NW, Washington, DC 20240; (202) 343-4662. Its work includes development of plans for use of water resources, groundwater recharge, construction and rehabilitation of water supply systems, and other water-related projects.

— *Coast Guard*, U.S. Department of Transportation, 2100 Second Street, SW, Washington, DC 20593; (202) 267-2229.

— *Council on Environmental Quality*, 722 Jackson Place, NW, Washington, DC 20503; (202) 395-5750. Advises the President on national environmental trends and policies.

— *Energy, U.S. Department of*, 2000 Independence Avenue, SW, Washington, DC 20585; (202) 586-5000. Provides framework for national energy policy, including energy conservation and regulatory programs; responsible for administering nuclear-weapons program and for developing new energy technologies.

— *Environmental Protection Agency*, 401 M Street, SW, Washington, DC 20460; (202) 382-2090. Its main divisions: Water Management; Hazardous Waste; Air, Pesticides, and Toxics; and Environmental Services. Other EPA offices include: *Office of Marine and Estuarine Protection*, (202) 475-8580; Clean Lakes Program, (202) 382-5700; Great Lakes National Program Office, 111 West Jackson, 10th Floor, Chicago, IL 60604; (312) 353-2117.

— *Farmers Home Administration*, U.S. Department of Agriculture, Federal Building, Suite 102, 101 South Main, Temple, TX 76501; (817) 774-1301. Provides credit for soil and water conservation, watershed protection, and other rural loans.

— *Fish and Wildlife Service*, U.S. Department of the Interior, Office of Public Affairs,

Eighteenth and C Streets, NW, Washington, DC 20240; (202) 208-5634. Responsibilities include listing and protecting endangered species and other fish and wildlife.
— *Forest Service*, U.S. Department of Agriculture, Public Affairs Office, P.O. Box 96090, Washington, DC 20090-6090; (202) 447-3760. Responsible for providing national leadership in forestry.
— *Interior, U.S. Department of*, Interior Building, 1849 C Street, NW, Washington, DC 20240; (202) 208-1100. The principal U.S. conservation agency, responsible for most of the federal public lands and resources, including water, minerals, fish and wildlife, national parks, and historical places.
— *National Advisory Committee on Oceans and Atmosphere*, 3300 Whitehaven Street, NW, Washington, DC 20235.
— *National Institute of Environmental Health Science*, U.S. Health and Human Services Department, P.O. Box 12233, Research Triangle Park, NC 27709; (919) 541-3212. As part of the National Institutes of Health, this institute studies the effects of chemical, biological, and physical factors in the environment on human health and well-being.
— *National Marine Fisheries Service*, U.S. Department of Commerce, Silver Spring Metro Center 1, 1335 East-West Highway, Silver Spring, MD 20910; (301) 427-2239. A component of NOAA, it is responsible for the protection of living marine resources.
— *National Ocean Policy Study*, U.S. Senate Commerce Committee, 527 Hart Senate Office Building, Washington, DC 20510.
— *National Ocean Service*, Marine Pollution Programs Office, Ocean and Coastal Resource Management, 1140 Rockville Pike, Rockville, MD 20852.
— *National Oceanic and Atmospheric Administration* (NOAA), U.S. Department of Commerce, Rockville, MD 20852; (301) 443-8910. Provides information on oceanic, atmospheric, solar, and space conditions; develops policy on coastal zone and ocean management.
— *Nuclear Regulatory Commission*, Office of Public Affairs, 1717 H Street, NW, Washington, DC 20555; (301) 492-0240. Responsible for ensuring that civilian uses of nuclear materials are consistent with public health, safety, and environmental quality.
— *Oceans and Fisheries Affairs Bureau*, U.S. Department of State, 2201 C Street, SW, Washington, DC 20520; (202) 647-2396.
— *Oceans and International Environmental and Scientific Affairs Bureau*, U.S. Department of State, 2201 C Street, SW, Washington, DC 20520; (202) 647-1555.
— *Office of Ocean Law and Policy*, U.S. Department of State, 2201 C Street, SW, Washington, DC 20520; (202) 647-9098.
— *President of the United States*, The White House, 1600 Pennsylvania Avenue, NW, Washington, DC 20500; (202) 395-7000. Has ultimate responsibility for the condition of the environment.
— *Soil Conservation Service*, U.S. Department of Agriculture, Public Information Office, P.O. Box 2890, Washington, DC 20013; (202) 447-4543. Administers a soil and conservation program in cooperation with landowners and other users. Assists in agricultural pollution control, environmental improvement, and rural community development.

Congressional Committees

U.S. House of Representatives
— *Committee on Agriculture*, Room 1301, Longworth House Office Building, Washington, DC 20515; (202) 225-2171. Responsible for forestry in general, protection of birds and animals in forest reserves, agricultural and industrial chemistry, and soil conservation. Subcommittee on Forests, Family Farms, and Energy.
— *Committee on Energy and Commerce*. Room 2125, Rayburn House Office Building, Washington, DC 20515; (202) 225-2927. Responsible for national energy policy, fossil-fuel, nuclear, and solar energy. Subcommittees: Energy and Power; Health and the Environment; Transportation and Hazardous Materials.
— *Committee on Interior and Insular Affairs*, Room 1324, Longworth House Office Building, Washington, DC 20515; (202) 225-2761. Responsible for forest reserves, national parks, irrigation, and Indian lands. Subcommittees: Energy and the Environment; Mining and Natural Resources;

National Parks and Public Lands; Water, Power, and Offshore Energy Resources.
— *Committee on Merchant Marine and Fisheries*, Room 1334, Longworth House Office Building, Washington, DC 20515; (202) 225-4047. Responsibility includes coastal zone management and international fishery. Subcommittees: Fisheries and Wildlife Conservation and the Environment; Oceanography and Great Lakes.
— *Committee on Public Works and Transportation*, Room 2165, Rayburn House Office Building, Washington, DC 20515; (202) 225-4472. Responsible for improvement of rivers and harbors, oil spills and other pollution of navigable waters, and flood control. Subcommittees: Surface Transportation; Water Resources.

U.S. Senate
— *Committee on Agriculture, Nutrition, and Forestry*, Room 328-A, Russell Building, Washington, DC 20510; (202) 224-2035. Responsibilities include agriculture, forestry, agricultural pests, pesticides. Subcommittees: Conservation and Forestry; Nutrition and Investigations.
— *Committee on Commerce, Science, and Transportation*, U.S. Senate SD-508, Washington, DC 20510; (202) 224-5115. Responsible for waterways, coastal management, marine fisheries, and oceans, including the National Ocean Policy Study.
— *Committee on Energy and Natural Resources*, Room SD-364, Dirksen Building, Washington, DC 20510; (202) 224-4971. Responsible for regulation, conservation, and extraction of minerals, petroleum, and nuclear energy, and for solar energy. Subcommittee on Water and Power.
— *Committee on Environmental and Public Works*, Room SD-458, Dirksen Building, Washington, DC 20510; (202) 224-6176. Responsible for environmental policy and research, ocean dumping, solid-waste disposal, and recycling. Subcommittees: Energy; Energy Research and Development; Environmental Protection; Mineral Resources Development and Production; Nuclear Regulation; Public Lands, National Parks, and Forests; Superfund, Ocean, and Water Protection; Toxic Substances, Environmental Oversight, Research and Development; Water Resources, Transportation, and Infrastructure.

Computer Networks

Networking by computer is an increasingly effective means of communication for environmentalists and conservationists. These networks offer a number of services including: **electronic mail,** which allows members to send private messages to one another and to members of many other networks around the world; **conferences**, providing users with access to information on a particular subject posted by conference managers and other users, affording discussion among members on the same general topic; **databases**, which allow computer users to consult compilations of information on particular topics; and **user directories**, which permit members to get lists of other members except those who request anonymity. The leading environmental computer network services include:
— *Earthnet Environmental Information Service*, a free global computer network consisting of "nodes" that carry some 30 conferences originating on various other computer nets. These include the *Bionews* conference, a read-only summary of environmental news items selected from *Econet* (see next listing). Its other conferences, such as the *Greenpeace* conference, are two-way computer E-mail discussion groups that you can read, reply to, and post your own messages on. Besides being a one-phone-call place to pick up environmental and conservation news and discussions, Earthnet has a number of sister boards that carry all the Earthnet conferences.

Earthnet can be reached via its main node in Long Island, New York, at (516) 321-4893 or in New York City at (212) 226-9045. Other nodes that carry Earthnet conferences are: Alternate Realty, Michigan, (713) 717-6472; Boardroom, North Carolina, (919) 831-0674; German Connection, San Antonio, Texas, (512) 532-4756; Jojac BBS Kettleby, Ontario, Canada, (416) 841-3701 or 939-2574/8660; North End Skyscraper, Oregon, (206) 756-9689; and Socialism On-Line, Watertown, Connecticut, (203) 274-4639.
— *EcoNet/Peacenet*, operated by the Institute for Global Communications (IGC), 18 De Boom Street, San Francisco, CA 94107, (415) 442-0220, is an international web of conferencing systems devoted to peace and the environment. It has 7,500 subscribers in the U.S. and 15,000 worldwide. Econet costs $15 for signing up plus a monthly fee of $10 allowing one free off-peak hour; a $10 fee

is charged for every hour of additional peak use or $5 for additional off-peak hours. If you are at a university, you can connect to IGC through *Internet*, the world's leading interuniversity network. *Note:* Just about every organization listed in this. Directory uses EcoNet.
— *Environet*, c/o Greenpeace, 139 Townsend Street, 4th Floor, San Francisco, CA 94107, (415) 512-9205, invites the public to participate free of charge in its computer bulletin board.
— *Save the Planet* is a free software program from Save the Planet Software, P.O. Box 45, Pitkin, CO 81241, (303) 641-5035. Containing many ideas for recycling and energy saving and "green shopping" suggestions aimed at reducing pollution and waste, its Save the Planet users' conference operates in conjunction with Econet.
— *The Well* (Whole Earth 'Lectronic Link), 27 Gate Five Road, Sausalito, CA 94965, (415) 332-4335 (voice), 332-6106 (modem—1200 baud), 332-7358 (modem—2400 baud), is operated by *Whole Earth Review*. It costs $10 a month plus a $2-an-hour on-line fee that does not include long-distance charges.

Prominent among dozens of other database service companies that provide on-line information on a wide variety of subjects are:
— *BRS*, 800 Westpark Drive, McLean, VA 22102; (800) 421-7248. Originally known as Bibliographic Retrieval Services, this is an on-line service with a wide range of databases in social sciences and the environment. As with the other companies listed below, the general practice is to charge users at a connect-hour or -minute rate that varies according to the category of customer (professional, corporate, etc.), time of day, and other factors. For more complete details on BRS and other similar services, write or check via the phone numbers provided here.
— *Dialog Information Service*, 3460 Hillview Avenue, Palo Alto, CA 94304; (800) 3-DIALOG. DIALOG provides access to more than 400 databases covering thousands of technical business publications, including some 1,100 journals, magazines, newspapers, and newsletters. Its special subject areas include energy and environment (with about 100 databases), law and government, and agriculture and nutrition.
— *Mead Data Central*, 9393 Springboro Pike, P.O. Box 933, Dayton, OH 45401; (513) 865-6900, (800) 227-4908, or, in Ohio, (800) 227-8379, and, in Canada, (800) 553-3685. A pioneer in the on-line information service, this company operates *Lexis* (for legal information), *Medis* (medicine), and *Nexis* (general news and business). With more than 320,000 subscribers, its main customers are lawyers, journalists, accountants, and other professionals.
— *WESTLAW*, 610 Opperman Drive, Eagan, MN 55123; (800) WESTLAW. Containing more than 3,000 databases, this is the computer-assisted legal research service from West Publishing Company, one of the nation's leading publishers of law books. WESTLAW provides coverage on environmental protection and conservation, including radioactive, solid, and toxic waste, clean air and water, and the impact of law on the environment. Databases include case law, statutes and regulations, administrative law, texts, periodicals, and other materials.

NOTES

1. Bill Gifford, "Inside the Environmental Groups," *Outside*, September 1990, p. 71. Gifford's report is hard-hitting, amusing, and generally on target.
2. Ibid., p. 74.
3. Ibid., p. 75.
4. Solicitation letter from Captain Paul Watson, April 1989.
5. Cited in Sea Shepherd brochure, 1989.
6. Sierra Club Legal Defense Fund, *Annual Report, 1988–1989*, p. 1.

B

PENDING LEGISLATION AND HOW YOU CAN INFLUENCE IT

In order to influence legislation you need to understand how the legislative process works. A bill's chance of passing depends on which of several committees it goes to (the chair of one may support it, the chair of another may not) and when it is introduced. Congress works on a cycle of two one-year sessions; a bill pending at the end of the first session carries over to the second, but one still pending at the close of the second session must be reintroduced and go through the whole process again. The first session of the 103rd Congress began in January 1993.

At any given time there are dozens of bills pending in the Senate and the House of Representatives. The best way to track specific bills in either House or Senate is through *Legislative and Information Bill Status* at (202) 225-1772. This helpful service is provided by the Office of the Clerk of the House of Representatives. You can also contact your own representative or senator(s), consult the *Congressional Record* or the *Federal Register*, or check with the Senate or House documents room. (For details, see *Resources* on page 364.)

For the general reader seeking a regular update on the state of legislation affecting the environment (along with a lot of other extremely valuable information) there is no better source than *Audubon Activist*, published ten times a year by the National Audubon Society (for details, see the Directory, page 339). Two more detailed but more expensive sources of information on pending legislation are the *Land Letter*, which comes out thirty four times a year and is published by the Conservation Fund, 1800 N. Kent Street, Suite 1120, Arlington, VA 22209, and the *Weekly Bulletin* of the Environmental and Energy Study Institute, 122 C Street, NW, Suite 700, Washington, DC 20001.

New Legislative Regulations

Getting a law on the books is only the first step. It usually falls to an agency such as the EPA or the BLM to translate the broad statutory language into specific rules and regulations. Federal agencies cannot announce new regulations without first publishing a proposal in the *Federal Register* for public comment. Although the details of many environmental regulations are highly technical, there is room for important contributions from the public.

As a practical matter, it is next to impossible for you as an individual to check every *Federal Register* (they come out daily) to see if it contains proposed regulations that you might be interested in. But the environmental groups do keep track of EPA's regulatory program, and the environmental

loose-leaf services (the *Environment Reporter* and the *Environmental Law Reporter*) flag proposed environmental regulations weekly. As a member of an environmental group, you can stay in touch with those working full-time on problems of concern to you and you can become aware of what EPA and other government agencies are working on.

How to Write an Effective Letter to Government Officials

1. *Address it correctly:*

a. *The President*
The White House
1600 Pennsylvania Avenue, NW
Washington, DC 20500

Dear Mr. President:

b. *The Honorable ....*
U.S. Senate
Washington, DC 20515

Dear Senator......................

c. *The Honorable ....*
U.S. House of Representatives
Washington, DC 20515

Dear Representative ....

2. *Write first to your own representative* and then to the chair or members of the committee dealing with the legislation you are interested in. If you write to a legislator other than your own, send a copy of your letter to your representative. He or she may become interested and offer support.
3. *Identify by name and number the bill that deals with the subject you are writing about* — e.g., "Grand Canyon Protection Act," H.R. 814, introduced by Representative George Miller, California.
4. *Keep it brief.* Politicians have short attention spans. Summarize your points and try to stick to one issue per letter and to confine your message to one page.
5. *Be personal.* Write in your own words. Don't send a form letter. Give your own reasons for supporting or opposing a piece of legislation.
6. *Follow up.* You will almost always get a reply, but it most likely will be a form letter. If you are not satisfied with it, write another letter or, if time is short, send a telegram or make a phone call.
7. *Be courteous.* Don't threaten. When appropriate, thank your legislator for taking positive action.
8. *Call:* The President, (202) 456-1111 or (202) 456-1414; U.S. Senate and House of Representatives, (202) 224-3121.
9. *Get others to write as well*; there is strength in numbers.

How You Can Get Involved in the Legislative Process

- Write a letter to your congressperson, referring to the bill by its name and number, and summarize your position and how you would like him or her to vote.
- Your best chance of influencing a congressional bill is *before* it goes to a conference committee — either during public hearings, at which testimony is given by scientists, representatives of all interested parties, and other witnesses, or at special meetings where the congressional staff consider alternative proposals. Your representative can give you information on legislation under review, which committees are responsible, and when hearings are going to take place.
- Even if your representative is not a member of the relevant committee, her or his aides can be very helpful in providing information on legislation.
- Find out when the appropriate committee of Congress is holding an *investigative* or an *oversight* hearing. Tell your representative that you want to attend and if you would like to testify.

* Your influence can be greater on state legislation than on federal because actions taken at the state level usually involve issues of local concern with which you are more familiar. State legislators are generally quite receptive to your input, and local media are usually more accessible than national.

The New Agenda

Congress is constantly considering new proposals and amendments to existing legislation. For example, critical amendments to the Clean Water Act and the Endangered Species Act that were unresolved during the 102nd Congress will be reconsidered in the new Congress. Paramount among the conservation and natural resource bills likely to dominate the agenda of the 103rd Congress are the following.

Water Resources and Wetlands

The debate over wetlands protection is expected to intensify as the Clean Water Act comes up for reauthorization. One particularly important proposal is a wetlands reform bill similar to that introduced by Representative Don Edwards (D-CA) in 1992. Designed to strengthen Section 404 and other sections of the CWA by regulating not only the filling but also the draining, flooding, and excavating of wetlands, it would give the EPA and the Fish and Wildlife Service greater roles in the Section 404 permitting process. Such a bill is strongly supported by conservationists. It is opposed by the so-called wise use movement, made up largely of ranchers, miners, and other development interests in the West. This movement and its supporters in Congress have pressed the argument that environmental regulations imposed on private property amount to a "taking" by the government and violate the Fifth Amendment of the U.S. Constitution.

In late October 1992 President Bush signed the Central Valley Project bill as part of a Western water bill. The measure ensures adequate clean water supplies to fourteen national and state wildlife refuges and makes restoration of threatened fish and wildlife species a priority. It also provides incentives for conservation through water pricing reforms. Supported by conservationists, fishermen, and major California businesses, the legislation was opposed by Central Valley farmers and California Governor Pete Wilson.

Ancient Forests

Although legislation protecting these forests was gridlocked in Congress during the Bush administration, important gains were made in the courts, in science, and in public opinion. Action on a forest bill stalled in June 1992 when House Speaker Thomas Foley (D-WA) pressured several House Interior Committee members to drop their support for a strongly conservationist bill sponsored by its chairperson, George Miller (D-CA). Later in the session some Northwestern members of Congress introduced amendments to open federal forests to more logging, but their efforts were rebuffed by a quick response from conservation and environmental activists.

The need for forest protection was affirmed in the Federal District Court in Seattle in 1991 when Judge William Dwyer issued an injunction banning logging across millions of acres of spotted owl habitat in national forests in Oregon and Washington until the Forest Service files an acceptable plan to protect the threatened owl.

In another positive development, just before the 102nd Congress adjourned, the Forest Service listed the marbled murrelet as a threatened species. The seabird nests in Pacific coastal forests that have been devastated by clear-cutting.

With less than 10 percent of the ancient forests remaining because of overlogging, the need for a solution to the ancient forest dispute is urgent. Bills similar to those proposed in 1992 by Representative George Miller and Senators Patrick Leahy (D-VT) and Brock Adams (D-WA) establishing an ancient forest reserve, banning raw log exports, and providing economic assistance to timber-dependent areas are expected to be introduced in the 103rd Congress.

Endangered Species

Although authorization for the Endangered Species Act (ESA) expired in September 1992, Congress voted to continue funding the act through fiscal 1993. Despite several bills attempting to weaken the ESA, there is considerable support in the House for a bill by Representative Gerry Studds (D-MA) that will broaden protection for endangered and threatened species and habitats, set deadlines for species recovery plans, and double funding to implement the act. Studds, who had 107 cosponsors of his bill in 1992, returns as chair of the Merchant Marine and Fisheries Committee, which has jurisdiction over the ESA.

Arctic Refuge

Despite approval by the Senate Environment and Public Works Committee for such legislation, the 102nd Congress failed to act on measures to protect the coastal plain of the Arctic National Wildlife Refuge (ANWR). The oil industry's efforts to open the refuge to oil drilling were blocked by conservationists who successfully stripped such a provision from the energy bill signed into law by President Bush in October 1992. But at the same time provisions to prohibit offshore oil drilling and to increase auto fuel-efficiency standards were removed by pro-oil legislators. In the 103rd Congress, wilderness protection legislation will be reintroduced but will continue to face fierce opposition from the powerful oil lobby.

Mining Reform

Legislation reforming the outdated 1872 Mining Law cleared the House Interior Committee in 1992 but did not get onto the floor of the House. It would mandate royalty payments, the restoration of mined lands, and an end to the arcane patenting system. Despite growing congressional support for the reform, there is strong opposition from the mining industry and its proponents in the House and Senate.

Grazing Fees

A provision to raise grazing fees on public lands was stripped from the Interior Appropriations bill in 1990, 1991, and 1992. Although the House voted to increase fees by one third to $2.56 per cow per month, Western senators prevailed in removing the increase from the final bill. Dogged opposition by Western senators is expected to continue throughout the 103rd Congress.

California Desert Protection

A bill to protect 7.1 million acres of southern California desert was passed by the House in 1992, but a similar measure introduced by retiring Senator Alan Cranston (D-CA) failed to make it through the Senate mainly because of opposition from Senator John Seymour (R-CA), who is no longer in the Senate. When this legislation is reintroduced in the 103rd Congress, it will have the support of both new California senators, Diane Feinstein and Barbara Boxer. It will, however, encounter fierce opposition from off-road vehicle users and from ranching and mining interests in the region.

Wilderness Protection

More than 17 million acres of wilderness — technically unprotected — remain in the northern Rocky Mountains, often referred to as "the last best place." In recent years Congress has dealt with the conservation-versus-development controversy by passing a series of individual state wilderness bills. Although these have provided protection to scattered areas, they have also allowed the Forest Service to sell off huge amounts of timber to private industry. In September 1992 a bill introduced by Representative Peter Kostmayer (D-PA) was the first attempt to manage national forest systems on a bioregional basis. The bill did not muster enough votes in the House and Kostmayer was

defeated in the November election. Conservationists are working hard to find a new sponsor for its reintroduction in the 103rd Congress.

In the last session of the 102nd Congress both houses passed Montana wilderness bills, but could not hammer out a compromise before they adjourned. The Senate version designated 1.2 million acres of federal lands as wilderness, while the House protected 1.5 million acres. Both bills left large tracts open for logging and other uses that would harm grizzly bear, elk, and gray wolf habitat. Montana's Audubon Council and other conservation groups support protection for at least 2.5 million acres when the bills are resurrected in the new Congress.

RESOURCES

1. U.S. Government

Distribution of Congressional Publications

Senate bills, reports, and documents are distributed through the **Senate Documents Room**, B-04, Hart Senate Office Building, Washington, DC 20510. House equivalents can be obtained from the **House Documents Room**, H-226, U.S. Capitol, Washington, DC 20515. Public laws are distributed by both document rooms.

U.S. Congress:

The *Congressional Record* publishes daily proceedings of Congress. Somewhat more generally accessible (through many public libraries, for example) is the *Federal Register*, which publishes rules and regulations before they are finalized in the *Code of Federal Regulations*. The *Register* also includes administrative notices and proposed rules and regulations.

To find out the status of a House or Senate bill, telephone (202) 225-1772.
To receive copies of bills, *write* (enclosing a self-addressed mailing label) to either House Documents Room, H-226, or Senate Documents Room, S-325, U.S. Capitol, Washington, DC 20510.
— *House Appropriations Committee*, H-218, U.S. Capitol, Washington, DC 20510.
— *Senate Appropriations Committee*, 118 Dirksen Senate Office Building, U.S. Capitol, Washington, DC 20510.

U.S. Government Printing Office, Superintendent of Documents, Washington, DC 20402, (202) 275-3030, sells copies of all U.S. laws, treaties, and implementing regulations. Copies of these documents may also be obtained from the main or regional offices of the federal agency responsible for enforcement.

The EPA Public Information Center, 401 M Street, SW, Washington, DC 20460, (202) 382-2080, will provide copies of laws for which EPA is the regulatory agency.

2. State Laws

Documents on state environmental legislation can be obtained from the individual state governments, most of which have departments of environmental protection or natural resources. Most state environmental laws are collected in the *Environmental Reporter*

GLOSSARY OF TERMS AND ABBREVIATIONS

A

Abiotic Nonliving.
Abyssal zone Bottom zone of the ocean characterized by dark water.
ACEC Area of Critical Environmental Concern.
Acre 1 acre = 4,840 square yards. 640 acres = 1 square mile.
Acre-Foot A volume equal to an acre covered with one foot of depth of water; equal to 325,900 gallons.
Aerobic Requiring oxygen to live. (Compare **Anaerobic.**)
Alluvial Pertaining to sand, mud, and other material deposited by a flowing stream.
Anadromous species Species of fish, such as salmon, that migrate from fresh water to salt water and back again to spawn.
Anaerobic Not requiring oxygen to live.
Ancient forest A forest in its late or climax stage of forest growth, usually taking 200 or more years to develop.
Annual A plant that completes its life cycle from seedling to mature seed-bearing plant in a single growing season and then dies. (Compare **Perennial.**)
Anoxia See **Hypoxia.**
Aquatic Pertaining to water.
Aquatic ecosystem A major ecosystem such as an ocean, river, lake, or pond that makes up the hydrosphere. (Compare **Biome.**)
Arid Dry, parched from heat.
Association A community consisting of a characteristic combination of species.
Atmosphere Layer of air surrounding the earth's surface.

B

Backfill Return soil or other material into a space from which it was removed.
Bacteria Smallest living organisms. With fungi, they comprise the decomposer level of the food web.
Barrier beach Gradually sloping land along a coastline usually having two rows of sand dunes that cushion the land behind them from the ravages of the sea.
Barrier island A long, narrow offshore island parallel to the coastline and consisting of sand, shells, or gravel.
Bathyl zone The cold, dark zone below the surface of the ocean in which some sunlight penetrates but not enough for photosynthesis.
Benthic zone The bottom of a body of water.
Bioaccumulation See **Bioconcentration.**

Biocentric Ecosystem-centered.
Bioconcentration The buildup of pollutants as smaller organisms are consumed by larger predators. Also known as **Biological amplification** or **Biomagnification.**
Biodegradable Capable of being broken down into simpler elements and compounds by bacteria or other decomposers.
Biodiversity The number of species of plants or animals in a community.
Biomass The total dry weight of all living organisms in a given area.
Biome Large land ecosystem such as a forest, grassland, or desert.
Biosphere The world in which life can exist, including sea, soil, and the lower atmosphere.
Biota All living things in a region — flora and fauna.
Biotic Living.
BLM Bureau of Land Management.
Boreal Northern.
Boundary currents Surface currents that flow north- or southward parallel and close to the continental margins.
Browser A herbivorous animal that feeds on browse — the leaves and twigs of woody plants.

C

Canopy The upper level of forest vegetation, which intercepts most of the sunlight, rain, and snow.
Carbon cycle The cyclic flow of carbon in various chemical forms throughout the biosphere.
Carnivores Animals that get their food by eating only other animals.
Carrying capacity The maximum number of species that can be supported indefinitely by a particular area.
Cellular respiration Process in cells of plants and animals whereby food molecules such as glucose combine with oxygen and break down into carbon dioxide.
Chaparral A type of vegetation dominated by shrubs with small, evergreen leaves, found in the Southwest and areas with a mediterranean climate.
Chemosynthesis Process whereby specialized bacteria and other organisms can convert chemicals obtained from the environment into chemical energy stored in food molecules without the presence of sunlight. (Compare **Photosynthesis.**)
Chlorophyll Green coloring material of plants, essential to the production of carbohydrates by photosynthesis.
Class-action suit Filing of a lawsuit by a group on behalf of a larger number of groups or citizens alleging similar damages but who do not need to be represented individually.
Clear-cutting The cutting of timber in which virtually all the trees in a forested area are removed.
Climax A relatively stable, mature stage of ecological development.
Climax ecosystem (Climax community) A mature system with a diverse array of species and capable of using energy and critical chemicals more efficiently than simpler, immature ecosystems.
 Climax forest A forest in the final stage of forest succession, in which species

composition shows no marked directional change with time.
Closed forest Terrain with an almost complete cover of trees. (Compare **Open forest**.)
Closed system A functional area that is mostly isolated by barriers that exclude certain outside influences.
Community An interacting group of different species occupying a particular area. For example, a Douglas fir-western hemlock community consists of these tree species together with many other plants, animals, and microorganisms.
Coniferous tree A tree or shrub whose seeds are borne in woody cones. There are about 500 to 600 species of living conifers.
Consumers Organisms that depend on other organisms for their food and energy. Generally divided into primary consumers (herbivores) and secondary (carnivores). (Compare **Decomposers, Producers**.)
Copepods Minute shrimplike crustaceans.
Crash A sharp decline in numbers of species after **overshoot** (see below).
Cultural eutrophication Overnourishment of an aquatic ecosystem with plant nutrients discharged by agriculture, industry, and other human activities. (See **Eutrophication**.)

D

Deciduous plants Plants that lose all their leaves during part of the year — e.g., oak and maple.
Decomposers Bacteria, mushrooms, fungi, and other organisms that obtain nutrients by breaking down matter in the wastes and dead bodies of other organisms into simpler chemicals.
Detritus Small particles of organic matter, largely derived from the breakdown of dead vegetation.
Detritus chain The part of the food web that is ultimately dependent on detritus as a source of food.
Diversity Variety; the number of species of plants or animals in a community.
Doldrums A belt of light, variable winds near the equator.
Drawdown Process by which the dominant species in an ecosystem uses up resources faster than they can be replaced.
Duff Decomposing organic matter found on the floor of a forest.

E

Ecological niche The part of a habitat occupied by an organism. The role played by an organism within an ecosystem.
Ecological succession Process whereby communities of plant and animal species are replaced in a given area over time.
Ecology The study of how living things relate to their natural environment.
Ecosystem A community of organisms and its physical habitat functioning as an ecological unit. Whether a fallen log or an entire watershed, an ecosystem includes resident organisms, nonliving components such as soil nutrients, inputs such as rainfall, and outputs such as organisms that disperse to other ecosystems.
Ecosystem services Services such as providing clean air and clean water performed by ecosystems.
Ecotone A transition zone between one ecosystem and another.

Effluent A substance, usually liquid, entering the environment from a **point source** (see below); refers generally to wastewater from an industrial or sewage treatment plant.

EIS, Environmental Impact Statement A study of the environmental impact that any project might or might not have.

Endangered species A wild species having so few individual survivors that it could soon become extinct in all or most of its natural range. (Compare **Threatened species.**)

Environment All the external conditions influencing the life of an organism or **population** (see below).

EPA U.S. Environmental Protection Agency.

Epiphytic Describing plants and fungi that grow on other plants but not as parasites. The only burdens an epiphytic plant (or epiphyte) places on its host are competition for sunlight and its weight on leaves and branches. Common epiphytes in Northwestern forests include green algae, liverworts, mosses, and ferns.

Erosion The process whereby earth and rock are loosened, dissolved, or worn away and moved by wind, water, and other means.

Estuarine zone Area near the coastline consisting of estuaries and coastal saltwater wetlands and extending to the edge of the continental shelf.

Estuary A partially enclosed body of water where river water meets and mixes with ocean water.

Eutrophic Characteristic of a body of water with an excessive supply of plant nutrients, mostly nitrates and phosphates.

Eutrophication The changes in a lake or stream that decrease oxygen supply and thus favor plant growth over animal life.

Extinction Complete disappearance of a species.

F

Family A group of closely related genera. For example, weasels and mink (genus *Mustela*), martens and fishers (genus *Martes*), and river otters (genus *Lutra*) are members of the family Mustelidae.

Feedback A signal sent into a self-regulating system eliciting a response.

Food chain Transfers of food energy in which one type of organism consumes another.

Food web A complex, interlocking series of food chains; an assemblage of organisms in an ecosystem, including plants, herbivores, and carnivores, showing the relationship of "who eats whom."

FWS U.S. Fish and Wildlife Service, managed by the Department of the Interior.

G

Gaia Mother Earth, the goddess of the Earth.

Gaia hypothesis The concept that the surface of the Earth is regulated by the activities of life. The hypothesis implies that if life were to be eliminated from Earth, its surface conditions would revert to those on Mars or Venus.

GAO U.S. General Accounting Office.

Genus (plural **genera**) A biological classification with one or a group of closely related species such as *Strix* (spotted and barred owls), *Tsuga* (western and mountain hemlocks), and *Oncorhynchus* (five northwestern species of Pacific salmon).

Grazer A herbivorous animal; in wildlife biology a herbivore feeding on grasses.

H **Habitat** The place where an organism or community of organisms lives.
Harvest Gathering a cultivated crop. In logging the term is used inappropriately as a euphemism for killing trees in virgin forests; in forestry, it is appropriate only when used regarding tree plantations, which are cultivated crops.
Herbivore An animal that eats plants or other photosynthetic organisms to get its food and energy. (See also **Carnivore.**)
HMP Habitat Management Plan
Holistic Relating to the study of complete living systems, rather than of their component parts in isolation.
Human ecology The study of the relations between a human community and its environment.
Hydrologic cycle The movement of water between the oceans, ground surface, and atmosphere by evaporation, precipitation, and the activity of living organisms.
Hydrosphere The water portion of the earth.
Hypoxia A deficiency of oxygen in tissue, blood, or a body of water.

I **Indicator species** A species whose presence indicates certain environmental conditions such as the relationship between old-growth forests and the northern spotted owl.
Intertidal zone See **Littoral zone.**
Invertebrates Organisms that lack a spinal column, including insects, mollusks, crustaceans, starfish, jellyfish, and many types of worms.
IPM Integrated pest management.

K **Keystone species** A species that influences other members of its community far out of proportion to its numbers.
Krill Tiny, shrimplike crustaceans eaten by whales.

L **Limnetic** Relating to the study of inland waters.
Lithosphere The crust of the earth.
Littoral zone The zone between mean high-water and mean low-water levels. Also known as the **Intertidal zone.**

M **Matter** Anything that has mass and occupies space. **Microhabitat** A small **habitat** within another one — such as the area in the air space under the snow, where environmental conditions differ from those in the surrounding area.
Microorganisms Bacteria, fungi, and other microscopic living things.
Monoculture Cultivation of a single species of plants or trees to the exclusion of other uses of the land.
Montane Pertaining to mountain conditions.
Multiple use Principle for managing a forest so that it is used for varied purposes, including timbering, wildlife preservation, recreation, and soil and water conservation.

N

Neap tide The lowest range of the tide, occurring near the times of the first and last quarter of the moon.
NPS National Park Service.
Nonpoint source Source of pollution where wastes are released from a diffuse number of points that are hard to control — e.g., stormwater and snowmelt runoff from land surfaces. (Compare **Point source.**)
Nutrient Element or compound needed for the growth, reproduction, and survival of a plant or animal.

O

Old-growth forest See **Ancient forest.**
Oligotropic A body of water with a low supply of nutrients.
Omnivores Creatures such as pigs, rats, cockroaches, or humans that eat plants and animals.
Open forest Land area covered only partially with trees.
Organic Derived from plants or animals.
Organism Any form of life.
Overgrazing Excessive grazing of rangeland by livestock to the point where it cannot be renewed, usually because of damage to the root system.
Overshoot The inevitable and irreversible consequence of continued **drawdown** (see above), when the use of resources in an ecosystem exceeds its carrying capacity and there is no way to recover or replace the loss.
Oxidation Chemical change involving the addition of oxygen, often increasing the action of decay-causing microorganisms.

P

Perennial A plant that lives for several years and usually produces seeds every year. (Compare **Annual.**)
Permafrost A layer of permanently frozen soil and other deposits, found in regions where the yearly average temperature is below freezing.
pH Numeric value indicating the relative acidity or alkalinity of a substance on a scale of 0 to 14, with a neutral point at 7. Acid solutions have pH values lower than 7.
Photosynthesis Process by which green plants in the presence of sunlight convert carbon dioxide and water into sugar. (Compare **Chemosynthesis.**)
Phytoplankton Small, often, single-celled plants that float in the water.
Pioneer community The first community that colonizes a site.
Plankton Small, free-floating organisms, including plants (phytoplankton) and animals (zooplankton).
Point source Source of pollution that involves discharge of wastes from a smokestack, sewage treatment plant, or other identifiable source. (Compare **Nonpoint source.**)
Population Group of individual organisms of the same kind (species).
Predator An organism that lives by eating an organism of another species.
Prey An animal captured as food by another animal.
Producers Organisms that use solar or chemical energy to manufacture their own food from inorganic materials, usually through photosynthesis. (Compare **Consumers, Decomposers.**)
Profundal zone Deep-water region of a body of water, which is not reached by sunlight.

Radiation Transmission through space of energy, such as heat or light, in the form of waves or rays.
Range Span of conditions that must be maintained for populations of a species to function normally.
Rangeland Uncultivated land that supports herbaceous or shrubby vegetation. It includes soil, water, atmosphere, vegetation, and animal life.
Riparian Occurring along the bank of a river or lake.
Riparian zone A narrow belt of land along a waterway, critical to the overall health of the surrounding environment.
Riprapping Process of lining a riverbank with rock.

Shoal An elevated area in the bottom of a body of water often exposed during low tides.
Silviculture The development, reproduction, and care of forest trees.
Sludge A semiliquid mixture of bacteria- and virus-laden organic matter, heavy metals, and synthetic organic chemicals removed from wastewater at a sewage-treatment plant.
Snag A standing dead tree. Once considered hazardous, they are now called "wildlife trees" in recognition of their crucial importance to wildlife and forest regrowth.
Solar energy Direct radiant energy from the sun and indirect forms of energy such as flowing or falling water and wind energy that are produced when solar energy interacts with the earth.
Species (singular or plural) A group of plants or animals having many common characteristics. Individuals belonging to the same species resemble each other and usually interbreed only with each other.
Structure Spatial and other arrangements of species within an ecosystem, taking into account the type and number of species, the biomass, life cycles, and spatial distribution.
Succession Replacement of one community by another. Changes over time in the species composition and structure of a community.
Succulent Cacti and other plants that store water and produce the food they need in the fleshy tissue of their stems and branches.
Sustainable development Development based on the use of renewable resources and that is ecologically sustainable over time.
Sustainable yield (sustainable capacity) Maximum extent to which a resource may be used without depletion. For example, trees may be taken from a forest provided the rate of cutting does not exceed the rate of their regrowth.
Swamp An area saturated with water for much of the time, but in which the soil surface is not deeply submerged. In a restricted sense, an area marked by woody vegetation; more widely used to include bogs, marshes, and other wetlands.
Symbiosis Two different species living together beneficially.
Synergy Interaction in which the combined effort is greater than the sum of the separate effects.
System Combination of things or parts forming a complex or unitary interacting whole.

Terrestrial Pertaining to land. (Compare **Aquatic**.)

Thermal stratification In a body of water, a succession of differentiated layers, each with a different temperature, the coldest at the bottom.

Threatened species A species considered to become endangered within the foreseeable future because of a decline in numbers.

Tidal flats Marshy or muddy areas covered and uncovered by the rise and fall of the tide; also called **tidal marshes**.

Tidal range Difference in height between consecutive high and low waters.

Transpiration The vaporization of water from plant tissues through pores called stomata.

Tree line Farthest limit of tree growth in northern regions and on mountains; the line beyond which living conditions are too harsh to allow growth of trees. Also known as **timber line**.

Trophic Pertaining to feeding.

Troposphere Innermost layer of the atmosphere, containing about 95 percent of the earth's air and extending 5 to 7 miles above the earth's surface.

Tundra A treeless plain of the arctic region dominated by mosses, lichens, and sedges.

Turbid Opaque.

Upwelling The process whereby water rises to a higher level, usually as a result of offshore currents or divergence.

USFS, also **FS** United States Forest Service.

Virgin forest Forest that has not been logged or burned by humans.

Wetland Transition area between terrestrial and aquatic systems where the water table is usually at or near the surface, or where the land is covered by shallow waters (e.g., swamps, marshes, bogs, intertidal mudflats, prairie potholes).

Weathering Process whereby bedrock is gradually broken down by physical and chemical exposure into small fragments that make up most of the soil's inorganic composition.

Wilderness A place where the earth and its community of life have not been seriously disturbed by humans.

Wilderness species Wild animal and plant species that flourish only in such relatively undisturbed vegetational communities such as large areas of mature forest, grassland, desert, and tundra.

Wildlife All undomesticated species of animals and plants.

Xerxes The first butterfly known to have become extinct as the result of human activity.

Xerophytes Plants with special characteristics such as small leaf size or thick corky bark enabling them to survive in very dry or very cold climates.

Yew Botanically known as *Taxus*, the yew is a member of the *Gymnospermae* family of much-branched trees and shrubs. The Pacific yew is threatened with extinction because 16,000 pounds of its bark are needed to produce 2.2 pounds of taxol, a potential anticancer drug.

Yield The volume or weight of theoutput or harvest of trees and other plants.

Zero Population Growth (ZPG) Condition in which the birth rate plus immigration equals the death rate plus emigration so that the population is no longer increasing.

Zone A layer encompassing a defined feature, structure, or property in water or other media.

Zooplankton Minute, often microscopic creatures that float in water, feeding on detritus, phytoplankton, and other zooplankton. One of the most common forms of zooplankton is the copepod, a small crustacean that serves as a major source of food for herring and other marine animals.

ACKNOWLEDGMENTS

Expressing our thanks to the many people who helped us with *This Land Is Your Land* is one of the most enjoyable aspects of putting the book together. To our literary agent Sarah Jane Freymann and the book's first editor John Michel, to Joseph Santoro the designer, and to its ultimate editor Wendy Wolf, each of whose enthusiasm, energy, and skills helped us complete the project, we are most grateful.

In between original proposal and finished book there developed an ever-expanding network of contributors, a veritable ecosystem of human effort and generosity. First, we acknowledge the early input of *This Land*'s scientific consultant, Dr. Richard A. Orson, botanist, ecologist, and geomorphologist, who helped shape the overall concept of the book and kept us on track throughout its progress. We also appreciate the technical and editorial readings generously supplied by the International Center for the Solution of Environmental Problems (especially from Dr. Joseph Goldman, ICSEP's director, and Bonnie McNairn), by Dr. Ed Grumbine, director of the University of California's Sierra Institute, Santa Cruz, Dr. Rodney Fujita, senior scientist, Environmental Defense Fund, and our two trusty research assistants, Mary Kay Carson and Fleur Templeton, both recruited from New York University's science and environmental reporting program.

On the government side we particularly want to acknoledge the invaluable help and support of; the late William Penn Mott, Jr., former director of the National Park Service; William Knapp, chief of the U.S. Fish and Wildlife Service division of Endangered Species and Habitat Conservation; Bill Radtkey, coordinator, Threatened and Endangered Species, Bureau of Land Management; Maureen Finnerty, superintendent of Olympic National Park; Valerie Meyer, librarian, Grand Canyon National Park; Joan Anzelmo, public affairs officer, Yellowstone National Park; Teresa M. Cherry, outdoor recreation planner, Great Dismal Swamp National Wildlife Refuge; Don Perkuchin, senior project leader, Okefenokee National Wildlife Refuge; Pamela Rizor, assistant manager, Arapaho National Wildlife Refuge; Rosa Wilson, photo librarian of the NPS; Jim Sanders, public affairs office, U.S. Forest Service; Victoria Flannagan, Protected Species Management, Florida Department of Natural Resources; Richard Wedepohl and Jana Suchy, both with the Wisconsin Department of Natural Resources; Betsy Peabody, Puget Sound Water Authority; and Susan Wilson Garms, marketing manager, Bureau of National Affairs Books.

From the environmental and conservation network we have received an avalanche of support and encouragement, especially from: Dr. Jan Beyea, senior scientist, National Audubon Society; Tensie Whelan, Audubon's vice-president for conservation information; Fred Baumgarten, editor, *Audubon Activist*; Randy Showstack, director of communications, American Rivers; Will Nixon, associate editor *E* magazine; Ron Geatz, media director, the Nature Conservancy; Ashok Gupta, senior energy analyst, Natural Resources Defense Council; Chris Van Daalen, codirector, Save America's Forests; Sally Ranney, executive director, American Wildlands; Ed Lewis, executive director, Greater Yellowstone Coalition; Richard Spotts, California representative, Defenders of Wildlife; Stephanie Kessler, Wyoming Outdoor Council; George Nickas, Utah Wilderness Association; Michael Herz, San Francisco BayKeeper; Greg Karras, director, clean bays and coastal waters program, Citizens for a Better Environment; Bill Davoren, executive director, San Francisco Bay Institute; Gar Smith, editor, *Earth Island Journal*; Richard Manning, author of *The Last Stand*; Karl Linn, Earth Island Institute urban habitat program; David Levine, the Learning Alliance; John Downey, Solar Coalition; Tom Turner, Sierra Club Legal Defense Fund; Paul Kemp, science and technology director, Coastal Coalition to Restore Coastal Louisiana; Joe Podgor, executive director, Friends of the Everglades; Jane Kerin-Moffat, coordinator, Long Island Sound Watershed Alliance; Jeff Babb, manager, Gray Ranch; Deborah Ortuno, director of development, Native Forest Council; Terrie Correll, the Desert Tortoise Council; Donna Beal, administrator, the Adirondack Council; Shannon Varner, citizen response coordinator, Chesapeake Bay Foundation; Holly Spousta, Massachusetts Audubon Society; Melissa Sargent; Rob Hayes; Susan Waters; Erika Jostad; and Erich Bollhorst, our indefatigable "in-house" researcher.

In a special category as guides to our understanding of ecosystems are: Winona LaDuke, executive director, White Earth Land Recovery Project; Mary Byrd Davis, publisher, and John Davis, editor of *Wild Earth* quarterly that they founded along with Dave Foreman and Reed Noss in spring 1991; Howie Wolke, coauthor with Foreman of the indispensable *Big Outside*; Jasper Carlton, executive director, the Biodiversity Legal Foundation; Steve Packard, Illinois director, the Nature Conservancy; Robert Stebbins, professor of zoology, University of California, Berkeley; Kristin Berry, desert tortoise authority, the Bureau of Land Management; John Berger, founder and director of Restoring the Earth; Paul and Julie Mankiewicz, codirectors of the Gaia Institute; William McDonough, architect; John Todd, founder, Oceans Arks International; Nancy Jack Todd, editor of *Annals of the Earth*; and the incomparable Lou Gold of Bald Mountain, Oregon.

There are many others who contributed to *This Land Is Your Land*, some of whom are credited on the dedication page in the front of the book or in the Resources and Notes. We greatly appreciate their help in making this book what we hope will be a useful guide to North America's endangered ecosystems. However, we acknowledge, in conclusion, that it is you the reader who will save and restore This Land.

INDEX

PRODUCTION NOTES

As in *Design for a Livable Planet,* a number of tools and pertinent new work procedures were utilized to write, design and produce this book. Whenever possible, we tried to avoid systems and materials that are harmful to the environment..

Through the "electronic office" namely, fax machines, modems, and desktop publishing, much time was conserved and material waste kept to a minimum. During the process of passing information between author, publisher, designer, and prepress, laser proofs and mailing envelopes were recycled, corrections were held to a minimum, and pages were processed directly from disk to film negatives, bypassing the outdated mechanical art process.

The following electronic and traditional tools aided in the realization of the marvelous product at hand:

Writing – Hardware/Northgate 386 and an Okidata OL830 laser printer. Software/WordPerfect 5.1.

Designing – Hardware/Macintosh IIfx and Quadra 950, Apple CD-ROM, Microtek 600Z scanner, and a LaserWriter IIg printer. Software/Microsoft Word 5.0, Caere OmniPage Pro 2.1, Quark XPress 3.11, Adobe Illustrator 3.2 and Photoshop 2.0.1, and Fractal Design Painter 2.0.

Typesetting – Hardware/Linotronic L330, RIP 4, and PLI SyQuest 44mb Removable Drive. Type/Helvetica Compressed, Univers Condensed, Stempel Garamond and Garamond #3 all from the Adobe Type Library.

Printing – Traditional page proofs and offset printing and, as expected, *this book was printed on recycled paper!*

DESIGN FOR A LIVABLE PLANET

HOW YOU CAN HELP CLEAN UP THE ENVIRONMENT

JON NAAR

COAUTHOR, DESIGN FOR A LIMITED PLANET

FOREWORD BY FREDERIC D. KRUPP
EXECUTIVE DIRECTOR, ENVIRONMENTAL DEFENSE FUND

DESIGN FOR A LIVABLE PLANET

The Indispensible Companion Book to
This Land Is Your Land

Winner of the American Library Association Award as Best Non-Fiction Book for Young Adults, *Design for a Livable Planet* has taken its place as the comprehensive source and guide to "What You Can Do" to clean up the environment.

"An absolutely first-rate, all-around guide to environmental action...provides in-depth background on ecological problems and detailed information on possible recommended individual and mass actions, as well as fascinating discussions of various advocacy groups. Encyclopedic in scope, but written in a clear personable manner...this stands out as the environmental handbook of greatest use for the general reader." — Kirkus Reviews

"Compelling. Design for a Livable Planet *is a particularly good read with snappy illustrations and hip typography."* — Boston Globe

"Ecology-energy expert Jon Naar's handy, all-round guide to individual and group action." — Time

"Good advice clearly presented, including lots of topics (VDT exposure, alternative energy, environmental law) you don't find in other ecology books." — USA Today

"Jon Naar presents his case clearly, with well-devised tables. He usefully lists ways to reduce energy use in the home." — New York Times Book Review

"Simply, the very best book on the environment. Brilliantly conceived, designed, and executed." — Joe Franklin, host and producer, "The Joe Franklin Show."

"Design for a Livable Planet *is refreshing and inspiring. Jon Naar has done a consummate job in preparing an action guide for consumers to live more lightly on the planet. I highly recommend it to teachers, students, and governmental and corporate decision-makers, as well as to the general public." — Representative Claudine Schneider, U.S. Congress*

"An outstanding book.... Jon Naar accomplishes the near impossible by fitting a host of environmental issues in one book and makes it highly readable, offering practical action steps that the average citizen can take to clean up the environment at home and at work. Great book!" — Will Collette, Organizing Director, Citizen's Clearinghouse for Hazardous Wastes

"Here is the definitive guide for the protection of our most valuable inheritance, our planet. It should be required reading for every elected official and policy maker in the country, as well as everybody concerned with the future of our precious and irreplaceable resources." — Mark Alan Siegel, New York State Assembly

"Jon Naar's Design for a Livable Planet *comes at exactly the right time. This book, with its emphasis on practical things you can do, is a valuable guide to help us envision and thus undertake change... he has given us plans that — if faithfully followed — will get us through the next few hundred years with the diversity of terrestrial life thriving and the planetary support systems intact." — Frederic D. Krupp, Executive Director, Environmental Defense Fund*

Photographs, diagrams, bibliography, glossary, and index, 340 pages
Introduction by Fred D. Krupp, Executive Director, Environmental Defense Fund

Design for a Livable Planet is available in bookstores, or call HarperCollins toll-free for credit card orders: 1-800-683-3080.
Hardcover: ISBN 0-06-05165-8, $25.95
Paperback: ISBN 0-06-096387-5, $14.00